The Handbook of Disgust Research

Philip A. Powell • Nathan S. Consedine
Editors

The Handbook of Disgust Research

Modern Perspectives and Applications

Springer

Editors
Philip A. Powell
School of Health and Related Research
University of Sheffield
Sheffield, United Kingdom

Nathan S. Consedine
Department of Psychological Medicine
University of Auckland
Auckland, New Zealand

ISBN 978-3-030-84485-1 ISBN 978-3-030-84486-8 (eBook)
https://doi.org/10.1007/978-3-030-84486-8

This Springer imprint is published by the registered company Springer Nature Switzerland AG
The registered company address is: Gewerbestrasse 11, 6330 Cham, Switzerland

To Annika and Ewan, my delightfully disgusting children.

—Philip A. Powell

To Carol who taught me how to be an academic and Ken who taught me how to think.

—Nathan S. Consedine

Contents

About the Editors

Philip A. Powell, PhD is a Research Fellow in the School of Health and Related Research at the University of Sheffield, United Kingdom. He is a Chartered Psychologist and Associate Fellow of the British Psychological Society and has degrees from the University of Sussex and the University of Sheffield. His doctoral thesis was on self-directed disgust in the context of depression. Current research interests include the role of emotions in health, well-being, and consumer behaviour; quality of life research; and methods in health economics. He has published his work widely in leading international journals and previously co-edited *The Revolting Self: Perspectives on the Psychological, Social, and Clinical Implications of Self-Directed Disgust* (Karnac Books 2015). He likes playing guitar, hiking, weight lifting, going on adventures with his family and dislikes coconut. He's a recent vegetarian.

Nathan S. Consedine, PhD works in health psychology in the Department of Psychological Medicine at the University of Auckland, New Zealand. His original training is in the basic understanding of emotion and emotion regulatory processes, evaluating how these factors may be linked to physical and mental health. He has held positions at Long Island University (Brooklyn) and Columbia University and returned to New Zealand in 2009 to begin testing specific ideas about the role of emotions in human behaviour in experimental and clinical designs. Current research interests include experimental studies of disgust, fear, and embarrassment, studies of medical compassion, and the links between emotion regulatory skills and health. He has published more than 150 scientific works and is an Associate Editor and reviewer for numerous international journals. He enjoys fishing, tennis, playing with his children, and listening to the sorts of music that his colleagues dislike. He's not a vegetarian but thinks he probably should be.

Contributors

Chris Askew, DPhil School of Psychology, University of Surrey, Guildford, UK

Charmaine Borg, PhD Department of Clinical Psychology and Experimental Psychopathology, University of Groningen, Groningen, Netherlands

Hannah K. Bradshaw, PhD Department of Psychology, Washington & Jefferson College, Washington, PA, USA

Elena Choleris, PhD Department of Psychology, University of Guelph, Guelph, ON, Canada

Nathan S. Consedine, PhD Department of Psychological Medicine, University of Auckland, Auckland, New Zealand

Graham C. L. Davey, PhD School of Psychology, University of Sussex, Brighton, UK

Vinayak Dev, MHealthPsych Department of Psychological Medicine, University of Auckland, Auckland, New Zealand

Jeffrey Gassen, PhD Department of Anthropology, Baylor University, Waco, TX, USA

Roger Giner-Sorolla, PhD School of Psychology, University of Kent, Canterbury, UK

Peter J. de Jong, PhD Department of Clinical Psychology and Experimental Psychopathology, University of Groningen, Groningen, Netherlands

Martin Kavaliers, PhD Department of Psychology, University of Western Ontario, London, ON, Canada
Department of Psychology, University of Guelph, Guelph, ON, Canada

Marco Tullio Liuzza, PhD Department of Medical and Surgical Sciences, "Magna Graecia" University of Catanzaro, Catanzaro, Italy

Tayla B. McCutcheon, BPsych(Hons) School of Psychology, UNSW Sydney, Sydney, Australia

Benjamin Oosterhoff, PhD Department of Psychology, Montana State University, Bozeman, MT, USA

Klaus-Peter Ossenkopp, PhD Department of Psychology, University of Western Ontario, London, ON, Canada

Paul G. Overton, PhD Department of Psychology, University of Sheffield, Sheffield, UK

Philip A. Powell, PhD School of Health and Related Research, University of Sheffield, Sheffield, UK

Gemma Reynolds, PhD Department of Psychology, Middlesex University, London, UK

Lisa M. Reynolds, PhD Department of Psychological Medicine, University of Auckland, Auckland, New Zealand

Sarah A. Rubenstein, BPsych(Hons) School of Psychology, UNSW Sydney, Sydney, Australia

Barış Sevi, MSc Department of Psychological Sciences, University of Connecticut, Storrs, CT, USA

Natalie J. Shook, PhD School of Nursing, University of Connecticut, Storrs, CT, USA

Jane Simpson, PhD Division of Health Research, Lancaster University, Lancaster, UK

Joshua M. Tybur, PhD Department of Experimental and Applied Psychology, Vrije Universiteit Amsterdam, Amsterdam, Netherlands

Lenny R. Vartanian, PhD School of Psychology, UNSW Sydney, Sydney, Australia

Ana B. Vivas, PhD Department of Psychology, CITY College, University of York Europe Campus, Thessaloniki, Greece

Part I
Introduction

Chapter 1
Researching the Revolting: An Introduction to the Handbook

Philip A. Powell and Nathan S. Consedine

The Psychology of Disgust: A Brief History

The first scientific exposition of disgust is broadly credited to the *Expression of the Emotions in Man and Animals*. Writing in the nineteenth century, Darwin notes:

> *Disgust is a sensation rather more distinct in its nature, and refers to something revolting, primarily in relation to the sense of taste, as actually perceived or vividly imagined; and secondarily to anything which causes a similar feeling, through the sense of smell, touch, and even of eyesight.* (Darwin 1872/2009, p. 250).

Vividly describing how disgust manifests in a way that functionally facilitates the ejection of things from the mouth—contracted nostrils, an open mouth with a retracted upper lip and lower lip protrusion—Darwin presaged the special relationship between disgust and oral incorporation, which would come to influence researchers' thinking over a century later.

For the next 100 years, however, scholarly work on disgust was confined to sporadic musings, largely reflecting currently dominant schools of thought. At the turn of the twentieth century, Sandor Ferenczi explained a child's *Disgust for Breakfast*, as "a displacement of an unconscious disgust for the maternal hand", which "prepared the breakfast with the same hands" that "played a part in those repudiated activities" in the "sexual relations of his parents" (Ferenczi 1919, p. 326). When behaviourism dominated, Abraham Maslow in *The "Emotion" of Disgust in Dogs*, ascribed dogs' rejection of consuming other dog flesh to odour (they ate it when rendered odourless) (Maslow 1932). In the 1940s, Andras Angyal (1941)

P. A. Powell (✉)
School of Health and Related Research, University of Sheffield, Sheffield, UK
e-mail: p.a.powell@sheffield.ac.uk

N. S. Consedine
Department of Psychological Medicine, University of Auckland, Auckland, New Zealand
e-mail: n.consedine@auckland.ac.nz

P. A. Powell, N. S. Consedine (eds.), *The Handbook of Disgust Research*,
https://doi.org/10.1007/978-3-030-84486-8_1

published *Disgust and Related Aversions,* characterising universal disgust elicitors as waste products, reflecting both the "danger of 'parasitic invasion'" (Angyal 1941, p. 407) and "remnants of archaic-animistic notions" (Angyal 1941, p. 402). Finally, as the study of human emotion was progressively legitimated, American psychologist Paul Ekman presented cross-cultural evidence on the universality of disgust expression and recognition (Ekman and Friesen 1971). These works and others would become the direct foundations for a series of more detailed contemporary theories of disgust that were to follow.

The Foundations of Modern Disgust Research

The building blocks of modern disgust research in psychology can be arguably traced back to a series of 1980s studies by Paul Rozin and his colleagues in the context of food rejection. Culminating in a seminal theoretical paper in *Psychological Review* (Rozin and Fallon 1987), these authors presented core ideas in the psychology of disgust, including the primacy of oral concerns, a role for "core" disease vector stimuli, a focus on animals and animal nature as potential disgust elicitors, and disgust's relationship to contamination. To the editors' knowledge, this was also the first work that attempted to systematically describe the acquisition of disgust throughout childhood.

Rozin's subsequent collaborations with Jon Haidt and Clark McCauley saw the development of an influential taxonomy of disgust (referred to as the "RHM model") that has influenced research for almost two decades (Haidt et al. 1994; Rozin et al. 2008). The RHM model charts an evolutionary (via both biological and sociocultural means) broadening of disgust's functioning from its roots in food rejection to several discrete classes of elicitor—"core disgust" responses to potentially infectious agents (e.g., waste products); "interpersonal disgust" responses to unknown others to reduce infection and protect the social order; "animal-nature disgust" responses to reminders of our baseness and animalistic nature (e.g., mortality); and "moral disgust" responses to certain divinity or purity violations (e.g., incest). In contemplating the diverse stimuli that appeared to elicit disgust, Rozin and colleagues noted that they "may have expanded to the point where the only thing they have in common is that decent people want nothing to do with them." (Rozin et al. 2008, p. 771).

Other early proponents of disgust research included Graham Davey and Peter de Jong, who began studying the links between disgust and mental health in the 90s and we are pleased to say have contributed to the current volume (Davey, Chap. 11, this volume; Borg and de Jong, Chap. 9, this volume). In perhaps the earliest paper on the topic, Davey and colleagues provided early evidence for a disease-avoidance model of animal phobias, showing that how easily people were disgusted was linked to fear of particular, non-harmful animals (e.g., maggots and slugs) (Matchett and Davey 1991). Peter de Jong too began his long career in disgust in the study of animal fears, exploring the role of evaluative learning in the link between disgust sensitivity and spider phobias (Merckelbach et al. 1993).

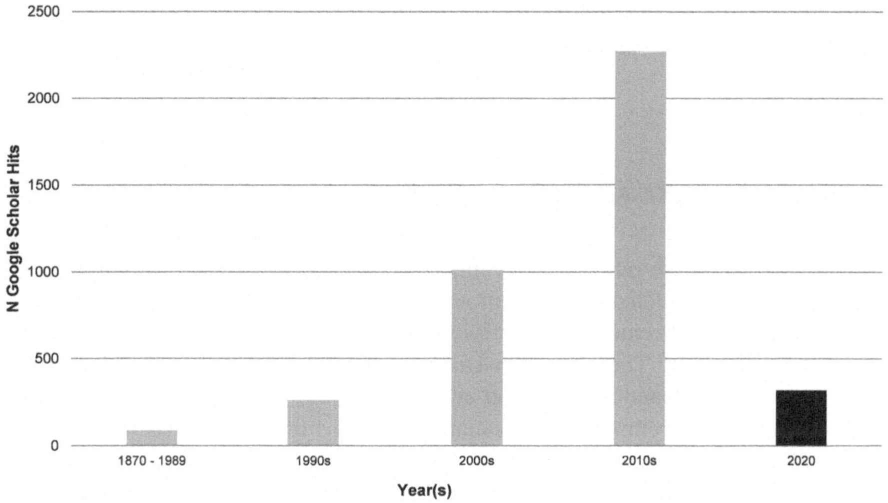

Fig. 1.1 Number of Google Scholar hits with "Disgust" in the title of article from 1870-2020, as of 10th April 2021

Up to this point, however, disgust research mostly remained within the purview of the academic and the academic clinician. This was soon to change. In the 2000s, social psychologists began to work with disgust, applying it to key societal problems. In politics, we discovered that conservatives may be more easily disgusted than liberals (Inbar et al. 2009). In group processes and intergroup relations, we learned that disgust sensitivity is implicated in negative intergroup attitudes (Navarrete and Fessler 2006). In physical health, we found that "disgust sensitive" people with an external bowel (colostomy) were less well adjusted (Smith et al. 2007). Key questions regarding the specificity of disgust's effects began to be considered, with researchers like Roger Giner-Sorolla—a contributor to this handbook (Giner-Sorolla, Chap. 8, this volume)—beginning to disentangle the role of disgust from other related emotions in moral judgment and behaviour (e.g., Gutierrez and Giner-Sorolla 2007). These papers and others became the modern building blocks for a whole host of fascinating research into disgust, a selection of which we are now pleased to share with you in the current volume.

Disgust research has since exploded. An estimated 2000+ scholarly papers were published throughout the 2010s with "disgust" in the title (see Fig. 1.1). Topics have been as wide ranging as sexual behaviour and dysfunction (Borg and De Jong 2012); biases in threat interpretation (Leathers-Smith and Davey 2011); bodily moral disgust (Russell and Giner-Sorolla 2013); disability and prosthesis use (Burden et al. 2018); disgust and fear modelling in children (Askew et al. 2014); cancer (Reynolds et al. 2014); politics and voting behaviour (Shook et al. 2017); pro-environmental decision-making (Powell et al. 2019); obesity stigma (Lieberman et al. 2012); stereotyping and prejudice (Vartanian et al. 2013); disgust and anticipatory nausea in rats (Cloutier et al. 2018); the potential moderation of disgust through mindfulness (Reynolds et al. 2015); body odour (Zakrzewska et al. 2019);

and self-directed disgust in mental health (Powell et al. 2013). This diverse body of work and much more was undertaken by the authors that have contributed to the current volume.

Modern Disgust Research in Its Fifth Decade

As data were acquired, other models, theories, and frameworks for thinking about disgust began to challenge the hegemony of the RHM model and its associated theoretical and measurement assumptions. Perhaps the most well-known of these new approaches is the Three-Domain Disgust (3DD) model of Josh Tybur, Debra Lieberman, and colleagues (Tybur et al. 2009, 2013). Tybur and collaborators critiqued the animal-reminder and social order functions of the RHM model as problematic and re-categorised disgust elicitors into pathogen avoidance, sexual, and moral domains. The emergence of this model has spurred a fascinating (and ongoing) discussion between proponents of the 3DD and RHM models on the role of biological and cultural evolution in shaping disgust (Rozin and Haidt 2013; Tybur 2013). We are pleased to feature Josh Tybur as an author in the present volume (Tybur, Chap. 6, this volume).

As is evident in Fig. 1.1, research into disgust shows no signs of abating. In 2020 alone there were as many papers published featuring the word "disgust" in the title as there were in the entirety of the 1990s. Thus, we are living through a continuing explosion in our understanding of the psychology of disgust and its relevance to the way we function as individuals, as groups, as consumers, and as a species. Perhaps surprisingly then, a general handbook on disgust research has been lacking, with notable exceptions being more focused volumes in clinical (Olatunji and McKay 2009) and moral (Strohminger and Kumar 2018) psychology. In any case, the curation of the current handbook is timely in bringing together international experts in disgust to share the latest findings, insights, and directions of disgust research. Research into disgust continues to be a hugely interesting and, at times, entertaining endeavour. Whether you find yourself spurting a can of *Liquid Ass*™ into a room to induce disgust in controlled experiments (Consedine, Chap. 2, this volume) or creating unique stimuli for surveys examining how disgust relates to people's willingness to pay for and consume novel food and drink, including those containing insects (Powell, Chap. 15, this volume), disgust is a field of research that we would recommend to anyone.

The Content of the Handbook

In this handbook, we are pleased to present a diverse set of works from some of the leading disgust researchers in the world. Contributors were specifically asked to avoid exhaustive, long-winded, and turgid reviews and to concentrate on key issues and the elements of disgust research they find most provocative and exciting. Rightly or wrongly, disgust researchers find the study of that which is icky deeply exciting and we wanted to be sure that this enthusiasm was evident. All chapters were peer-reviewed by two independent contributors to the current volume, which has helped to assure their quality.

Theoretical Issues in Disgust

The first half of the book focuses on theoretical issues in disgust. In the next chapter, Nathan Consedine presents an up-to-date perspective on methods to measure and induce disgust in psychological research. In addition to demonstrating the considerable design creativity in disgust research, his chapter reminds us that measurement is complex and not independent of our underlying theoretical assumptions (Consedine, Chap. 2, this volume). In the subsequent chapter, Hannah Bradshaw and Jeffrey Gassen chart the evolution of disgust, detailing the emergence and shaping of disgust as a consequence of our evolutionary arms race with pathogens and introducing its role as the affective component of the "behavioural immune system" (Bradshaw and Gassen, Chap. 3, this volume). Following this, Martin Kavaliers and colleagues provide an interesting analogue on the role and expression of pathogen and toxin disgust in a non-human animal: the rodent (Kavaliers et al., Chap. 4, this volume). In Chap. 5, Gemma Reynolds and Chris Askew explore the development and acquisition of disgust as a result of direct and indirect learning mechanisms, including its role in the development of fear in children (Reynolds and Askew, Chap. 5, this volume). In Chap. 6, Josh Tybur elucidates people's individual differences in sensitivity to disgust, how we measure and interpret these, their implications, and the latest developments in the field (Tybur, Chap. 6, this volume). In the penultimate chapter of this section, Marco Tullio Liuzza takes us on a tour of the role olfaction plays in disgust and disease avoidance, with a focus on gender differences, aging, and sexual behaviour (Liuzza, Chap. 7, this volume). In the section's final chapter, Roger Giner-Sorolla discusses the paradox of moral disgust, including the oxymoronic nature of the term and how the apparent paradox may be resolved through careful theorising and research (Giner-Sorolla, Chap. 8, this volume).

Disgust and Its Applications

The second half of the handbook then shifts its focus to consider how disgust may be implicated in key applied issues. In the first chapter of this section, Charmaine Borg and Peter de Jong discuss work on sexual disgust as the antithesis of sexual arousal, charting the ostensibly opposing evolutionary goals of avoiding infection risk while engaging in functional and/or pleasurable sexual behaviour (Borg and de Jong, Chap. 9, this volume). Next up, Lenny Vartanian and colleagues concentrate on the role disgust plays in prejudice, reviewing the evidence, mechanisms, and unanswered questions regarding this link (Vartanian et al., Chap. 10, this volume). In Chap. 11, Graham Davey provides a commanding overview of the mechanistic links between disgust and mental health problems, with reference to three decades of research (Davey, Chap. 11, this volume). In Chap. 12, Paul Overton and collaborators provide a neural model of the "disgust system" and relate potential deficits in this system to a wide range of neurological disorders, including epilepsy and Parkinson's disease (Overton et al., Chap. 12, this volume). In Chap. 13, Lisa Reynolds and Vinayak Dev discuss the particular relevance of disgust as a health-based emotion and explore the implications disgust has for adaptation to chronic health conditions (Reynolds and Dev, Chap. 13, this volume). In the penultimate chapter in the section on applications, Natalie Shook and colleagues broaden our lens to review how disgust moved from playing a role in disease (avoidance) to playing a role in Western democracies, cataloguing the current state and future directions for disgust in politics (Shook et al., Chap. 14, this volume). In the last chapter on applications, Philip Powell maintains this contemporary focus, offering a comprehensive overview of how disgust can both constrain and motivate consumer behaviour, disgust's links to sustainability, and why certain consumption habits elicit repugnance (Powell, Chap. 15, this volume). The volume concludes with an afterword by the editors in which we touch on the continuing relevance disgust has for the contemporary issues confronting humanity—pandemic threat, climate change and sustainability, and the increasingly divided nature of societal exchange. Collated under the shadow of the COVID pandemic, editing this work has been challenging but it has nonetheless been a labour of love. We hope you enjoy it.

References

Angyal, A. (1941). Disgust and related aversions. *The Journal of Abnormal and Social Psychology, 36*(3), 393. https://doi.org/10.1037/h0058254

Askew, C., Çakır, K., Põldsam, L., & Reynolds, G. (2014). The effect of disgust and fear modeling on children's disgust and fear for animals. *Journal of Abnormal Psychology, 123*(3), 566. https://doi.org/10.1037/a0037228

Borg, C., & De Jong, P. J. (2012). Feelings of disgust and disgust-induced avoidance weaken following induced sexual arousal in women. *PloS ONE, 7*(9), e44111. https://doi.org/10.1371/journal.pone.0044111

Burden, N., Simpson, J., Murray, C., Overton, P. G., & Powell, P. A. (2018). Prosthesis use is associated with reduced physical self-disgust in limb amputees. *Body Image, 27*, 109-117. https://doi.org/10.1016/j.bodyim.2018.08.001

Cloutier, C. J., Zevy, D. L., Kavaliers, M., & Ossenkopp, K.-P. (2018). Conditioned disgust in rats (anticipatory nausea) to a context paired with the effects of the toxin LiCl: Influence of sex and the estrous cycle. *Pharmacology Biochemistry and Behavior, 173*, 51-57. https://doi.org/10.1016/j.pbb.2018.08.008

Darwin, C. (1872/2009). *The Expression of the Emotions in Man and Animals*. Harper Perennial.

Ekman, P., & Friesen, W. V. (1971). Constants across cultures in the face and emotion. *Journal of Personality and Social Psychology, 17*(2), 124. https://doi.org/10.1037/h0030377

Ferenczi, S. (1919). Disgust for breakfast. *Further Contributions to Psychoanalysis*, 326.

Gutierrez, R., & Giner-Sorolla, R. (2007). Anger, disgust, and presumption of harm as reactions to taboo-breaking behaviors. *Emotion, 7*(4), 853. https://doi.org/10.1037/1528-3542.7.4.853

Haidt, J., McCauley, C., & Rozin, P. (1994). Individual differences in sensitivity to disgust: A scale sampling seven domains of disgust elicitors. *Personality and Individual Differences, 16*(5), 701-713. https://doi.org/10.1016/0191-8869(94)90212-7

Inbar, Y., Pizarro, D. A., & Bloom, P. (2009). Conservatives are more easily disgusted than liberals. *Cognition & Emotion, 23*(4), 714-725. https://doi.org/10.1080/02699930802110007

Leathers-Smith, E., & Davey, G. C. (2011). The disgust threat interpretation bias is not moderated by anxiety & disgust sensitivity. *Journal of Experimental Psychopathology, 2*(1), 63-76. https://doi.org/10.5127/jep.007410

Lieberman, D. L., Tybur, J. M., & Latner, J. D. (2012). Disgust sensitivity, obesity stigma, and gender: Contamination psychology predicts weight bias for women, not men. *Obesity, 20*(9), 1803-1814. https://doi.org/10.1038/oby.2011.247

Maslow, A. (1932). The "emotion" of disgust in dogs. *Journal of Comparative Psychology, 14*(3), 401. https://doi.org/10.1037/h0074405

Matchett, G., & Davey, G. C. (1991). A test of a disease-avoidance model of animal phobias. *Behaviour Research and Therapy, 29*(1), 91-94. https://doi.org/10.1016/s0005-7967(09)80011-9

Merckelbach, H., de Jong, P. J., Arntz, A., & Schouten, E. (1993). The role of evaluative learning and disgust sensitivity in the etiology and treatment of spider phobia. *Advances in Behaviour Research and Therapy, 15*(4), 243-255. https://doi.org/10.1016/0146-6402(93)90011-P

Navarrete, C. D., & Fessler, D. M. (2006). Disease avoidance and ethnocentrism: The effects of disease vulnerability and disgust sensitivity on intergroup attitudes. *Evolution and Human Behavior, 27*(4), 270-282. https://doi.org/10.1016/j.evolhumbehav.2005.12.001

Olatunji, B. O., & McKay, D. (2009). *Disgust and its disorders: Theory, assessment, and treatment implications*. American Psychological Association.

Powell, P. A., Jones, C. R., & Consedine, N. S. (2019). It's not queasy being green: The role of disgust in willingness-to-pay for more sustainable product alternatives. *Food Quality and Preference, 78*, 103737. https://doi.org/10.1016/j.foodqual.2019.103737

Powell, P. A., Simpson, J., & Overton, P. G. (2013). When disgust leads to dysphoria: A three-wave longitudinal study assessing the temporal relationship between self-disgust and depressive symptoms. *Cognition & Emotion, 27*(5), 900-913. https://doi.org/10.1080/02699931.2013.767223

Reynolds, L. M., Lin, Y. S., Zhou, E., & Consedine, N. S. (2015). Does a brief state mindfulness induction moderate disgust-driven social avoidance and decision-making? An experimental investigation. *Journal of Behavioral Medicine, 38*(1), 98-109. https://doi.org/10.1007/s10865-014-9582-5

Reynolds, L. M., McCambridge, S. A., Bissett, I. P., & Consedine, N. S. (2014). Trait and state disgust: An experimental investigation of disgust and avoidance in colorectal cancer decision scenarios. *Health Psychology, 33*(12), 1495. https://doi.org/10.1037/hea0000023

Rozin, P., & Fallon, A. E. (1987). A perspective on disgust. *Psychological Review, 94*(1), 23. https://doi.org/10.1037/0033-295X.94.1.23

Rozin, P., & Haidt, J. (2013). The domains of disgust and their origins: Contrasting biological and cultural evolutionary accounts. *Trends in Cognitive Sciences, 17*(8), 367-368. https://doi.org/10.1016/j.tics.2013.06.001

Rozin, P., Haidt, J., & McCauley, C. R. (2008). Disgust. In M. Lewis, J. M. Haviland-Jones & L. F. Barrett (Eds.), *Handbook of* emotions (3rd *ed*., pp. 757-776). The Guilford Press.

Russell, P. S., & Giner-Sorolla, R. (2013). Bodily moral disgust: what it is, how it is different from anger, and why it is an unreasoned emotion. *Psychological Bulletin, 139*(2), 328. https://doi.org/10.1037/a0029319

Shook, N. J., Oosterhoff, B., Terrizzi Jr, J. A., & Brady, K. M. (2017). "Dirty politics": The role of disgust sensitivity in voting. *Translational Issues in Psychological Science, 3*(3), 284. https://doi.org/10.1037/tps0000111

Smith, D. M., Loewenstein, G., Rozin, P., Sherriff, R. L., & Ubel, P. A. (2007). Sensitivity to disgust, stigma, and adjustment to life with a colostomy. *Journal of Research in Personality, 41* (4), 787-803. https://doi.org/10.1016/j.jrp.2006.09.006

Strohminger, N., & Kumar, V. (2018). *The moral psychology of disgust*. Rowman & Littlefield.

Tybur, J. M. (2013). *Cultural and biological evolutionary perspectives on disgust—oil and water or peas and carrots?* http://www.epjournal.net/blog

Tybur, J. M., Lieberman, D., & Griskevicius, V. (2009). Microbes, mating, and morality: individual differences in three functional domains of disgust. *Journal of Personality and Social Psychology, 97*(1), 103. https://doi.org/10.1037/a0015474

Tybur, J. M., Lieberman, D., Kurzban, R., & DeScioli, P. (2013). Disgust: evolved function and structure. *Psychological Review, 120*(1), 65. https://doi.org/10.1037/a0030778

Vartanian, L. R., Thomas, M. A., & Vanman, E. J. (2013). Disgust, contempt, and anger and the stereotypes of obese people. *Eating and Weight Disorders-Studies on Anorexia, Bulimia and Obesity, 18*(4), 377-382. https://doi.org/10.1007/s40519-013-0067-2

Zakrzewska, M., Olofsson, J. K., Lindholm, T., Blomkvist, A., & Liuzza, M. T. (2019). Body odor disgust sensitivity is associated with prejudice towards a fictive group of immigrants. *Physiology & Behavior, 201*, 221-227. https://doi.org/10.1016/j.physbeh.2019.01.006

Part II
Theoretical Issues in Disgust

Chapter 2
This Is How We Do It: Inductions, Methods, and Measurement in Disgust Research

Nathan S. Consedine

As indicated by the collection of works in this volume, research now suggests that disgust—a "forgotten" emotion—regarding objects, animals, foods, and the behaviour of others is an important influence on how people feel and act. Disgust predicts health behaviours, indices of mental health, and political attitudes. We have evidence from cross-sectional studies, prospective designs, and experimental studies, but what do we really know? More technically it seems, we know that self-report indices of a construct we think is probably disgust is correlated with certain outcomes. We know that when we show people images or videos we think are disgusting (or others have told us are disgusting), they score on our scales (and sometimes behave) in particular ways. Thus, in a very real sense, our confidence in the conclusions of those working with disgust is predicated on our confidence in the ways in which they induce, manipulate, and measure this "thing" we call disgust.

In contributing to what is now a fast-moving area of study, the following chapter will review and critically evaluate the common methodologies social scientists use in the conduct of disgust research. The purpose of this review is to "take stock" in how we study disgust, evaluate how our measures do (and do not) fit with our theories, and use these observations to provide guidance for ongoing methodological thought and development in the area. The chapter is broadly organised into two major sections—a section on the methods we use to induce (and measure) disgust in the laboratory and a section on how we study what appear to be relatively stable differences across people in the general tendency to get disgusted (or experience disgust as difficult). Observations, criticisms, and future directions are provided throughout the text. Note that *cross-population methods* such as those used in the study of the behavioural immune system are not covered here (see Tybur et al. 2014 for a recent overview).

N. S. Consedine (✉)
Department of Psychological Medicine, University of Auckland, Auckland, New Zealand
e-mail: n.consedine@auckland.ac.nz

P. A. Powell, N. S. Consedine (eds.), *The Handbook of Disgust Research*,
https://doi.org/10.1007/978-3-030-84486-8_2

The Measurement of Dispositional/Trait Disgust via Self-Report

Overview and Description

As is common in the early studies of individual differences, most empirical work on disgust has concentrated on linking self-reported trait measures of responses to potentially disgusting stimuli to attitudinal, behavioural, and/or mental health outcomes. The list of outcomes associated with trait disgust as measured in this way is truly remarkable but would (in brief) include food preferences (e.g., Powell et al. 2019), taste sensitivities (Schienle and Schlintl 2020), cancer screening frequency (Reynolds et al. 2018), cancer treatment side effects (Dev et al. 2020), the stigmatisation of people with cancer (Azlan et al. 2020), and mating strategies (Al-Shawaf et al. 2015), as well as attitudes towards immigrants, Muslims (Brenner and Inbar 2015), homosexuals (Kiss et al. 2020), and obese individuals (Vartanian et al. 2016; see also Vartanian et al., Chap. 10, this volume). Measures are slowly being adjusted for new groups such as children (Rottman et al. 2019), with recent adaptations to adult instruments (Muris et al. 2012) as well as novel child-specific measures (Viar-Paxton et al. 2015).

So far, so good. At a dispositional level, the tendency to get disgusted looks like it's "a thing," a stable aspect of functioning that is associated with outcomes in theoretically sensible ways. Our confidence in these findings is, however, only as good as the evidence and interpretations regarding the measures of dispositional disgust we employ. It is also worth remembering that disgust propensity and sensitivity (the "experience" of disgust) are latent constructs insofar as they may exist but cannot be directly measured. All self-report (and many other indices of disgust responding) must thus be thought of as representing what are sometimes termed "manifest" or "indicator" variables. They are thus indices of the latent disgust construct but inevitably measure it with a degree of error. Put simply, measures of a construct like disgust do not only measure this construct but also capture variation of other kinds. This places very real constraints on how confident we can be.

Below, the major measures of dispositional or trait disgust are briefly considered. Given the availability of good reviews in this area (Tybur, Chap. 6, this volume; Tybur and Karinen 2018), rather than exhaustively compile prior psychometric data (or advocate for one measure versus another), this information is used to illustrate the considerations that researchers might contemplate when choosing an instrument. In parallel, it serves as a reminder that the way in which we measure (or induce) disgust, and the confidence we have in interpretations, is interwoven with our notions regarding what we think disgust is, how it is instantiated or manifest in psychological (and other processes), and what we think it is for.

In addition to early trait discrete emotion measures measuring the tendency to feel a range of emotions (e.g., the Differential Emotions Scale [DES]; Izard et al. 1974), the last three decades have seen a steady growth in the availability of disgust-specific trait measures. Most are what we might term "domain" based, maximising scale content validity and interpretability by assessing disgust as it arises in response to different classes of elicitor. In general, content in these measures have been developed based on a combination of self-reported disgust in response to theoretically viable elicitors (rather than theory per se), although some have placed a greater emphasis on sensory (e.g., Schienle et al. 2020) or theoretical considerations (e.g., Curtis and de Barra 2018). Established measures are described more fully below.

The Disgust Scale (DS; Haidt et al. 1994) was the first of the more focal and detailed indices. Developed on the basis of self-reported elicitors, the scale is a mixture of true–false and disgust-rating questions grouped into seven domains (food, animals, body products, sex, body-envelope violations, death, and hygiene). It has been adapted several times (see Tybur et al. 2014 for details), with each variant employing slightly different component and scoring procedures. The Disgust Scale–Revised (DS-R; Olatunji et al. 2007) removed seven items (including most of the sexual items) resulting in a more robust three-component—core, animal-reminder, and contamination disgust—structure that became the most widely used measure for several years. Perhaps in part because of structural concerns and the omission of moral disgust from the DS-R, the Three-Domain Disgust Scale was later developed (TDDS; Tybur et al. 2009). For this measure, participants use a 1 (not at all disgusting) to 7 (extremely disgusting) scale to rate 21 items from three domains—pathogen, sexual, and moral disgust.

Sitting slightly off to one side of the domain-based measures is a measure based around a different model of how disgust may manifest in dispositional functioning. Originally presented in 2006, the Disgust Sensitivity and Propensity Scale (DPSS; van Overveld et al. 2006), and the subsequent adaptation in the Disgust Propensity and Sensitivity Scale-Revised (DPSS-R; van Overveld et al. 2010), measures how easily an individual is disgusted (their disgust *propensity*) as well as how unpleasant they find the experience of disgust (disgust *sensitivity*).

Evaluating Measures of Dispositional Disgust: Things to Consider

Across the different measures, face validity and internal reliabilities are generally acceptable, as are studies of convergent and criterion validity (e.g., Karinen et al. 2021; Rozin et al. 1999), and test-retest reliability (e.g., Polak et al. 2019). There are relatively few studies of test-retest reliability. Available data suggest varied stability ranging from .48 for moral disgust to .79 for sexual disgust across 12 weeks (Olatunji et al. 2012), although timeframes are quite short (e.g., Jones et al. 2018). Nonetheless, studies to date seem to suggest that the items we typically think of as indexing disgust sensitivity are reliably grouped together and that scores are consistent across assessments. Conversely, perhaps the most serious challenges for trait disgust measurement lie in ongoing questions regarding issues of structural and discriminant validity. These issues are briefly considered below.

As a starting point, it is worth noting that trait measurements vary in the number of underlying components they suggest; between two (the DPSS-R) and five (the new German 5-FES, Eickmeier et al. 2019) components have been suggested. In considering what this variation might mean for the study of disgust, it is initially worth recalling that an instrument's structure reflects the empirical regularities across self-reported data points while the labelling of the resultant groupings reflects particular theoretical interests. Empirically, however, we are talking about groupings (subscales) of different disgust items that covary in some systematic way. The groupings that result from component analyses thus mostly tell us that people who are more (or more frequently) disgusted by a certain stimulus are also likely to report being more disgusted by another. Nothing more and nothing less.

There are, however, many reasons that individual disgust items will covary—response sets, social desirability, participant ability to discriminate, item phrasing and proximity, and semantic overlaps, to name but a few—that have nothing to do with the content itself (Podsakoff et al. 2012). Thus, although it is often forgotten, using self-report to generate the content of measurement (and data reduction to generate the structure) can lead to the development of measures that then "echoes" through the subsequent structuring of our theories—a form of what we might term *measurement primacy*. The "animal reminder" items in the DS-R, for example, shows clear reference to Rozin's earlier theoretical work, but might well be labelled and interpreted differently in the context of the TDDS. More broadly, while measurement considerations are clearly central to the scientific endeavour, they are not (and should not be) the only basis upon which to theorise about disgust. Think Freud's unconscious, plate tectonics, and, potentially, string theory—hugely useful theories even if unequivocal proof was initially (or remains) lacking.

The measure put forth by Curtis and de Barra (2018) provides an example of how difficult reconciling theoretical and psychometric considerations can be in disgust research. Their theoretical model suggested that elicitors should be grouped in a manner corresponding to six distinct pathogen transmission pathways. Factor analyses did not support the model, perhaps suggesting that the disgust system(s) are not structured in this way. Indeed, I am not suggesting that they are. What I am suggesting is that *even if* the underlying disgust systems *were* structured in this way, there is no guarantee that this structure would be reflected in the structure of a self-report measure (see also Armstrong et al. 2021 on this and related issues). Clearly, we need reliability, factorial generalisability, and construct validity in our measures. Equally clearly, however, we should not forget that measures are, first and foremost, a tool and a frequently imperfect one at that. Choose wisely and avoid being bewitched by internal reliabilities and structure.

Relatedly, it is worth noting that while most trait measures are *content* based, content groupings are not the only possible way to structure elements of trait disgust. The DPSS-R, for example, is purported to index how easily an individual is disgusted (*propensity*) as well as how unpleasant they find it (*sensitivity*). Although the distinction is theoretically plausible and appears to have utility in clinical samples, like the content-based measures (Campbell et al. 2020), studies of the DPSS variants' structure have returned somewhat mixed results (e.g., three versus the expected two components for the DPSS-12 in Goetz et al. 2013), with at least some evidence suggesting shorter versions may be more robust (Fergus and Valentiner 2009).

A second concern with the measurement of trait disgust regards separating it from other aspects of dispositional functioning. Although there are exceptions (see Tybur, Chap. 6, this volume), measures of trait disgust are reliably associated with other "negative" aspects of personality functioning (Oosterhoff et al. 2018; Tybur and Karinen 2018). In one meta-analysis, for example, disease avoidance traits (including both trait disgust and germ aversion) were associated with neuroticism, conscientiousness, openness to experience, and extraversion (Oosterhoff et al. 2018). Other studies have linked disgust sensitivity with political conservatism, anxiety, trait negative affectivity, depression, orderliness, and intuitive thinking, to name but a few. Perhaps more problematically, emotions tend to covary (Consedine and Moskowitz 2007). Both trait and state disgust covary with multiple emotions including fear, anxiety, embarrassment, and anger (McCambridge and Consedine 2014; Muris et al. 2008; Russell and Giner-Sorolla 2013). The question, of course, is how serious these associations are and how should we think about what they mean.

The magnitude of the correlations between disgust and other indices varies across studies. Some data suggest moderate correlations between indices of trait disgust and neuroticism (e.g., Druschel and Sherman 1999), while others suggest correlations less than .20 between disgust/germ aversion and personality factors (Oosterhoff et al. 2018). However, even if we are willing to tolerate this level of covariation as either (a) reflecting shared method characteristics or (b) inconsequential (both possibilities), there is nonetheless reason to be cautious. First, the problem may be more pronounced in particular domains, notably in studies of moral judgment where anger and disgust are often quite closely associated (e.g., Olatunji et al. 2012; see also Giner-Sorolla, Chap. 8, this volume) and in health (below).

Second, there is a sense in which contrasting indices of trait disgust with global personality or affective measures risks being a straw man test of what the problem actually is; demonstrating that trait disgust is only modestly correlated with *global* aspects of personality functioning is not a particularly convincing test of discriminant validity. Predictively, it may be that some specific (rather than general) emotional factor is conflating findings, but that its effects are "hidden" within the noise of a general measure of negative affectivity or neuroticism. Certainly in studies of emotion, our studies psychometrically "bundle" potential affective confounds within measures of general negative affectivity and then demonstrate that either that (a) disgust is only weakly associated with negative affectivity or (b) that disgust is a better predictor of outcomes (e.g., Olatunji et al. 2011). My suggestion here is that rather than uncritically assume our measures of disgust are separate from other measures or bundle multiple affective constructs together as some form of "control" variable, we should be identifying specific and likely confound candidates and carefully measuring them.

This issue has been grappled with for some time in studies of avoidance in healthcare. The problem here is that the same stimuli that elicit disgust in health settings (e.g., a stool collection procedure, some sort of bodily insertion) also concurrently elicit embarrassment and fear (and perhaps shame). Of equal import, all of these emotions appear to have evolved, at least in part, because they promote avoidance (Consedine and Moskowitz 2007). Being sure that it is disgust specifically that is responsible for avoidance in such scenarios is thus exceedingly difficult, both conceptually and in terms of measurement. In one study, for example, Reynolds et al. (2018) found that fear and embarrassment regarding bowel screening were correlated with indices of faecal and insertion screening disgust at levels between .50 and .80 and with trait negative affects at approximately between .25 and .40; all emotion indices were correlated with screening outcomes. Thus, while these particular data reflect feelings about very specific stimuli (rather than general dispositions), they also show how multiple, closely-related emotional responses can potentially underpin the same outcome.

In making this comment, I am by no means suggesting that our measures of trait disgust are not dispositional. They probably are. Recent evidence suggests that self-other agreement is quite high (r's of .36, .46, and .66 for moral, pathogen, and sexual disgust sensitivity, respectively), implying that self-reported sensitivities are reliably detected by others (Karinen et al. 2021). I am, nonetheless, suggesting that we must not become complacent in our thinking about the problem posed by discriminant validity where emotional responding is concerned. A significant number of studies simply measure elements of trait disgust and correlate scores with outcome without considering the very real conceptual and practical implications that walk hand-in-hand with the associations between disgust and other emotional responses. Evidence from studies of morality and health suggest an ongoing need to test the discriminant validity of dispositional measures of disgust from other specific elements of trait emotionality (rather than general aspects).

Domain Specific Disgust Measurement

As disgust measurement has matured and researchers have begun to more systematically pin down how disgust relates to outcomes in particular domains, the literature has also seen the parallel emergence of domain specific disgust measures. In some ways this development is the logical extension of the "content" or domain-based measurement that predominates, the assumption being that disgust regarding particular stimuli should (on balance) better predict behaviour or attitudes to that stimuli than more general measures. To the extent that emotions likely evolved to motivate behaviour regarding the elicitor (Consedine and Moskowitz 2007), the possibility that disgust sensitivities in particular domains are differentially associated with the motivation to avoid specific classes of elicitor is important for both theoretical and practical reasons.

In this line, measures have been developed to assess disgust specifically regarding *food* using either verbal self-report (Hartmann and Siegrist 2018) or pictures (Ammann et al. 2018) and there are multiple measures in colorectal cancer. Several groups have used the Emotional Barriers to Bowel Screening scale (EBBS, used by

Davis et al. 2017, and Reynolds et al. 2018) which provides separate indices of faecal and insertion disgust. Others have developed the ICK-C scale (used by O'Carroll et al. 2015, and Chambers et al. 2016) or, more recently, a two component (core and interpersonal disgust) measure of colostomy specific disgust (Jin et al. 2020). Interestingly, early evidence suggests some utility to more specific measures. In one study, for example, disgust regarding having something inserted into the body (but not faecal disgust, embarrassment, or fears about outcome) predicted the odds of having had an invasive bowel test in the last 5 years (Reynolds et al. 2018). Although such findings may reflect covariation created by shared measurement characteristics or language, they also stand in contrast to experimental findings which tend to suggest a relatively domain neutral effect to disgust inductions. The question of how *precise* the avoidant, risk-minimising adaptations comprising the disgust system are remains an area of active inquiry. Do indices of disgust regarding specific elicitors actually index a construct that is distinct from general sensitivities and are they a better predictor of outcomes? Answering such questions has the potential to substantially advance our thinking not only about measurement but also about the nature of disgust itself.

Concluding Remarks

Overall, while these comments may raise as many questions as they answer, the measurement of trait disgust has come no small distance—quiet confidence in the increasing maturity of measurement is warranted. Further, the sorts of considerations we should be giving weight to in choosing a measure are becoming increasingly clear. In addition to considering basic psychometric indices, researchers should use contextual theory and their specific research foci to determine the instrument best-suited to testing particular hypotheses (see Tybur et al. 2014). Further, it seems increasingly likely that particular measures, tapping into particular sensitivities or content, will be better suited to predicting (and perhaps explaining) variation in particular types of attitudinal and behavioural outcomes.

Inducing and Measuring Disgust in the Laboratory

Overview and Characterisation

In part because disgust is thought to produce avoidance in both physical (Reynolds et al. 2019) and mental (Deacon and Olatunji 2007) health contexts, experimental studies in which disgust is validly and discriminantly *induced* are critical to the advancement of the field. The question, of course, is how to go about inducing and measuring disgust in ways that are best suited to answering the complex questions disgust research entails. Perhaps because nearly all of the pathogens that disgust evolved to counteract are too small to be directly observed, humans rely on indirect (and imperfect) sensory cues for their potential existence (Rottman et al. 2019). Textures, smells, noises, tastes, and visual cues that map onto aspects of the stimuli that historically posed a contamination threat are thus all potential induction modalities for the disgust researcher. More to the point in terms of the current chapter, and as in emotion inductions in general (see Siedlecka and Denson 2019), each means of inducing disgust has a mixture of conceptual and design advantages, some of which can be considered in advance (see Table 2.1).

Although mixed manipulations have been undertaken, most inductions occur via a single sensory modality, typically via visual (Culpepper et al. 2018) or olfactory routes (e.g., Reynolds et al. 2014; Tybur et al. 2011; see also Liuzza, Chap. 7, this volume). Cover stories designed to increase attention or post-hoc checks are sometimes used. Newer technologies have been incorporated, including use of virtual reality visual paradigms (e.g., Ammann et al. 2020) or pseudo-word stimuli (e.g., Darcy and Fontaine 2020) to elicit disgust. Other studies have used olfactory inductions (e.g., McCambridge and Consedine 2014; Reynolds et al. 2014; Tybur et al. 2011) or autobiographical narrative/imagination-based inductions that have people imagine contact with disease threats, such as mucus (e.g., White et al. 2013). Few studies have manipulated disgust through auditory, gustatory, or tactile means (although see Oum et al. 2011).

Content in visual stimuli includes wounds, rotten foods, vomit, faeces (e.g., Ammann et al. 2020), bodily products (Shenhav and Mendes 2014), watching a video of people eating animal eyeballs (Aldao and Wisco 2015), and similarly appetising entertainments. Pictures drawn from the International Affective Picture System (IAPS; Lang et al. 1999) are used, as are sets of images devised by Curtis et al. (2004), Haberkamp et al. (2017), and Culpepper et al. (2018). Other researchers use idiosyncratically developed sets with content varying from depictions of "immoral" sexuality, insects, bodily products, or disgusted faces. Film stimuli are similarly diverse and have included dissections, faeces or vomit, and painful injuries (e.g., Vianna and Tranel 2006). Sometimes visual cues are juxtaposed with information about disease threats (e.g., Ackerman et al. 2009).

Evaluating Disgust Induction Modalities: Things to Consider

As might be expected, different induction modalities offer distinct blends of advantages and disadvantages to researchers. Compared to other induction modalities, visual stimuli are advantaged insofar as they are available in standardised and previously validated sets, can be directly assessed for face validity, permit high control over conceptual content (e.g., contamination versus sexual stimuli), are remotely deliverable, and are potentially capable of being "dosed" and/or sustaining a disgust response for a period of time (see Simpson et al. 2006 on the issue of elicitation modality and time course). Conversely, however, designs employing visual stimuli are likely weaker in terms of participant blindness and demand.

Table 2.1 Preliminary assessment of design strengths and weaknesses of disgust induction modalities

Modality		Efficacy	Availability	Standardisation or control	Participant blindness	Experimenter blindness	Can be dosed?
Visual	Pictures	++	++	++	–	++	Probably
	Video	++	++	++	–	++	Probably
Olfactory		++	++	++	+	–	Probably
Audio		?	?	++	?	?	?
Tactile		?	?	++	?	?	?
Gustatory		+	?	++	?	?	?
Mental	Imaginary	++	++	+	–	++	Possibly
	Narrative/recall	++	++	– –	– –	++	?

Note: ++ = strong; + = modest; – = poor; – – = weakest; ? = unknown; comments on dosing are speculative

Ratings demonstrating the effect of the experimental manipulation are typically made regarding either stimuli themselves (i.e., how do you feel about these pictures?) or about feelings during stimuli exposure (i.e., how did you feel when you were looking at the pictures?), increasing the odds that participants become aware of and respond to experimental expectation. Put simply, it may be difficult to prevent participants gradually coming to realise that disgust responses are central to the study when stimuli are obviously curated to elicit disgust. Attempting to hide this fact via cover stories and/or burying disgust terms in longer lists of affective adjectives are the most common attempts to mitigate this problem but how successful they are is unknown. Despite these limitations, visual elicitation of disgust is the most common modality and, at least according to some writers, the most effective way to manipulate disgust in a controlled manner (Siedlecka and Denson 2019).

Induction of disgust via olfaction is increasingly common. In olfactory induction studies, the manipulation is often covert or not explained/justified to participants who, at least in theory, are less likely to suspect that odours form part of the experiment. Studies of this kind include Schnall et al. (2008) in which olfactant was sprayed into bags lining an outdoor trash receptacle, a technique also used in several laboratory studies (e.g., Fong and Sündermann 2020; McCambridge and Consedine 2014; Perone et al. 2020; Reynolds et al. 2014, 2015; Rottman and Kelemen 2012; Tybur et al. 2011). The most commonly used olfactant is a novelty spray called "Liquid Ass" ™ which is non-hazardous but has a distinct faecal aroma. Three sprays of this product appears to be the modal "dose" but variations in space, ventilation, and participant proximity to the olfactant necessitate pilot testing.

Compared to visual inductions, inducing disgust via odour has several distinct advantages. With careful titration of the dose (pilot testing is a must!), it increases the odds of maintaining participant blindness, though (for obvious reasons) experimenter blindness can be difficult to maintain. Depending on the specific research foci, it may also be that there are advantages to eliciting disgust "directly" rather than requiring (as is the case with visual inductions) some cognitive mediation; detecting likely pathogen presence via olfaction may be taken as indexing the presence of a disease threat with an immediacy that is lacking in visual images. Finally, it is possible that olfactory manipulations offer a more precise manipulation in which disgust alone is impacted (e.g., McCambridge and Consedine 2014). Reynolds et al. (2014), for example, used an olfactory manipulation and specifically tested for effects on other avoidance-producing emotions; there were no effects on either fear or embarrassment.

However, in contrast to visual inductions, olfactory inductions may see the experimenter lose control over the duration of the exposure (via dissipation or, potentially, olfactory habituation processes) as well as the "content" of the disgust elicitor, something that may be important when grappling with questions in specific domains (discussed below). In addition, the extent to which odours elicit disgust may be conflated with individual differences in olfactory sensitivity (see e.g., Chan et al. 2020), allergies, viral infections, and/or smoking history is not entirely clear. Olfactory methods will likely require more extensive pilot testing, facilities must often be ventilated between participants, designs require the blocking of experimental and control sessions (to prevent cross-condition contamination), and, of course, dedicated space in which experimental odours will not interfere with the research of others. Trust me on this one.

Autobiographical recall of disgusting events seems to reliably induce disgust (Lane et al. 1997), as do imaginary exposures. Some studies use vignettes, asking participants to narrate descriptions of incest or contact with bodily waste (e.g., Ottaviani et al. 2013). However, while they may reliably induce disgust, there are some limitations. Perhaps most prominently, demand is almost certainly higher than in other modalities for the simple reason that participants are specifically asked to recall a disgusting event. Control is also lower insofar as different people may choose to recall different events that will vary widely in content as well as in severity. Finally, it is not clear that participants are actually disgusted in that moment versus *having being disgusted* when the event actually occurred. No disease or contamination threat is currently present and the risk is that we end up correlating a mental representation of historical/expected feelings with another representational outcome (e.g., an attitude). This possibility noted, Siedlecka and Denson's (2019) review suggests both autobiographical and imagery methods may impact physiology, although whether physiological changes are specific to disgust is not yet clear.

In contrast to developing literatures regarding these methods, studies using audio or gustatory pathways to induce disgust are almost non-existent. According to Siedlecka and Denson's (2019) review of emotion induction methods, certain noises (burping, flatulence, and vomiting sounds) *may* increase disgust, but the review only references a single study (Marzillier and Davey 2005) in which noises were *combined* with imagery and recall. Gustatory and tactile inductions are similarly uncommon; anecdotal data from experienced disgust researchers suggest that such noises are often experienced as humorous, at least by university students. One study reported using consumption of a bitter drink to increase disgust (Adolph and Pause 2012) and tactile stimuli with wet or biological consistency appear more disgusting than dry/inanimate consistencies.

Overall, while it has been claimed that visual elicitation is the best technique for disgust elicitation while situational manipulations (including olfaction) are the worst (Siedlecka and Denson 2019), my own sense is that the issue is far from settled. A recent meta-analysis of 50 moral judgment studies (Landy and Goodwin 2015), for example, found that olfactory inductions had a stronger effect than visual disgust inductions. More broadly, core questions that have deep implications for how we think about disgust are yet to be tested. Does induction modality matter in terms of how disgust manifests in the different response subsystems? Can different modalities be "dosed" differently or do they have different time courses? What about habituation? How long can we elicit disgust in controlled ways without habituation effects? Does the *content* of visual disgust inductions matter in terms of the effects on attitudes or behaviour? Quite clearly, we do not know. What seems clear is that the choice of induction modality is a complex trade-off between how important control, blindness, resources, the content of elicitation, and the specific questions at stake are.

Stepping back slightly further, we can see that disgust induction studies have some interesting but potentially troublesome commonalities. Typically, in such approaches, visual or olfactory stimuli validated as being disgusting are presented (sometimes with a cover story and sometimes without), the manipulation is verified via statistical analyses of self-reported disgust, and the effects on psychological and behavioural processes tested. So, for example, in one recent study (Stewart et al. 2020), participants used a 100-point visual analogue scale to indicate how disgusted they were by images of dog faeces, vomit, or a cold sore. The resultant differences between experimental and control conditions was taken as evidence that the "stimuli that were used in this study powerfully induced the experience of disgust with a Cohen's d of 1.91" (Stewart et al. 2020, p. 6).

To this point, the causal-experimental logic seems quite reasonable. However, while reporting and interpretations of this kind are *technically* accurate, because no other emotional responses were measured, we cannot be certain that any effects on outcomes (in Stewart et al. 2020, on "scrupulosity") is caused by disgust per se rather than by other negative emotions that were inadvertently also elicited. Recent works have highlighted the difficulties inherent to validly measuring state emotion in laboratory studies (Weidman et al. 2017). In addition to issues with construct and face validity and use of single item measures (see Weidman et al. 2017), covariation and co-elicitation of negative emotions are common (McCambridge and Consedine 2014; Siedlecka and Denson 2019) making it difficult to determine which of several covarying negative emotions is causally responsible for effects (Consedine and Moskowitz 2007; Reynolds et al. 2018; Suls and Bunde 2005). Common emotion inductions such as eating chocolate might induce happiness, but also guilt. Similarly, consuming a bitter beverage might induce surprise in addition to disgust (Siedlecka and Denson 2019).

While there are exceptions (e.g., McCambridge and Consedine 2014), this issue pervades experimental emotions work. It may, however, be of particular relevance to studies of disgust because (a) avoidance or a negative judgment of some kind is so often the outcome of interest and (b) so many emotions (including disgust) evolved because they facilitate this type of function. Avoidant behaviour following a disgust induction can easily reflect the influence of other inadvertently induced emotions. Suffice to say that researchers must start more systematically measuring the effects of experimental manipulations in a way that permits the exclusion of competing (affective) explanations. This requires greater thought than merely obscuring target affect(s) and maximising blindness by hiding disgust terms amongst additional adjectives. Instead, conceptually viable confounds should be specifically targeted, measured, and proven to be irrelevant to the target effects of the manipulations.

Relatedly at a methods level, it is not yet clear which class of induction modality best discriminantly induces disgust rather than a mixed emotional response and/or whether induction modality matters for the magnitude or duration of the induction. Visual stimuli might initially appear have the advantage, at least at some levels, of being able to discriminantly induce disgust rather than other emotions, but, as has been discussed above, this is not entirely clear. Less debatably, because stimuli are comparatively simple, pictorial stimuli may offer a greater degree of control over content; we can examine the effects of disgust regarding particular domains of stimuli. Video stimuli offer a greater degree of complexity and naturalism, but seem likely to induce a more differentiated emotional response. These observations noted, we simply do not know and, until we do, the above comments stand: we must carefully choose modalities and stimuli as well as measure the confounding presence of other emotional responses that are likely or conceptually important.

Measuring Disgust Outside of Self-Report

Disgust researchers have recognised the inherent limitations of self-report and, in many cases quite creatively, adapted existing methods or developed new methods to index disgust more objectively. A short list of alternate measurements would include salivary responses (Proctor and Carpenter 2007), facial coding (Ekman and Friesen 1979), facial electromyography (Chapman et al. 2009), electrogastrography (Meissner et al. 2011), and fMRI (Jabbi et al. 2008), which are periodically used. Obviously, the choice of index depends on the research question(s). However, objective indices are often used primarily to increase certainty regarding elicitation.

Perhaps surprisingly, however, the disgust face is not reliably produced when feeling disgust (e.g., covertly filmed students smelling urine did not necessarily show disgust, Gilbert et al. 1987) and neural localisation of emotional responses remains unlikely (Chapman and Anderson 2012). As importantly, such indices are design expensive, requiring expertise, time, and specialised equipment. A simpler (and cheaper) solution is to more routinely incorporate the measurement of behaviour into disgust study design. Disgust was selected in evolution because it promoted rejection and withdrawal from pathogens, yet it is these same behaviours that are often the most problematic. Studies have tested willingness to engage with apparently contaminated or disgusting objects such as maggots, dirty clothing, ostensibly used stomas (e.g., Reynolds et al. 2014; Stevenson et al. 2010) or indexed the desire to maintain seating distance (Reynolds et al. 2015; Vartanian et al. 2016).

Again, however, interpretative caution is needed. Behavioural avoidance or rejection can reflect emotions other than disgust (Rottman et al. 2019; Stevenson et al. 2010) and key questions remain unclear. For example, most experimental studies rely on disgust inductions that are *incidental* to the target of any subsequent judgment or behaviour. That is to say, we disgust our participants (say, using images of mutilation) and then assess behaviours or processes that are completely unrelated—the disgust is not necessarily "about" the issues that are being investigated. So, for example, people become more opposed to gay marriage when disgusted by an odour (Adams et al. 2014). It may be that disgust acts relatively non-specifically in terms of increasing the avoidance of potential threats, but it could also be that we're not testing this question properly. According to Tybur et al. (2014), while tactile cues to pathogens (e.g., touching something wet or viscous; Oum et al. 2011) might lead to a stronger "pull away" responses than, say, olfactory cues, key questions remain unanswered. Are all methods eliciting

(a) the same basic response and/or (b) are different induction modalities activating a response that is differentially manifest in different components of the disgust response?

Concluding Remarks

Although studies of disgust are stretching their slightly sticky fingers into an increasingly wide range of research crevices, there is a very real sense in which our field remains young. Yes, the tendency to be disgusted appears trait like and relatively stable but questions regarding how important the differences between possible classes of core, sexual, contamination, and moral elicitors remain. Yes, we can reliably induce disgust and index disgust in controlled experimental settings but whether the eliciting modality matters (and for which outcomes) is a mystery to this point. Other methodological questions (e.g., regarding the dosing of disgust induction and response) are yet to be seriously asked. Methods and measurement are not truly distinct from our theories and there is a very real need to critically consider the implications the former has on the latter. Methods themselves must be scrutinised in the same way that we would examine, test, replicate, and verify any other phenomenon. Too often, we latch onto a measure or method that is widely used and has reasonable face validity and interpret is as "disgust" without thought or reflection. Ultimately, we must decide whether we want to evaluate and develop our tools or whether we are willing to let empirical regularities and covariation "drive" our thinking about this most intriguing of emotions.

References

Ackerman, J. M., Becker, D. V., Mortensen, C. R., Sasaki, T., Neuberg, S. L., & Kenrick, D. T. (2009). A pox on the mind: Disjunction of attention and memory in the processing of physical disfigurement. *Journal of Experimental Social Psychology*, 45, 478-485.

Adams, T. G., Stewart, P. A., Blanchar, J. C. (2014). Disgust and the politics of sex: Exposure to a disgusting odorant increases politically conservative views on sex and decreases support for gay marriage, *PLoS ONE, 9(5)*, e95572.

Adolph, D., & Pause, B. M. (2012). Different time course of emotion regulation towards odors and pictures: Are odors more potent than pictures? *Biological Psychology, 91,* 65-73.

Aldao, E., & Wisco, B. E. (2015). Motivational conflict influences the timing of emotions and their regulation. *Motivation & Emotion, 39,* 943-952.

Al-Shawaf, L. Lewis, D. M. G., & Buss, D. M. (2015). Disgust and mating strategy. *Evolution & Human Behaviour, 36,* 199-205.

Ammann, J., Hartmann, C., Peterhans, V., Ropelato, S., Siegrist, M. (2020). The relationship between disgust sensitivity and behaviour: A virtual reality study on food disgust. *Food Quality & Preference, 80*, 103833.

Ammann, J., Hartmann, & Siegrist, M. (2018). Development and validation of the Food Disgust Picture Scale. *Appetite, 125,* 367-379.

Armstrong, T., Wilbanks, D., Leong, D., & Hsu, K. (2021). Beyond vernacular: measurement solutions to the lexical fallacy in disgust research. PsyArXiv. https://doi.org/10.31234/osf.io/a8u5m

Azlan, H. A., Overton, P. G., Simpson, J., & Powell, P. A. (2020). Disgust propensity has a causal link to the stigmatization of people with cancer. *Journal of Behavioural Medicine, 43,* 377-390.

Brenner, C. J., & Inbar, Y. (2015). Disgust sensitivity predicts political ideology and policy attitudes in the Netherlands. *European Journal of Social Psychology, 45,* 27-38.

Campbell, R. L., Bynion, T-M., Forte, J., Feldner, M. T., & Adams Jr, T. G. (2020). Specificity of disgust in the prediction of behavioural avoidance of possible contaminants. *Cognitive Therapy & Research, 44,* 386-392.

Chan, K. Q., van Dooren, R., Holland, R. W., & van Knippenberg, A. (2020). Disgust lowers olfactory threshold: a test of the underlying mechanism. *Cognition & Emotion.* Advance online publication. https://doi.org/10.1080/02699931.2019.1660145.

Chambers, J. A., Callander, A. S., Grangeret, R. & O'Carroll, R. E. (2016). Attitudes towards the Faecal Occult Blood Test (FOBT) versus the Faecal Immunochemical Test (FIT) for colorectal cancer screening: perceived ease of completion and disgust. *BMC Cancer, 16,* 96.

Chapman, H. A., Kim, D. A., Susskind, J. M., & Anderson, A. K. (2009). In bad taste: evidence for the oral origins of moral disgust. *Science, 323,* 1222-1226.

Chapman, H. A., & Anderson, A. K. (2012). Understanding disgust. *Annals of the New York Academy of Sciences, 1251,* 62-76.

Consedine, N. S., & Moskowitz, J. (2007). The role of discrete emotions in health: A critical review. *Applied and Preventive Psychology, 12,* 59-75.

Culpepper, P. D., Havlícek, J., Leongómez, J. D., & Craig Roberts, S. (2018). Visually activating pathogen disgust: A new instrument for studying the behavioral immune system. *Frontiers in Psychology, 9 (AUG),* 1397.

Curtis, V., Aunger, R., & Rabie, T. (2004). Evidence that disgust evolved to protect from risk of disease. *Proc. R. Soc. B* 271, S131-S133.

Curtis, V., & de Barra, M. (2018). The structure and function of pathogen disgust. *Philosophical Transactions of the Royal Society B: Biological Sciences,* 373(1751), 20170208.

Darcy, I., & Fontaine, N. M. G. (2020). The Hoosier Vocal Emotions Corpus: a validated set of North American pseudo-words for evaluating emotion processing. *Behaviour Research Methods, 52,* 901-917.

Davis, M., Oaten, M., Occhipinti, S., & Stevenson, R. J. (2017). An investigation into the emotion of disgust as an affective barrier to intention to screen for colorectal cancer. *European Journal of Cancer Care, 26,* e12582.

Deacon, B., & Olatunji, B. O. (2007). Specificity of disgust sensitivity in the prediction of behavioural avoidance in contamination fear. *Behavior Research & Therapy, 45,* 2110-2120.

Dev, V., Consedine, N. S., & Reynolds, L. M. (2020). The "Ick" factor: an unrecognized affective predictor of chemotherapy side effects. *Annals of Behavioral Medicine.* Advance online publication. https://doi.org/10.1093/abm/kaaa055.

Druschel, B. A., & Sherman, M. F. (1999). Disgust sensitivity as a function of the Big Five and gender. *Personality and Individual Differences, 26,* 739-748.

Eickmeier, K., Hoffmann, L., & Banse, R. (2019). The 5-factor disgust scale: Development and validation of a comprehensive measure of disgust propensity. *European Journal of Psychological Assessment, 35*(3), 403-413.

Ekman, P., & Friesen, W. V. (1979). Measuring facial movement. *Environmental Psychology & Nonverbal Behaviour, 1*(1), 56-75.

Fergus, T. A., & Valentiner, D. P. (2009). The Disgust Propensity and Sensitivity Scale-Revised: An examination of a reduced-item version. *Journal of Anxiety Disorders, 23*(5), 703-710.

Fong, Z. H., & Sündermann, O. (2020). Modulating disgust in mental contamination: Experimental evidence for the role of disgust. *Journal of Behavior Therapy and Experimental Psychiatry, 68,* 101567.

Gilbert, A. N., Fridlund, A. J., & Sabini, J. (1987). Hedonic and social determinants of facial displays to odors. *Chemical Senses, 12*(2), 355-363.

Goetz, A. R., Cougle, J. R., & Lee, H.-J. (2013). Revisiting the factor structure of the 12-item Disgust Propensity and Sensitivity Scale—Revised: Evidence for a third component. *Personality and Individual Differences, 55*(5), 579-584.

Haberkamp, A., Glombiewski, J. A., Schmidt, F., & Barke, A. (2017). The DIsgust-RelaTed-images (DIRTI) database: validation of a novel standardized set of disgust pictures. *Behaviour, Research, & Therapy,* 89, 86–94.

Haidt, J., McCauley, C., & Rozin, P. (1994). Individual-differences in sensitivity to disgust: A scale sampling seven domains of disgust elicitors. Personality and Individual Differences, 16, 701-713.

Hartmann, C., & Siegrist, M. (2018). Development and validation of the Food Disgust Scale. *Food Quality & Preference, 63,* 38-50.

Izard, C. E., Dougherty, F. E., Bloxom, B. M., & Kotsch, W. E. (1974). The Differential Emotions Scale: A methods of measuring the subjective experience of discrete emotions. Unpublished manuscript.

Jabbi, M., Bastiaansen, J., & Keysers, C. (2008). A common anterior insula representation of disgust observation, experience, and imagination shows divergent functional connectivity pathways. *PLoS One, 3*(8), e2939.

Jin, Y., Ma, H., Zhang, Y., & Jimenez-Herrera, M. (2020). Development and psychometric evaluation of the colostomy disgust scale in patients with colostomy. *European Journal of Cancer Care, 29,* e13323.

Jones, B. C., Hahn, A. C., Fisher, C. I., Wang, H., Kandrik, M., Lee, A. J., Tybur, J.M., & DeBruine, L. M. (2018). Hormonal correlates of pathogen disgust: testing the compensatory prophylaxis hypothesis. *Evolution and Human Behavior, 39,* 464-467.

Karinen, A. K., Tybur, J. M., & de Vries, R. E. (2021). The disgust traits: self-other agreement in pathogen, sexual, and moral disgust sensitivity and their independence from HEXACO personality. *Emotion.* Advance online publication. https://doi.org/10.1037/emo0000795

Kiss, M. J., Morrison, M. A., & Morrison, T. G. (2020). A meta-analytic review of the association between disgust and prejudice toward gay men. *Journal of Homosexuality, 67,* 674-676,

Landy, J. F., & Goodwin, G. P. (2015). Does incidental disgust amplify moral judgment? A meta-analytic review of experimental evidence. *Perspectives on Psychological Science, 10*(4), 518-536.

Lang, P. J., Bradley, M. M., & Cuthbert, B. B. (1999). International Affective Picture System (IAPS): *Instruction manual and affective ratings.* Center for Research in Psychophysiology, University of Florida.

Lane, R., Reiman, E., Ahern, G., Schwartz, G., & Davidson, R. (1997). Neuroanatomical correlates of happiness, sadness, and disgust. *American Journal of Psychiatry, 154,* 926-933.

McCambridge, S. A., & Consedine, N. S. (2014). For whom the bell tolls: Experimentally-manipulated disgust and embarrassment cause anticipated sexual health avoidance among some people. *Emotion, 14*(2), 407-415.

Marzillier, S., & Davey, G. (2005). Anxiety and disgust: evidence for a unidirectional relationship. *Cognition & Emotion, 19,* 729-750.

Meissner, K., Muth, E. R., & Herbert, B. M. (2011). Bradygastric activity of the stomach predicts disgust sensitivity and perceived disgust intensity. *Biological Psychology, 86,* 9-16.

Muris, P., Mayer, B., Huijding, J., & Konings, T. (2008). A dirty animal is a smelly animal! Effects of disgust-related information on fear beliefs in children. *Behaviour Research & Therapy, 46,* 137-144.

Muris, P., Huijding, J., Mayer, B., Langkamp, M., Reyhan, E., & Olatunji, B. (2012). Assessment of disgust sensitivity in children with an age-downward version of the Disgust Emotion Scale. *Behaviour Therapy, 43,* 876-886.

O'Carroll, R. E., Chambers, J., Brownlee, Libby, G., & Steele, R. (2015). Anticipated regret to increase uptake of colorectal cancer screening (ARTICS): a randomised controlled trial. *Social Science & Medicine, 142,* 118–27.

Olatunji, B. O., Adams, T., Ciesielski, B., David, B., Sarawgi, S., & Broman-Fulks, J. (2012). The Three Domains of Disgust Scale: factor structure, psychometric properties, and conceptual limitations. *Assessment, 19*(2), 205-225.

Olatunji, B. O., Ebesutani, C., David, B., Fan, Q., & McGrath, P. B. (2011). Disgust proneness and obsessive compulsive symptoms in a clinical sample: structural differentiation from negative affect. *Journal of Anxiety Disorders, 25,* 932-938.

Olatunji, B. O., Williams, N. L., Tolin, D. F., Sawchuk, C. N., Abramowitz, J. S., Lohr, J. M., & Elwood, L. (2007). The Disgust Scale: Item analysis, factor structure, and suggestions for refinement. *Psychological Assessment, 19,* 281-297.

Oosterhoff, B., Shook, N. J., & Iyer, R. (2018). Disease avoidance and personality: a meta-analysis. *Journal of Research in Personality, 77,* 47-56.

Ottaviani, C., Mancini, F., Petrocchi, N., Medea, B., & Couyoumdjian, A. (2013). Autonomic correlates of physical and moral disgust. *International Journal of Psychophysiology, 89,* 57-62.

Oum, R. E., Lieberman, D., & Aylward, A. (2011). A feel for disgust: Tactile cues to pathogen presence. *Cognition and Emotion, 25*(4), 717-725.

Perone, P., Becker, D. V., & Tybur, J. M. (2020). Visual disgust elicitors produce an attentional blink independent of contextual and trait-level pathogen avoidance. *Emotion.* Advance online publication. https://doi.org/10.1037/emo0000751.

Podsakoff, P. M., MacKenzie, S. B., & Podsakoff, N. P. (2012). Sources of method bias in social science research and recommendations on how to control it. *Annual Review of Psychology, 63,* 539-569.

Proctor, G. B., & Carpenter, G. H. (2007). Regulation of salivary gland function by autonomic nerves. *Anatomic Neuroscience: Basis and Clinical, 133,* 3-18.

Polak, J., Landova, E., & Frynta, D. (2019). Undisguised disgust: a psychometric evaluation of a disgust propensity measure. *Current Psychology, 38,* 608-617.

Powell, P. A., Jones, C. R., & Consedine, N. S. (2019). It's not queasy being green: The role of disgust in willingness-to-pay for sustainable alternatives. *Food Quality & Preference, 78,* 103737.

Reynolds, L. M., Bissett, I., & Consedine, N. S. (2018). Emotional predictors of bowel screening: the avoidance-promoting role of fear, embarrassment, and disgust. *BMC Cancer, 18*(1), 518.

Reynolds, L. M., Lin, Y-S., Zhou, E., & Consedine, N. S. (2015). Does a state mindfulness induction moderate disgust-driven social avoidance and decision-making? An experimental investigation. *Journal of Behavioral Medicine, 38* (1), 98-109.

Reynolds, L. M., McCambridge, S. A., Bissett, I. P., & Consedine, N. S. (2014). Trait and state disgust: An experimental investigation of disgust and avoidance in colorectal cancer decision scenarios. *Health Psychology, 33*(12), 1495-1506.

Reynolds, L. M., Powell, P. A., Lin, Y. S., Ravi, K., Chung, K., & Consedine, N. S. (2019). Fighting the flinch: Experimentally induced compassion makes a difference in healthcare providers. *British Journal of Health Psychology, 24*(4), 982-998.

Rozin, P., Haidt, J., McCauley, C., Dunlop, L., & Ashmore, M. (1999). Individual differences in disgust sensitivity: comparisons and evaluations of paper-and-pencil versus behavioural measures. *Journal of Research in Personality, 33,* 330-351.

Rottman, J., DeJesus, J. M., & Greenebaum, H. (2019). Developing disgust: theory, measurement and application. In V. LoBue, K. Perez-Edgar, & K. Buss (Eds.), *Handbook of emotional development* (pp. 283-309). Springer.

Rottman, J., & Kelemen, D. (2012). Aliens behaving badly: children's acquisition of novel purity-based morals. *Cognition, 124 (3),* 356-360.

Russell, P. S., & Giner-Sorolla, R. (2013). Bodily moral disgust: what it is, how it is different from anger, and why it is an unreasoned emotion. *Psychological Bulletin, 139,* 328-351.

Schienle, A., & Schlintl, C. (2020). The association between quinine hydrochloride sensitivity and disgust proneness in children and adults. *Chemosensory Perception, 13,* 78-83.

Schienle, A., Zorjan, S., & Wabnegger, A. (2020). A brief measure of disgust propensity. *Current Psychology*, 1-7.

Schnall, S., Haidt, J., Clore, G. L., & Jordan, A. H. (2008). Disgust as embodied moral judgment. *Personality and Social Psychology Bulletin, 34,* 1096-1109.

Shenhav, A., & Mendes, W. B. (2014). Aiming for the stomach and hitting the heart: Dissociable triggers and sources for disgust reactions. *Emotion, 14*, 301-309.

Siedlecka, E., & Denson, T. F. (2019). Experimental methods for inducing basic emotions: a qualitative review. *Emotion Review, 11,* 87-97.

Simpson, J., Carter, S., Anthony, S. H., & Overton, P. G. (2006). Is disgust a homogenous emotion? *Motivation & Emotion, 30,* 31-41.

Stevenson, R. J., Oaten, M. J., Case, T. I., Repacholi, B. M., & Wagland, P. (2010). Children's responses to adult disgust elicitors: development and acquisition. *Developmental Psychology, 46*(1), 165-177.

Stewart, P. A., Adams, T. G., & Senior, C. (2020). The effect of trait and state disgust on fear of God and sin. *Frontiers in Psychology, 11,* 51.

Suls, J., & Bunde, J. (2005). Anger, anxiety, and depression as risk factors for cardiovascular disease: The problems and implications of overlapping affective dispositions. *Psychological Bulletin, 131*(2), 260-300.

Tybur, J. M., Bryan, A. D., Magnan, R. E., & Hooper, A. E. C. (2011). Smells like safe sex: olfactory pathogen primes increase intentions to use condoms. *Psychol. Sci.* 22, 478-480.

Tybur, J. M., & Karinen, A. K. (2018). Measurement and theory in disgust sensitivity. In V. Zeigler-Hill & T. K. Shackelford (Eds.) *Handbook of Personality and Individual Differences* (pp. 159-179). Sage Publications.

Tybur, J. M., Lieberman, D. L., & Griskevicius, V. (2009). Microbes, mating, and morality: individual differences in three functional domains of disgust. *Journal of Personality & Social Psychology, 97,* 103-122.

Tybur, J. M., Pollet, T. V., & Frankenhuis, W. E. (2014). Behavioral immune system methods: surveying the present to shape the future. *Evolutionary Behavioural Sciences, 8,* 274–283.

van Overveld, M., de Jong, P. J., & Peters, M. L. (2010). The Disgust Propensity and Sensitivity Scale—Revised: its predictive value for avoidance behaviour. *Personality & Individual Differences, 49,* 706-711.

van Overveld, W. J. M., de Jong, P. J., Peters, M. L., Cavanagh, K., & Davey, G. C. L. (2006). Disgust propensity and disgust sensitivity: Separate constructs that are differentially related to specific fears. *Personality and Individual Differences, 41,* 1241-1252.

Vartanian, L. R., Trewartha, T., & Vanman, E. J. (2016). Disgust predicts prejudice and discrimination toward individuals with obesity. *Journal of Applied Social Psychology, 46,* 369-375.

Vianna, E., & Tranel, D. (2006). Gastric myoelectrical activity as an index of emotional arousal. *International Journal of Psychophysiology, 61,* 70-76.

Viar-Paxton, M. A., Ebesutani, C., Kim, E. H., Ollendick, T., Young, J., & Olatunji, B. O. (2015). Development and initial validation of the Child Disgust Scale. *Psychological Assessment, 27,* 1082-1096.

Weidman, A. C., Steckler, C. M., & Tracy, J. L. (2017). The jingle and jangle of emotion assessment: Imprecise measurement, casual scale usage, and conceptual fuzziness in emotion research. *Emotion, 17*(2), 267-295.

White, A. E., Kenrick, D. T., & Neuberg, S. L. (2013). Beauty at the ballot box: Disease threats predict preferences for physically attractive leaders. *Psychological Science*, 24, 2429-2436.

Chapter 3
The Evolution of Disgust, Pathogens, and the Behavioural Immune System

Hannah K. Bradshaw and Jeffrey Gassen

When considering factors that posed a threat to survival throughout evolutionary history, things like lions, tigers, bears, inclement weather, and environmental catastrophes readily come to mind. However, the most potent and ubiquitous selection pressure shaping human evolution arises from microorganisms (i.e., parasites, viruses, bacteria) that are typically too small to be seen by the naked eye. For instance, evidence from small scale forager-horticulturalist societies demonstrates that infectious diseases are responsible for over half of all deaths (Gurven et al. 2007). As such, humans and other animals have developed a repertoire of emotional, cognitive, behavioural, and physiological mechanisms that function as protection against the persistent threat posed by infection and illness.

In the current chapter, we first provide an overview of past literature supporting the reasoning that key aspects of our psychology—disgust and the behavioural immune system—represent adaptations to the selection pressures posed by infectious microorganisms. Although disgust has been shown to underlie moral reasoning, we limit our focus here to implications directly concerning infection and illness. We end by presenting innovative ideas bearing on disgust and the behavioural immune system, which incorporate perspectives from psychology, ecology, and psychoneuroimmunology.

H. K. Bradshaw (✉)
Department of Psychology, Washington & Jefferson College, Washington, PA, USA
e-mail: hbradshaw@washjeff.edu

J. Gassen
Department of Anthropology, Baylor University, Waco, TX, USA
e-mail: Jeffrey_Gassen@baylor.edu

P. A. Powell, N. S. Consedine (eds.), *The Handbook of Disgust Research*,
https://doi.org/10.1007/978-3-030-84486-8_3

Disgust: An Infection-Avoidance Adaptation

Characterised by a feeling of revulsion, the emotion of disgust is a major player in our evolved strategies to minimise the risk of disease and illness. Evidence supporting this proposition emerges from different research methodologies exploring the various elicitors of disgust. Qualitative, cross-cultural work focusing on what people find disgusting shows that responses centre around potential vectors for disease and infectious microorganisms: bodily excretions (e.g., faeces), decay and rotten food, certain animals (e.g., cockroaches, lice, rats), and certain people (e.g., sick or unhygienic people; Curtis and Biran 2001). Other research including over 40,000 participants across a wide variety of cultures illustrates that people evaluate visual stimuli containing cues of disease to be more disgusting than similar stimuli without disease cues (Curtis et al. 2004). Cues of infection detected by other sensory systems similarly elicit disgust. For instance, people commonly find the smell of decay, faeces, and vomit to be disgusting (Croy et al. 2011; Glass et al. 2014), and tactile cues of moisture, which are indicative of pathogen presence, also elicit disgust (Oum et al. 2011; Saluja and Stevenson 2019). Together, these findings demonstrate remarkable consistency in the cues that elicit disgust, indicating that disgust is intimately tied to threats of disease and contamination.

Additional research finds that disgust is associated with distinct changes in perception and behaviour. For instance, when disgust is experimentally induced, people exhibit greater sensitivity to touch (Hunt et al. 2017) and smells (Chan et al. 2019). Such perceptual changes, which occur in the context of disgust, might function to better facilitate the detection of cues of contamination. Disgust is further related to increased behavioural avoidance of potential contaminants (Campbell et al. 2019; Dorfan and Woody 2011; Olatunji et al. 2014) and greater likelihood of hygienic behaviour after exposure to potential contaminants (e.g., handwashing; Porzig-Drummond et al. 2009). The results of this work are consistent with the reasoning that disgust represents an adaptation to the selection pressure posed by disease and infection, by motivating avoidance of potential contaminants (Curtis et al. 2011; Tybur et al. 2013).

Of importance to note here is that accurate knowledge regarding the transmission of illness and disease is not a prerequisite for feeling disgust towards or being motivated to avoid contaminants. Lay beliefs that disease and infection are spread by contact predate modern theories about disease transmission (Karamanou et al. 2012). For example, the set of rules for dealing with leprosy outlined in Leviticus emphasise quarantining and isolating individuals with the disease, presumably to prevent transmission (*King James Bible* 1769/ 2008, Leviticus 13). Moreover, disgust and hygienic practices that serve to reduce infection risk exist across history and cultures, long before scientists gained knowledge of bacteria and viruses (Curtis 2007). Our ancestors who felt disgust towards potential contaminants would have been better able to avoid disease and infection than those who did not, thus facilitating survival and reproduction (though, the trade-off between anti-pathogen defences and reproduction is not straightforward; Borg and de Jong, Chap. 9, this volume). The disease-avoidance advantage of disgust would have led to such traits being more frequent in subsequent populations. Recent research finds approximately half of the variation in disgust sensitivity is due to underlying genetic factors (Sherlock et al. 2016), which suggests that disgust sensitivity is, to some degree, heritable. Obviously, it is impossible to invent a time machine to study the presence of disgust and the survival advantage it may have conferred in ancestral times. However, the presence of disgust across history, its functional correlates, and the cross-cultural ubiquity of disgust elicitors point to the existence of disgust as an evolved mechanism against disease.

More Broadly: The Behavioural Immune System

Disgust, however, is just the tip of the iceberg when it comes to evolved disease avoidance mechanisms. Within the last decade, research has established that, much like other animals, humans possess a suite of perceptual, cognitive, and behavioural components, which strategically function to defend against the threat of illness and infection. Together, these defences are referred to as the *behavioural immune system* (Schaller and Park 2011). Below, we discuss the evidence related to each of these components.

Beforehand, it bears noting that there is some disagreement in the literature regarding the relationship between disgust and the behavioural immune system. Some argue that these constructs are functionally identical (Lieberman and Patrick 2014). According to this perspective, given that disgust and the behavioural immune system rely on the same input (i.e., cues of contamination), involve the same computational processes (e.g., estimating potential cost of infection), and produce functionally similar outputs (e.g., avoidance), the distinction between the two is merely semantic. Others claim that, while disgust and the behavioural immune system can operate in concert, disgust is not a necessary prerequisite for the activities of the behavioural immune system (Schaller 2014). That is, people engage in prophylactic (e.g., handwashing) or culturally normative (e.g., hygiene) behaviour that pre-emptively serves to mitigate the threat of infection without consciously experiencing disgust. This disagreement is, as of yet, unsettled; future research is needed to delineate the extent to which disgust is or is not intrinsic to the activity of the behavioural immune system.

Perception and Cognition

The behavioural immune system is theorised to guide the detection of cues indicative of illness and infection. Research in this vein has explored whether individuals can detect the presence of illness in others through various sensory modalities. For instance, in one study, participants were shown faces of targets that had either been injected with lipopolysaccharide (LPS), a bacterial stimulus that triggers an immune response, or a placebo, and were instructed to rate whether the target was sick or healthy (Axelsson et al. 2018). The results provided evidence that people identified the individuals injected with LPS as sick above chance level. Other research has used similar acute illness (vs. placebo) inductions to examine how people evaluate the health of targets based on smell (Olsson et al. 2014) or videos of their gait patterns (Sundelin et al. 2015). Compared to targets who were injected with the placebo, the scent and gait patterns of sick targets were rated to be less healthy by those injected with the immune trigger. The findings of this work demonstrate that our sensory systems are functionally tuned to detect cues of sickness and infection in conspecifics.

Other research has examined how the behavioural immune system influences key aspects of cognition, such as attention and memory. For instance, visual cues of contaminants (e.g., vomit) draw and hold people's attention (van Hooff et al. 2014; Vogt et al. 2011). Not only do people pay more attention towards potential sources of infection, they also exhibit increased memory for these stimuli. Research in this vein finds that images that elicit disgust are remembered better than neutral images or images that elicit fear (Chapman 2018). Moreover, people exhibit better memory for items that were touched by purportedly sick (vs. healthy) targets (Bonin et al. 2019; Fernandes et al. 2017).

Increased attention to, and memory for, sources of infection may help facilitate disease avoidance while individuals navigate their social world. For example, avoiding potential sources of infection first requires identifying them. Accordingly, being especially attuned to cues connoting disease risk—particularly when those cues are subtle—is key to appropriately activating emotional reactions, like disgust, that promote avoidance behaviours. Additionally, heightened memory for disgusting stimuli may help individuals avoid second encounters with situations conferring disease risk.

Behaviour and Sociality

As may be ascertained from the name, behaviour plays a major role in the behavioural immune system. Objects or other people containing cues of contamination are intuitively avoided. For instance, people exhibit greater behavioural avoidance—measured using an approach-avoid task—when viewing videos featuring contamination threats, as compared to videos eliciting threats of violence or danger (Newhagen 1998). This pattern of behaviour is found in human cultures removed from Western society and even in nonhuman animals. For instance, Apicella et al. (2018) demonstrate that Hadza and Tannese adults and children exhibit rejection of food that has come into contact with contaminants (Apicella et al. 2018). Nonhuman primates display similar aversion towards, and avoidance of, contaminated food (Sarabian et al. 2017, 2018).

Other research has found that experimentally activating concerns of disease can influence hygiene behaviour. For instance, people wash their hands more when in the presence of a disgusting odour (Pellegrino et al. 2016). Disease concerns are additionally shown to reduce sociality. When disease concerns are made salient, people report lower extraversion, exhibit greater behavioural avoidance of human targets, and report less interest in affiliation (Mortensen et al. 2010; Sacco et al. 2014). Such shifts in sociality in the face of disease threats make adaptive sense, as contact with human targets pose a risk for disease transmission.

Smoke Detector Principle and Functional Flexibility

Although features of the behavioural immune system help detect cues of disease and serve to facilitate avoidance of sources of infection and illness, its responses are necessarily imperfect. Given the high cost of illness and infection across evolutionary time, our evolved disease-detection mechanisms are extremely sensitive and can be activated even in the absence of actual disease threats. For example, recent work presented participants with the sounds of coughs and sneezes from targets who reported currently experiencing an infectious illness (e.g., cold, flu) and from targets who were coughing and sneezing for reasons unrelated to illness (e.g., consumption of spices; Michalak et al. 2020). The study found that participants could not differentiate between the two sources, misidentifying non-contagious sounds of illness as posing an infection threat. Other research on this topic finds that people respond similarly towards contagious others as they do towards those who have non-contagious physical anomalies (Kouznetsova et al. 2012; Park et al. 2013; Ryan et al. 2012). For instance, people display similar behavioural avoidance towards objects touched by a target with influenza and a target with a facial birthmark (Ryan et al. 2012). Together this work demonstrates that the processes of the behavioural immune system can lead to psychologically categorising non-infectious others and objects as infectious and prompt behavioural avoidance, even when avoidance is unnecessary. As such, the responses of the behavioural immune system can themselves impose costs on one's ability to successfully navigate the challenges inherent in survival and reproduction (e.g., finding shelter, food, mates, friends, etc.).

Like other evolved threat management systems, the behavioural immune system is characterised by *functional flexibility*, such that the intensity of responses varies based on the costs and benefits of disease avoidance in a given context (Ackerman et al. 2018; Tybur and Lieberman 2016; Schaller and Park 2011). That is, the activities of the behavioural immune system are shown to be particularly amplified in contexts where benefits of disease avoidance more strongly outweigh the costs, such as when one is especially vulnerable to illness and infection (for a review see Schaller and Park 2011). For instance, the association between disease and physical anomalies is stronger for those who are chronically concerned about infection or when infection concerns are made salient (Miller and Maner 2012; Park et al. 2007).

Additionally, those who are chronically concerned about their health even evaluate relatively healthy individuals as being less healthy (Hedman et al. 2016).

On the other hand, behavioural immune system activity is downregulated in circumstances where avoidance restricts one's ability to meet other goals. For instance, taking care of children requires one to interact with potential contaminants (e.g., changing diapers, cleaning up vomit, etc.). Research finds that women who are parents of children report less disgust towards the smell of a dirty diaper from their baby, compared to a dirty diaper from an unrelated child (Case et al. 2006). Along similar lines, sexual intercourse involves close physical contact exposure to bodily fluids. When sexually aroused, both men and women report decreased disgust towards sex-related stimuli (Borg and de Jong, Chap. 9, this volume; Borg and De Jong 2012; Stevenson et al. 2011a). However, it is worth noting that one online study found that sexual arousal did not blunt women's disgust (Zsok et al. 2017). Other research finds that people who are food deprived exhibit less facial disgust in response to unpalatable food images (Hoefling et al. 2009). Together, this research demonstrates that responses of the behavioural immune system are contextually dependent, varying based on state vulnerability and salient goals.

Social Circumstances and Individual Differences in Disgust Sensitivity

Just as disgust varies transiently based on contextual factors that influence the relative costs of engaging in behavioural disease avoidance (e.g., when one is food deprived), recent research suggests that more stable social characteristics of individuals that affect these costs might also impact disgust sensitivity. That is, individuals whose social circumstances increase the costs associated with behavioural disease avoidance should demonstrate lower disgust sensitivity. In contrast, individuals whose social circumstances diminish the costs associated with behavioural avoidance should exhibit higher disgust sensitivity. For example, because disgust promotes avoidance of people or stimuli that elicit disgust, it could be costly because it reduces the number of options one has for social partners, romantic partners, or food options. Therefore, disgust should be higher for individuals who are better able to withstand the costs of losing opportunities in these domains compared those to who may have fewer options. In other words, individuals who can afford to be more selective should tend to exhibit greater disgust than those who cannot.

One potential important determinant of an individual's ability to withstand the cost of elevated disgust responses is one's social status. This is because the wealth and prestige of having high social status endows an individual with a greater number of choices when it comes to romantic partners (Bereczkei et al. 1997; Stringhini et al. 2012; Von Rueden and Jaeggi 2016), housing (Dwyer 2007), and foods (Drewnowski and Specter 2004), compared to those of lower status. With these insights in mind, one recent set of studies examined relationships between social status and disgust sensitivity (Bradshaw et al. n.d.). This work demonstrated that individuals reporting a higher social status also reported having greater disgust sensitivity. A set of experiments further revealed both that participants expected high status targets to exhibit greater disgust sensitivity than low status targets, and in reverse, that targets described as displaying disgust were rated as higher status than those not displaying disgust.

Together, this research suggests that social status and disgust are closely linked and sets the stage for future research to examine how additional social factors that influence one's ability to withstand the costs of disgust may impact its expression. Moreover, there are also a number of additional social and cultural factors related to socioeconomic status that might influence disgust sensitivity. Therefore, additional research is needed to determine the causal factors involved in this link. For example, individuals of high and low socioeconomic often differ in parenting behaviours (Hoff et al. 2002) and emotional styles (Manstead 2018), which might impact attitudes towards disgust-eliciting stimuli.

Relationships between the Biological Immune System and Behavioural Immune System

While a large body of work has demonstrated the far-reaching effects of the behavioural immune system on psychology and behaviour, research into how behavioural pathogen defence mechanisms intersect with the activities of the immune system is still in its nascent stage (Ackerman et al. 2018; Gassen et al. 2018; Gassen and Hill 2019; Murray et al. 2019). A leading hypothesis from the evolutionary psychology literature proposes that the biological and behavioural immune systems may operate in a compensatory manner (see compensatory prophylaxis hypothesis; Fessler et al. 2005; Fleischman and Fessler 2011), and some recent research in the field of psychoneuroimmunology has provided initial support for this possibility (e.g., Bradshaw et al. n.d.; Gassen et al. 2018). On the other hand, separate research suggests that—in some situations—disgust and biological immune activation may co-occur (e.g., Rubio-Godoy et al. 2007; Stevenson et al. 2011b). Here, we review previous theory and research on relationships between biological and behavioural pathogen defence mechanisms. We propose that progress in this domain has been limited by an over-reliance on indirect hypothesis testing (e.g., inferring immune status from hormone levels) and a failure to fully appreciate the complexity of immune function. We conclude by presenting future directions for disgust research that may help unravel complex relationships between immunity and behaviour that have evolved in the context of the host-pathogen arms race.

The Compensatory Prophylaxis Hypothesis

The compensatory prophylaxis hypothesis postulates that behavioural disease avoidance mechanisms have evolved to complement the activities of the biological immune system in protecting an individual from infection. However, because prophylactic disease avoidance behaviours bear energetic and social costs (as briefly outlined in the previous section), the extent to which individuals engage in these behaviours should not be constant, but should rather increase when the biological immune system is impaired and one is particularly vulnerable to the threat of infectious disease (Fessler et al. 2005; Fleischman and Fessler 2011). Support for this hypothesis includes research demonstrating that pregnant women reported elevated disgust sensitivity during the first trimester (compared to the second and third trimesters), a period when the maternal immune system is believed to be suppressed to prevent the maternal immune system from attacking the foetus (Fessler et al. 2005). Additional support for the compensatory prophylaxis hypothesis comes from research finding that, in naturally-cycling women, higher levels of salivary progesterone—a hormone that suppresses certain types of immune responses—were associated with greater disgust sensitivity, contamination-related rumination, and self-grooming behaviours (Fleischman and Fessler 2011; but see Jones et al. 2018).

Some recent research has provided additional support for compensatory prophylaxis by examining the relationship between inflammation—a crucial component of the immune system—and the activities of the behavioural immune system. For example, one study found that healthy individuals with lower levels of basal inflammation (both *in vivo* and *in vitro*) reported greater germ aversion than those with higher levels (Gassen et al. 2018). Other research shows that participants administered aspirin—a non-steroidal anti-inflammatory drug—reported feeling more negative towards pictures of disgusting stimuli than those administered a placebo (Bradshaw et al. n.d.). In both cases, these findings are consistent with the compensatory prophylaxis hypothesis, as they suggest that lower levels of inflammation are associated with increased sensitivity to disease-connoting contexts and stimuli.

Beyond the Compensatory Prophylaxis Hypothesis

Research finds that the biological and behavioural immune systems do not always operate in a compensatory manner, however. Instead, some studies have shown that experimentally activating the behavioural immune system via exposure to disgust-connoting stimuli can also elicit an immune response (Ersche et al. 2014; Schaller et al. 2010; Stevenson et al. 2011b, 2012). In other words, both the biological and behavioural immune systems appear to increase in concert when an individual is directly faced with stimuli that may confer infectious disease risk. For example, one experiment found that exposure to disgusting visual stimuli (relative to neutral and negative, but not disgusting stimuli) led to increased oral inflammation (Stevenson et al. 2011b). Other work demonstrated that inducing disgust through a similar procedure led to both an increase in a marker of oral inflammation, as well as elevated body temperature (Stevenson et al. 2012).

The idea that the biological and behavioural immune systems may operate in parallel when an individual faces direct exposure to stimuli connoting infection risk is not mutually exclusive with insights from the compensatory prophylaxis hypothesis. Instead, it is possible, and likely, that interactions between these systems are highly context-dependent and moderated by individual-level factors, such as one's control over pathogen contact (see e.g., Bradshaw et al. n.d.), energetic status (see e.g., moderation by hormone leptin; Han et al. 2003; Iikuni et al. 2008), age (Fessler et al. 2005; Franceschi et al. 2007), and sex (Klein et al. 2015; Tybur et al. 2011a). However, a more comprehensive theoretical framework that accounts for each of these previous findings is needed to help guide predictions about when, how, and for whom relationships between the biological and behavioural immune systems follow a given pattern.

Current Limitations of Research Examining Links between Immune Function and Disgust

While research on disgust and the behavioural immune system has exploded in the last two decades, sparse effort has been devoted to integrate insights from immunology and related fields into this programme of study (with the exception of Schaller et al. 2010; and Stevenson et al. 2011b, 2012). Accordingly, there has been little progress towards elucidating the biological-behavioural immune system interface. This is particularly troublesome because the structure and function of evolved emotional, cognitive, and behavioural pathogen defence tools cannot be fully understood without knowing how they relate to our body's primary defence against infectious illness.

We argue that the first major limitation of research examining links between the biological and behavioural immune systems, to date, has been an over-reliance on indirect methods for measuring immune function. A strength of evolutionary psychology is its focus on considering ultimate explanations for behaviour. However, developing and testing evolutionary theories involves designing empirical research to test putative mechanisms. Without the appropriate methodology to do so, direct relationships and causality become difficult to interpret and both significant and null results can present more questions than answers. This has often been the case with extant research on relationships between the biological and behavioural immune systems. For example, studies testing the compensatory prophylaxis hypothesis have largely relied on menstrual cycle phase, pregnancy trimester, or levels of progesterone as a proxy for the activities of the immune system (e.g., Fessler et al. 2005;

Fleischman and Fessler 2011; Jones et al. 2018). However, each of these factors have complex, pleiotropic effects on immune function (i.e., rather than uniformly impairing immunity; Hall and Klein 2017; Hughes 2012; Lorenz et al. 2015; Mor and Cardenas 2010; Saito et al. 2010), and considered in isolation, likely provide a poor index of one's current infection risk.

Further limiting the utility of these measures is the fact that their effects on immunity are moderated by individual-level factors, such as whether or not one is sexually active (e.g., hormones and cycle phase; Prasad et al. 2014) and one's health/diet prior to and during pregnancy (e.g., Dellschaft et al. 2015; Sen et al. 2013). Therefore, it is vital that the activities of the immune system are directly measured when testing hypotheses about the biological-behavioural immune system interface. This also extends to measuring the behavioural immune system. Direct measurements of the behavioural immune system (e.g., approach-avoid tasks, behavioural tasks) should be prioritised to improve the external validity of such research.

Measuring immune function, however, is easier said than done. The immune system is unbelievably complex, and classes of immune responses can be categorised in a variety of ways, whether it be innate versus adaptive immunity (Hoebe et al. 2004), cellular versus humoral immunity (Xu et al. 2004), Th_1 versus Th_2 versus Th_{17} responses (Romagnani 1992), or pro-inflammatory versus anti-inflammatory (Dinarello 1997), among many more. The second limitation of research examining relationships between the biological and behavioural immune systems is that literature has largely ignored this intricacy, leaving gaps in our understanding of how these systems truly interact. For example, the immune system is often described as being broadly "elevated" or "diminished" (e.g., Bradshaw et al. n.d.; Fessler et al. 2005; Fleischman and Fessler 2011; Gassen et al. 2018; Jones et al. 2018), but these states likely only reflect a specific facet of immune function, such as inflammatory activity, not the immune system as a whole. Moreover, even lower levels of inflammation do not necessarily indicate diminished immune function. In fact, lower systemic levels of inflammation within the normal range may actually indicate the opposite (see e.g., Cohen et al. 1999, 2012).

While on the surface it may appear daunting, appreciating the complexity of immune function is necessary for developing informed hypotheses and designing appropriate tests of relationships between the biological and behavioural immune system. Viewed more optimistically, the complexity of immunity provides a wealth of opportunities to generate novel and nuanced hypotheses about how specific aspects of immune function relate to behavioural pathogen avoidance mechanisms. For example, we might find that disgust is related to the function of innate immune cells, such as natural killer cells, but not adaptive immunity, like the production of antibodies by B cells. Further, we might find that relationships between behavioural immune system activity and different facets of immunity differ by age, sex, or ecology. Each of these shapes both disgust responses, as well as immune responses. Recent advancements in immunology that increased the affordability and accessibility of technologies like flow cytometry and transcriptomics now make it possible for more scientists—even social scientists—to test these and other hypotheses.

Future Directions for Disgust Research

In the previous sections, we presented evidence suggesting that disgust and the activities of the behavioural immune system represent a suite of evolved cognitive, psychological, and behavioural mechanisms that function to reduce exposure to potential sources of infection. We also briefly summarised theory and research on how these mechanisms might interact with the biological immune system. Below, we build on this past research and integrate insights from evolutionary biology, ecology, and psychoneuroimmunology to present new evolution-informed hypotheses about how environmental-, host-, and pathogen-level factors may interact to influence both the activities of the behavioural immune system and the course of the arms race between host sensory systems and harmful microorganisms.

Different Behavioural Responses to Different Classes of Pathogens

The behavioural immune system is believed to be relatively domain general, promoting avoidance of a broad range of stimuli connoting infection risk (Schaller 2011). However, it is possible that, like the activities of the biological immune system, behavioural pathogen avoidance strategies may be fine-tuned to the type of disease threat encountered and its primary vector (Abbas and Janeway 2000; Everett and McFadden 1999; Girardin et al. 2002; Romagnani 1991; Sher et al. 2003; Wang et al. 2016). For instance, the threat posed by intestinal parasites differs from that posed by viruses and bacteria. As such, we may expect that visual exposure to threats of intestinal parasites may elicit a different reaction (e.g., disgust plus gastrointestinal distress) than would viewing pictures of vectors of viruses or bacteria (e.g., general disgust).

The threat posed by ectoparasites (e.g., lice, ticks, fleas) illustrates the utility of specialised behavioural responses to specific classes of pathogen threats. Ectoparasites are common vectors for infectious diseases, and, given that they attack the host via attaching to external surfaces (i.e., the skin), the traditional avoidance defence of the behavioural immune system offers little protection against this class of threat. Kupfer and Fessler (2018) suggest that humans and other animals possess specifically tailored behavioural defences (e.g., scratching and grooming) that function to counter the threat posed by ectoparasites. Consistent with this reasoning, initial research finds parasites and pathogens that are primarily transmitted via the skin elicit emotional responses and behavioural reactions—such as a skin-crawling sensation—that are distinct from general disgust (Blake et al. 2017). A separate study found that exposure to live maggots—also ectoparasites—led to increased tactile sensitivity relative to exposure to control stimuli (i.e., rice) presented in the same container (Hunt et al. 2017). Future research is needed to examine whether exposure to stimuli connoting risk for infection by different classes of pathogens evokes different emotional, cognitive, and behavioural responses, as well as the role that learning plays in such processes.

The fact that different classes of pathogen threats elicit distinct behavioural responses may provide some clarification for recent work regarding the lack of a relationship between regional variability in infectious disease prevalence and disgust sensitivity (Tybur et al. 2016). If disgust sensitivity is an evolved solution to the threat of infection, one may expect individuals who reside in areas where the threat of infectious disease is relatively high to exhibit greater disgust sensitivity than those in areas where the threat of infectious disease is lower. That no such relationship has been found, at first blush, appears to suggest that disgust sensitivity does not serve its alleged adaptive function. However, this research used an index of infectious disease prevalence that did not differentiate between diseases caused by different classes of pathogens. This index is necessarily simplistic, as specific classes of pathogens vary as a function of certain environmental features (e.g., temperature and precipitation; Guernier et al. 2004). As such, it is possible that relationships between disgust sensitivity and pathogen richness or prevalence may emerge when considering pathogen classes and their most common vectors separately. For example, individuals living in regions especially rich in viral-induced illnesses (but not macroparasites) may be more attuned to infection cues in conspecifics—a primary vector for viral transmission—than those living in regions where macroparasites pose more of a burden. Further, given that unique immune responses to different types of pathogens also have distinct metabolic and oxidative costs (e.g., see defence vs. tolerance: Allen and Sutherland 2014; Medzhitov et al. 2012; Wang et al. 2019), the selection pressure that pathogen density poses on behavioural disease avoidance mechanisms likely varies based on types of micro- or macro-organisms most commonly encountered in a given region.

An additional consideration, here, also arises when taking into account the relationship between pathogen prevalence and immune system activity. That is, the Tsimane, who live in a pathogen-dense ecology, have elevated levels of many immune parameters compared to industrialised populations, such as immunoglobulins and certain cells types including eosinophils (involved in parasite defences), B cells (involved in adaptive immunity), and natural killer cells (involved in innate immunity; Blackwell et al. 2016). Given recent evidence of a negative relationship between inflammation and behavioural immune system activity (e.g., Bradshaw et al. n.d.; Gassen et al. 2018), disgust sensitivity may actually be lower in regions rich in pathogens that elicit systemic inflammatory responses (e.g., intracellular pathogens). More research is needed to understand how various types of pathogen threats might differentially influence both the biological and behavioural immune system. Research in this vein, moreover, will hopefully provide opportunities for cross-disciplinary and cross-cultural collaboration.

Co-evolution between Pathogens and Host Sensory Systems

Humans and other animals are in a never-ending evolutionary arms race with pathogens. This is evident in the myriad strategies disease-causing microorganisms have for subverting and down-regulating host immune responses (Lucas et al. 2001; Olivier et al. 2005). Animals also use various sensory modalities to detect and avoid cues connoting infection risk, chief among them being olfaction (Poirotte et al. 2017; Tybur et al. 2011b). Little attention has been paid to the possibility that our ability to smell and avoid contamination may exert a selection pressure on microorganisms to make themselves and their vectors less aversive, or "smelly" and disgust-inducing (for more on olfaction and disgust see Liuzza, Chap. 7, this volume).

With food in particular, we often know almost instantaneously (via visual or olfactory inspection) whether our food is undercooked or spoiled and harbouring bacteria that may do us harm. However, research finds that these "smelly" bacteria that render your food inedible are not typically the types that cause illness in humans (Gram et al. 2002). In contrast, some of the most dangerous bacteria living in your food can only be detected in the lab. For example, *Listeria monocytogenes*, which infects approximately 1600 Americans each year, cannot be smelled, tasted, or seen (Farber and Peterkin 1991). The same is true for many of the other top causes of foodborne illness, including *Salmonella* (nontyphoidal), *Campylobacter* (multiple species), and *Escherichia coli* (O157) (Centre for Disease Control 2018). Although wildly speculative, is it possible that a taxonomy of bacterial "smelliness" would reveal a history of host olfaction-pathogen co-evolution that has endowed disease-causing bacteria with the ability to downregulate production of compounds hosts find aversive so that they can live right under our noses?

Similar reasoning applies when considering the evolution of infectious microorganisms that are spread primarily through social contact. Such pathogens, according to Nesse and Williams (1994), may downregulate virulence in order to facilitate their spread. While illness and infection are typically accompanied by a feeling of lassitude, motivating the host to conserve energy (Schrock et al. 2020), certain viruses may have evolved longer asymptomatic periods before the onset of symptoms. Take, for example, the recent COVID-19 pandemic. Individuals infected with COVID-19 are most infectious prior to the onset of any symptoms (He et al. 2020). Moreover, because viral infections induce inflammation (Kawai and Akira 2006), and inflammation has been linked to greater impulsivity (Gassen et al. 2019a), these viruses may also have evolved to manipulate their hosts' social behaviour to increase transmissibility. This reasoning, however, is entirely speculative. Future research would benefit from examining how various diseases might exploit the behaviour of hosts.

Disease Seasonality and Host Defences

One important parameter that has been essentially neglected in the behavioural immune system literature is the seasonality of disease. Rather than being constant, infectious diseases vary cyclically across seasons. For instance, vector-borne diseases and parasitic infections occur at higher levels during the warm and rainy seasons (Altizer et al. 2006; Amin 2002), and in temperate climates, viral respiratory infections predictably increase during the winter months (Moriyama et al. 2020). These seasonal peaks in infection are thought to be driven by various environmental factors that influence pathogen transmissibility and host susceptibility (Altizer et al. 2006; Dowell 2001; Fisman 2007). For example, contexts of relatively low temperature and humidity not only increase the transmissibility of the influenza virus but can also increase the host's susceptibility to infection (Moriyama et al. 2020).

Furthermore, research finds that the activities of the immune system also exhibit seasonal variation (Demas and Nelson 1998a, 1998b; Dopico et al. 2015; Gassen et al. 2019b). In the wild, the stressful conditions that accompany winter—such as shorter days, thermoregulatory demands, and reduced food availability—together compromise immune function and leave animals more susceptible to infectious illnesses (Nelson 2004; Demas and Nelson 1998a, 1998b). Winter, however, is reliably predicted annually by the shortening of days. To combat the effects of seasonal stress on infection risk, research finds that animals, even in the lab when exposed artificially to shortened days, tend to upregulate investment in immune function as winter approaches (Nelson 2004; Demas and Nelson 1998a, 1998b). More recent research finds that humans, too, exhibit more robust immune responses as the days shorten (Dopico et al. 2015; Gassen et al. 2019a).

Given that seasonality both influences pathogen transmissibility and the activities of the immune system, disgust and the behavioural immune system may also exhibit seasonal variation. For example, people may display greater disgust and behavioural avoidance in response to cues of viral infection (e.g., coughs, sneezing) during the winter months than the summer months. Behavioural immune system activity might further be influenced by changes in ambient environmental temperature. Foodborne bacteria, for instance, is more likely to grow in relatively high ambient temperature (Han et al. 2016). As such, people may exhibit more disgust towards questionable food sources when in environments with a high (vs. low) ambient temperature. To the best of our knowledge, there has been no research examining the relationship between activities of the behavioural immune system and seasonality or temperature. While these possibilities are, at the moment, mere speculations, they represent a potentially fruitful direction for researchers seeking to more fully understand how environmental factors might influence our disease avoidance psychology.

Conclusions

Diseases have been present throughout human history. Although medical advances, such as vaccines, have increased our ability to withstand the threat of illness and infection, viruses and bacteria evolve rapidly, and, despite our best efforts, are likely to pose a threat to our survival for the foreseeable future. The strength of this selection pressure has shaped both our body and our psychology to counter the persistent threat of infection. Disgust and the behavioural immune system represent aspects of our psychology that are perhaps best understood through an evolutionary lens. An evolutionary perspective bearing on how disease threats have shaped behaviour can not only provide an explanation for why we find certain things more disgusting than others, but can also help provide insights into individual-and context-based differences in disgust sensitivity. The research on disgust is still in its infancy, and there is much left to learn. Future research into the nature of disgust will continue to benefit from applying evolutionary thinking, as well as integrating theory and methods from fields like immunology, psychoneuroimmunology, and behavioural ecology, to further uncover the complex bio-behavioural tools that protect the body from the omnipresent threat of infectious illness.

References

Abbas, A. K., & Janeway, C. A. (2000). Immunology: improving on nature in the twenty-first century. *Cell*, 100(1), 129-138.

Ackerman, J. M., Hill, S. E., & Murray, D. R. (2018). The behavioral immune system: Current concerns and future directions. *Social and Personality Psychology Compass*, *12*(2), e12371.

Allen, J. E., & Sutherland, T. E. (2014). Host protective roles of type 2 immunity: parasite killing and tissue repair, flip sides of the same coin. *Seminars in Immunology*, *26*(4), 329-340.

Altizer, S., Dobson, A., Hosseini, P., Hudson, P., Pascual, M., & Rohani, P. (2006). Seasonality and the dynamics of infectious diseases. *Ecology Letters*, *9*(4), 467-484.

Amin, O. M. (2002). Seasonal prevalence of intestinal parasites in the United States during 2000. *The American Journal of Tropical Medicine and Hygiene*, *66*(6), 799-803.

Apicella, C. L., Rozin, P., Busch, J. T., Watson-Jones, R. E., & Legare, C. H. (2018). Evidence from hunter-gatherer and subsistence agricultural populations for the universality of contagion sensitivity. *Evolution and Human Behavior*, *39*(3), 355-363.

Axelsson, J., Sundelin, T., Olsson, M. J., Sorjonen, K., Axelsson, C., Lasselin, J., & Lekander, M. (2018). Identification of acutely sick people and facial cues of sickness. *Proceedings of the Royal Society B: Biological Sciences*, *285*(1870), 20172430.

Bereczkei, T., Voros, S., Gal, A., & Bernath, L. (1997). Resources, attractiveness, family commitment; reproductive decisions in human mate choice. *Ethology*, *103*(8), 681-699.

Blackwell, A. D., Trumble, B. C., Maldonado Suarez, I., Stieglitz, J., Beheim, B., Snodgrass, J. J., Kaplan, H. & Gurven, M. (2016). Immune function in Amazonian horticulturalists. *Annals of Human Biology*, *43*(4), 382-396.

Blake, K. R., Yih, J., Zhao, K., Sung, B., & Harmon-Jones, C. (2017). Skin-transmitted pathogens and the heebie jeebies: evidence for a subclass of disgust stimuli that evoke a qualitatively unique emotional response. *Cognition and Emotion*, *31*(6), 1153-1168.

Bonin, P., Thiebaut, G., Witt, A., & Méot, A. (2019). Contamination is "good" for your memory! Further evidence for the adaptive view of memory. *Evolutionary Psychological Science*, *5*(3), 300-316.

Borg, C., & De Jong, P. J. (2012). Feelings of disgust and disgust-induced avoidance weaken following induced sexual arousal in women. *PloS One*, *7*(9).

Bradshaw, H. K., Gassen, J., Prokosch, M. L., Boehm, G. W., & Hill, S. E. (under review). Control over pathogen exposure and basal immunological activity influence pathogen-avoidance motivation.

Campbell, R. L., Bynion, T. M., Forte, J., Feldner, M. T., & Adams, T. G. (2019). Specificity of Disgust in the Prediction of Behavioral Avoidance of Possible Contaminants. *Cognitive Therapy and Research*, 1-7.

Case, T. I., Repacholi, B. M., & Stevenson, R. J. (2006). My baby doesn't smell as bad as yours: The plasticity of disgust. *Evolution and Human Behavior*, *27*(5), 357-365.

Centre for Disease Control. (2018). Burden of Foodborne Illness: Findings. Retrieved June 14, 2020, from https://www.cdc.gov/foodborneburden/2011-foodborne-estimates.html.

Chan, K. Q., van Dooren, R., Holland, R. W., & van Knippenberg, A. (2019). Disgust lowers olfactory threshold: a test of the underlying mechanism. *Cognition and Emotion*, 1-7.

Chapman, H. A. (2018). Enhanced recall of disgusting relative to frightening photographs is not due to organisation. *Cognition and emotion*, *32*(6), 1220-1230.

Cohen, S., Doyle, W. J., & Skoner, D. P. (1999). Psychological stress, cytokine production, and severity of upper respiratory illness. *Psychosomatic Medicine*, *61*(2), 175-180.

Cohen, S., Janicki-Deverts, D., Doyle, W. J., Miller, G. E., Frank, E., Rabin, B. S., & Turner, R. B. (2012). Chronic stress, glucocorticoid receptor resistance, inflammation, and disease risk. *Proceedings of the National Academy of Sciences*, *109*(16), 5995-5999.

Croy, I., Olgun, S., & Joraschky, P. (2011). Basic emotions elicited by odors and pictures. *Emotion*, *11*(6), 1331–1335.

Curtis, V. A. (2007). Dirt, disgust and disease: a natural history of hygiene. *Journal of Epidemiology & Community Health*, *61*(8), 660-664.

Curtis, V., & Biran, A. (2001). Dirt, disgust, and disease: Is hygiene in our genes?. *Perspectives in Biology and Medicine*, *44*(1), 17-31.

Curtis, V., Aunger, R., & Rabie, T. (2004). Evidence that disgust evolved to protect from risk of disease. *Proceedings of the Royal Society of London. Series B: Biological Sciences*, *271* (suppl_4), S131-S133.

Curtis, V., De Barra, M., & Aunger, R. (2011). Disgust as an adaptive system for disease avoidance behaviour. *Philosophical Transactions of the Royal Society B: Biological Sciences*, *366*(1563), 389-401.

Dellschaft, N. S., Ruth, M. R., Goruk, S., Lewis, E. D., Richard, C., Jacobs, R. L., Curtis, J. M. & Field, C. J. (2015). Choline is required in the diet of lactating dams to maintain maternal immune function. *British Journal of Nutrition*, *113*(11), 1723-1731.

Demas, G. E., & Nelson, R. J. (1998a). Exogenous melatonin enhances cell-mediated, but not humoral, immune function in adult male deer mice (Peromyscus maniculatus). *Journal of Biological Rhythms*, 13(3), 245-252.

Demas, G. E., & Nelson, R. J. (1998b). Photoperiod, ambient temperature, and food availability interact to affect reproductive and immune function in adult male deer mice (Peromyscus maniculatus). *Journal of Biological Rhythms*, 13(3), 253-262.

Dinarello, C. A. (1997). Role of pro-and anti-inflammatory cytokines during inflammation: experimental and clinical findings. *Journal of Biological Regulators and Homeostatic Agents*, *11*(3), 91-103.

Dopico, X. C., Evangelou, M., Ferreira, R.C., Guo, H., Pekalski, M. L., Smyth, D. J., Cooper, N., Burren, O. S., Fulford, A. J., Hennig, B. J. & Prentice, A. M. (2015). Widespread seasonal gene expression reveals annual differences in human immunity and physiology. *Nature Communications*, 6(1), 1-13.

Dorfan, N. M., & Woody, S. R. (2011). Danger appraisals as prospective predictors of disgust and avoidance of contaminants. *Journal of Social and Clinical Psychology*, *30*(2), 105-132.

Dowell, S. F. (2001). Seasonal variation in host susceptibility and cycles of certain infectious diseases. *Emerging Infectious Diseases*, *7*(3), 369.

Drewnowski, A., & Specter, S. E. (2004). Poverty and obesity: the role of energy density and energy costs. *The American Journal of Clinical Nutrition*, *79*(1), 6-16.

Dwyer, R. E. (2007). Expanding homes and increasing inequalities: US housing development and the residential segregation of the affluent. *Social Problems*, *54*(1), 23-46.

Ersche, K. D., Hagan, C. C., Smith, D. G., Abbott, S., Jones, P. S., Apergis-Schoute, A. M., & Döffinger, R. (2014). Aberrant disgust responses and immune reactivity in cocaine-dependent men. *Biological Psychiatry*, *75*(2), 140-147.

Everett, H., & McFadden, G. (1999). Apoptosis: an innate immune response to virus infection. *Trends in Microbiology*, 7(4), 160-165.

Farber, J. M., & Peterkin, P. I. (1991). Listeria monocytogenes, a food-borne pathogen. *Microbiology and Molecular Biology Reviews*, *55*(3), 476-511.

Fernandes, N. L., Pandeirada, J. N., Soares, S. C., & Nairne, J. S. (2017). Adaptive memory: The mnemonic value of contamination. *Evolution and Human Behavior*, *38*(4), 451-460.

Fessler, D. M., Eng, S. J., & Navarrete, C. D. (2005). Elevated disgust sensitivity in the first trimester of pregnancy: Evidence supporting the compensatory prophylaxis hypothesis. *Evolution and Human Behavior*, *26*(4), 344-351.

Fisman, D. N. (2007). Seasonality of infectious diseases. *Annu. Rev. Public Health*, *28*, 127-143.

Fleischman, D. S., & Fessler, D. M. (2011). Progesterone's effects on the psychology of disease avoidance: Support for the compensatory behavioral prophylaxis hypothesis. *Hormones and Behavior*, *59*(2), 271-275.

Franceschi, C., Capri, M., Monti, D., Giunta, S., Olivieri, F., Sevini, F., Panourgia, M. P., Invidia, L., Celani, L., Scurti, M. & Cevenini, E. (2007). Inflammaging and anti-inflammaging: a systemic perspective on aging and longevity emerged from studies in humans. *Mechanisms of Ageing and Development*, *128*(1), 92-105.

Gassen, J., & Hill, S. E. (2019). Why inflammation and the activities of the immune system matter for social and personality psychology (and not only for those who study health). *Social and Personality Psychology Compass*, *13*(6), e12471.

Gassen, J., Prokosch, M. L., Makhanova, A., Eimerbrink, M. J., White, J. D., Proffitt Leyva, R. P. P., Peterman, J. L., Nicolas, S. C., Reynolds, T. A., Maner, J. K., & McNulty, J. K. (2018). Behavioral immune system activity predicts downregulation of chronic basal inflammation. *PloS One*, *13*(9), e0203961.

Gassen, J., Leyva, R. P. P., Mengelkoch, S., White, J. D., Peterman, J. L., Prokosch, M. L., Bradshaw, H. K., Eimerbrink, M. J., Corrigan, E. K., Cheek, D. J., & Boehm, G. W. (2019a). Day length predicts investment in human immune function: Shorter days yield greater investment. *Psychoneuroendocrinology*, 107, 141-147.

Gassen, J., Prokosch, M. L., Eimerbrink, M. J., Leyva, R. P. P., White, J. D., Peterman, J. L., Burgess, A., Cheek, D. J., Kreutzer, A., Nicolas, S. C., & Boehm, G. W. (2019b). Inflammation predicts decision-making characterized by impulsivity, present focus, and an inability to delay gratification. *Scientific Reports*, *9*(1), 1-10.

Girardin, S. E., Sansonetti, P. J., & Philpott, D. J. (2002). Intracellular vs extracellular recognition of pathogens–common concepts in mammals and flies. *Trends in Microbiology*, 10(4), 193-199.

Glass, S. T., Lingg, E., & Heuberger, E. (2014). Do ambient urban odors evoke basic emotions?. *Frontiers in psychology*, *5*, 340.

Gram, L., Ravn, L., Rasch, M., Bruhn, J. B., Christensen, A. B., & Givskov, M. (2002). Food spoilage—interactions between food spoilage bacteria. *International Journal of Food Microbiology*, *78*(1-2), 79-97.

Guernier, V., Hochberg, M. E., & Guégan, J. F. (2004). Ecology drives the worldwide distribution of human diseases. *PLoS Biol*, *2*(6), e141.

Gurven, M., Kaplan, H., & Supa, A. Z. (2007). Mortality experience of Tsimane Amerindians of Bolivia: regional variation and temporal trends. *American Journal of Human Biology: The Official Journal of the Human Biology Association*, *19*(3), 376-398.

Hall, O. J., & Klein, S. L. (2017). Progesterone-based compounds affect immune responses and susceptibility to infections at diverse mucosal sites. *Mucosal Immunology*, *10*(5), 1097-1107.

Han, Z., Yan, J. Q., Luo, G. G., Liu, Y., & Wang, Y. L. (2003). Leptin receptor expression in the basolateral nucleus of amygdala of conditioned taste aversion rats. *World Journal of Gastroenterology: WJG*, *9*(5), 1034.

Han, N., Mizan, M. F. R., Jahid, I. K., & Ha, S. D. (2016). Biofilm formation by Vibrio parahaemolyticus on food and food contact surfaces increases with rise in temperature. *Food Control, 70*, 161-166.

He, X., Lau, E.H., Wu, P., Deng, X., Wang, J., Hao, X., Lau, Y.C., Wong, J. Y., Guan, Y., Tan, X., & Mo, X. (2020). Temporal dynamics in viral shedding and transmissibility of COVID-19. *Nature Medicine, 26*(5), 672-675.

Hedman, E., Lekander, M., Karshikoff, B., Ljótsson, B., Axelsson, E., & Axelsson, J. (2016). Health anxiety in a disease-avoidance framework: Investigation of anxiety, disgust and disease perception in response to sickness cues. *Journal of Abnormal Psychology, 125*(7), 868.

Hoebe, K., Janssen, E., & Beutler, B. (2004). The interface between innate and adaptive immunity. *Nature Immunology, 5*(10), 971-974.

Hoefling, A., Likowski, K. U., Deutsch, R., Häfner, M., Seibt, B., Mühlberger, A., Weyers, P., & Strack, F. (2009). When hunger finds no fault with moldy corn: Food deprivation reduces food-related disgust. *Emotion, 9*(1), 50.

Hoff, E., Laursen, B., Tardif, T., & Bornstein, M. (2002). Socioeconomic status and parenting. *Handbook of parenting Volume 2: Biology and ecology of parenting, 8*(2), 231-52.

Hughes, G. C. (2012). Progesterone and autoimmune disease. *Autoimmunity Reviews, 11*(6-7), A502-A514.

Hunt, D. F., Cannell, G., Davenhill, N. A., Horsford, S. A., Fleischman, D. S., & Park, J. H. (2017). Making your skin crawl: The role of tactile sensitivity in disease avoidance. *Biological Psychology, 127*, 40-45.

Iikuni, N., Kwan Lam, Q. L., Lu, L., Matarese, G., & Cava, A. L. (2008). Leptin and inflammation. *Current Immunology Reviews, 4*(2), 70-79.

Jones, B. C., Hahn, A. C., Fisher, C. I., Wang, H., Kandrik, M., Lee, A.J., Tybur, J. M., & DeBruine, L. M. (2018). Hormonal correlates of pathogen disgust: Testing the Compensatory Prophylaxis Hypothesis. *Evolution and Human Behavior, 39*(2), 166-169.

Karamanou, M., Panayiotakopoulos, G., Tsoucalas, G., Kousoulis, A. A., & Androutsos, G. (2012). From miasmas to germs: a historical approach to theories of infectious disease transmission. *Infez Med, 20*(1), 58-62.

Kawai, T., & Akira, S. (2006). Innate immune recognition of viral infection. *Nature Immunology, 7* (2), 131-137.

King James Bible. (2008). Oxford University Press (Original work published 1769).

Klein, S. L., Marriott, I., & Fish, E. N. (2015). Sex-based differences in immune function and responses to vaccination. *Transactions of the Royal Society of Tropical Medicine and Hygiene, 109*(1), 9-15.

Kouznetsova, D., Stevenson, R. J., Oaten, M. J., & Case, T. I. (2012). Disease-avoidant behaviour and its consequences. *Psychology & Health, 27*(4), 491-506.

Kupfer, T. R., & Fessler, D. M. (2018). Ectoparasite defence in humans: relationships to pathogen avoidance and clinical implications. *Philosophical Transactions of the Royal Society B: Biological Sciences, 373*(1751), 20170207.

Lieberman, D., & Patrick, C. (2014). Are the behavioral immune system and pathogen disgust identical? *Evolutionary Behavioral Sciences, 8*(4), 244–250.

Lorenz, T. K., Heiman, J. R., & Demas, G. E. (2015). Sexual activity modulates shifts in TH1/TH2 cytokine profile across the menstrual cycle: an observational study. *Fertility and Sterility, 104* (6), 1513-1521.

Lucas, M., Karrer, U. R. S., Lucas, A., & Klenerman, P. (2001). Viral escape mechanisms–escapology taught by viruses. *International Journal of Experimental Pathology, 82*(5), 269-286.

Manstead, A. S. (2018). The psychology of social class: How socioeconomic status impacts thought, feelings, and behaviour. *British Journal of Social Psychology, 57*(2), 267-291.

Medzhitov, R., Schneider, D. S., & Soares, M. P. (2012). Disease tolerance as a defense strategy. *Science*, 335(6071), 936-941.

Michalak, N. M., Sng, O., Wang, I. M., & Ackerman, J. (2020). Sounds of sickness: Can people identify infectious disease using sounds of coughs and sneezes?. *Proceedings of the Royal Society B*, *287*(1928), 20200944.
Miller, S. L., & Maner, J. K. (2012). Overperceiving disease cues: The basic cognition of the behavioral immune system. *Journal of Personality and Social Psychology*, *102*(6), 1198.
Mor, G., & Cardenas, I. (2010). The immune system in pregnancy: a unique complexity. *American Journal of Reproductive Immunology*, *63*(6), 425-433.
Moriyama, M., Hugentobler, W. J., & Iwasaki, A. (2020). Seasonality of respiratory viral infections. *Annual Review of Virology*, *7*.
Mortensen, C. R., Becker, D. V., Ackerman, J. M., Neuberg, S. L., & Kenrick, D. T. (2010). Infection breeds reticence: The effects of disease salience on self-perceptions of personality and behavioral avoidance tendencies. *Psychological Science*, *21*(3), 440-447.
Murray, D. R., Prokosch, M. L., & Airington, Z. (2019). PsychoBehavioroimmunology: connecting the behavioral immune system to its physiological foundations. *Frontiers in Psychology*, *10*, 200.
Nelson, R. J. (2004). Seasonal immune function and sickness responses. Trends in immunology, 25 (4), 187-192.
Nesse, R. M., & Williams, G. C. (1994). *Why we get sick: The new science of Darwinian medicine*. New York: Times Books.
Newhagen, J. E. (1998). TV news images that induce anger, fear, and disgust: Effects on approach-avoidance and memory. *Journal of Broadcasting & Electronic Media*, *42*(2), 265-276.
Olatunji, B. O., Ebesutani, C., Haidt, J., & Sawchuk, C. N. (2014). Specificity of disgust domains in the prediction of contamination anxiety and avoidance: A multimodal examination. *Behavior Therapy*, *45*(4), 469-481.
Olivier, M., Gregory, D. J., & Forget, G. (2005). Subversion mechanisms by which Leishmania parasites can escape the host immune response: a signaling point of view. *Clinical Microbiology Reviews*, *18*(2), 293-305.
Olsson, M. J., Lundström, J. N., Kimball, B. A., Gordon, A. R., Karshikoff, B., Hosseini, N., Sorjonen, K., Olgart Höglund, C., Solares, C., Soop, A.. & Axelsson, J. (2014). The scent of disease: human body odor contains an early chemosensory cue of sickness. *Psychological Science*, *25*(3), 817-823.
Oum, R. E., Lieberman, D., & Aylward, A. (2011). A feel for disgust: tactile cues to pathogen presence. *Cognition and Emotion*, *25*(4), 717-725.
Park, J. H., Schaller, M., & Crandall, C. S. (2007). Pathogen-avoidance mechanisms and the stigmatization of obese people. *Evolution and Human Behavior*, *28*(6), 410-414.
Park, J. H., Van Leeuwen, F., & Chochorelou, Y. (2013). Disease-avoidance processes and stigmatization: Cues of substandard health arouse heightened discomfort with physical contact. *The Journal of Social Psychology*, *153*(2), 212-228.
Pellegrino, R., Crandall, P. G., & Seo, H. S. (2016). Using olfaction and unpleasant reminders to reduce the intention-behavior gap in hand washing. *Scientific Reports*, *6*, 18890.
Poirotte, C., Massol, F., Herbert, A., Willaume, E., Bomo, P. M., Kappeler, P. M., & Charpentier, M. J. (2017). Mandrills use olfaction to socially avoid parasitized conspecifics. *Science Advances*, *3*(4), e1601721.
Porzig-Drummond, R., Stevenson, R., Case, T., & Oaten, M. (2009). Can the emotion of disgust be harnessed to promote hand hygiene? Experimental and field-based tests. *Social Science & Medicine*, *68*(6), 1006-1012.
Prasad, A., Mumford, S. L., Louis, G. M. B., Ahrens, K. A., Sjaarda, L. A., Schliep, K. C., Perkins, N. J., Kissell, K. A., Wactawski-Wende, J., & Schisterman, E. F. (2014). Sexual activity, endogenous reproductive hormones and ovulation in premenopausal women. *Hormones and Behavior*, *66*(2), 330-338.
Romagnani, S. (1991). Human TH1 and TH2 subsets: doubt no more. *Immunology Today*, 12(8), 256-257.

Romagnani, S. (1992). Induction of TH1 and TH2 responses: a key role for the 'natural' immune response?. *Immunology Today*, *13*(10), 379-381.

Rubio-Godoy, M., Aunger, R., & Curtis, V. (2007). Serotonin–A link between disgust and immunity?. *Medical Hypotheses*, *68*(1), 61-66.

Ryan, S., Oaten, M., Stevenson, R. J., & Case, T. I. (2012). Facial disfigurement is treated like an infectious disease. *Evolution and Human Behavior*, *33*(6), 639-646.

Sacco, D. F., Young, S. G., & Hugenberg, K. (2014). Balancing competing motives: Adaptive trade-offs are necessary to satisfy disease avoidance and interpersonal affiliation goals. *Personality and Social Psychology Bulletin*, *40*(12), 1611-1623.

Saito, S., Nakashima, A., Shima, T., & Ito, M. (2010). Th1/Th2/Th17 and regulatory T-cell paradigm in pregnancy. *American Journal of Reproductive Immunology*, *63*(6), 601-610.

Saluja, S., & Stevenson, R. J. (2019). Perceptual and cognitive determinants of tactile disgust. *Quarterly Journal of Experimental Psychology*, *72*(11), 2705-2716.

Sarabian, C., Ngoubangoye, B., & MacIntosh, A. J. (2017). Avoidance of biological contaminants through sight, smell and touch in chimpanzees. *Royal Society Open Science*, *4*(11), 170968.

Sarabian, C., Belais, R., & MacIntosh, A. J. (2018). Feeding decisions under contamination risk in bonobos. *Philosophical Transactions of the Royal Society B: Biological Sciences*, *373*(1751), 20170195.

Schaller, M. (2011). The behavioural immune system and the psychology of human sociality. *Philosophical Transactions of the Royal Society B: Biological Sciences*, *366*(1583), 3418-3426.

Schaller, M. (2014). When and how disgust is and is not implicated in the behavioral immune system. *Evolutionary Behavioral Sciences*, *8*(4), 251.

Schaller, M., & Park, J. H. (2011). The behavioral immune system (and why it matters). *Current Directions in Psychological Science*, *20*(2), 99-103.

Schaller, M., Miller, G. E., Gervais, W. M., Yager, S., & Chen, E. (2010). Mere visual perception of other people's disease symptoms facilitates a more aggressive immune response. *Psychological Science*, *21*(5), 649-652.

Schrock, J. M., Snodgrass, J. J., & Sugiyama, L. S. (2020). Lassitude: The emotion of being sick. *Evolution and Human Behavior*, *41*(1), 44-57.

Sen, S., Iyer, C., Klebenov, D., Histed, A., Aviles, J. A., & Meydani, S. N. (2013). Obesity impairs cell-mediated immunity during the second trimester of pregnancy. *American Journal of Obstetrics and Gynecology*, *208*(2), 139-e1.

Sher, A., Pearce, E., & Kaye, P. (2003). Shaping the immune response to parasites: role of dendritic cells. *Current Opinion in Immunology*, 15(4), 421-429.

Sherlock, J. M., Zietsch, B. P., Tybur, J. M., & Jern, P. (2016). The quantitative genetics of disgust sensitivity. *Emotion, 16*(1), 43-51.

Stevenson, R. J., Case, T. I., & Oaten, M. J. (2011a). Effect of self-reported sexual arousal on responses to sex-related and non-sex-related disgust cues. *Archives of Sexual Behavior*, *40*(1), 79-85.

Stevenson, R. J., Hodgson, D., Oaten, M. J., Barouei, J., & Case, T. I. (2011b). The effect of disgust on oral immune function. *Psychophysiology*, *48*(7), 900-907.

Stevenson, R. J., Hodgson, D., Oaten, M. J., Moussavi, M., Langberg, R., Case, T. I., & Barouei, J. (2012). Disgust elevates core body temperature and up-regulates certain oral immune markers. *Brain, Behavior, and Immunity*, *26*(7), 1160-1168.

Stringhini, S., Berkman, L., Dugravot, A., Ferrie, J. E., Marmot, M., Kivimaki, M., & Singh-Manoux, A. (2012). Socioeconomic status, structural and functional measures of social support, and mortality: The British Whitehall II Cohort Study, 1985–2009. *American Journal of Epidemiology*, *175*(12), 1275-1283.

Sundelin, T., Karshikoff, B., Axelsson, E., Höglund, C. O., Lekander, M., & Axelsson, J. (2015). Sick man walking: Perception of health status from body motion. *Brain, Behavior, and Immunity*, *48*, 53-56.

Tybur, J. M., & Lieberman, D. (2016). Human pathogen avoidance adaptations. *Current Opinion in Psychology*, *7*, 6-11.

Tybur, J. M., Bryan, A. D., Lieberman, D., Hooper, A. E. C., & Merriman, L. A. (2011a). Sex differences and sex similarities in disgust sensitivity. *Personality and Individual Differences, 51* (3), 343-348.

Tybur, J. M., Bryan, A. D., Magnan, R. E., & Hooper, A. E. C. (2011b). Smells like safe sex: Olfactory pathogen primes increase intentions to use condoms. *Psychological Science, 22*(4), 478-480.

Tybur, J. M., Lieberman, D., Kurzban, R., & DeScioli, P. (2013). Disgust: Evolved function and structure. *Psychological Review, 120*(1), 65.

Tybur, J. M., Inbar, Y., Aarøe, L., Barclay, P., Barlow, F. K., De Barra, M., Becker, D. V., Borovoi, L., Choi, I., Choi, J. A., & Consedine, N. S. (2016). Parasite stress and pathogen avoidance relate to distinct dimensions of political ideology across 30 nations. *Proceedings of the National Academy of Sciences, 113*(44), 12408-12413.

van Hooff, J. C., van Buuringen, M., El M'rabet, I., de Gier, M., & van Zalingen, L. (2014). Disgust-specific modulation of early attention processes. *Acta Psychologica, 152*, 149-157.

Vogt, J., Lozo, L., Koster, E. H., & De Houwer, J. (2011). On the role of goal relevance in emotional attention: Disgust evokes early attention to cleanliness. *Cognition and Emotion, 25* (3), 466-477.

Von Rueden, C. R., & Jaeggi, A. V. (2016). Men's status and reproductive success in 33 nonindustrial societies: Effects of subsistence, marriage system, and reproductive strategy. *Proceedings of the National Academy of Sciences, 113*(39), 10824-10829.

Wang, A., Luan, H. H., & Medzhitov, R. (2019). An evolutionary perspective on immunometabolism. *Science*, 363(6423).

Xu, R., Johnson, A. J., Liggitt, D., & Bevan, M. J. (2004). Cellular and humoral immunity against vaccinia virus infection of mice. *The Journal of Immunology, 172*(10), 6265-6271.

Zsok, F., Fleischman, D. S., Borg, C., & Morrison, E. (2017). Disgust trumps lust: women's disgust and attraction towards men is unaffected by sexual arousal. *Evolutionary Psychological Science, 3*(4), 353-363.

Chapter 4
Pathogen and Toxin Disgust in Rodents

Martin Kavaliers, Klaus-Peter Ossenkopp, and Elena Choleris

What is disgust and how is it expressed, elicited, and regulated? Before examining the expression and neurobiology of disgust in rodents and other non-human animals it is important to be clear about the concept of disgust. As noted throughout this handbook, investigations with humans have provided a variety of categorisations of the possible functions and origins of disgust (Stevenson et al. 2019). Recent approaches contend that human disgust is a continuation of pathogen and toxin/contamination avoidance behaviours that are ubiquitous in non-human animals (e.g., Curtis 2014; Kavaliers et al. 2019b; Stevenson et al. 2019; see also Bradshaw and Gassen, Chap. 3, this volume). From an evolutionary perspective, disgust responses (or analogues of what we call 'disgust' in humans) have a key role in the avoidance of pathogens, contaminants, and toxins across species.

Although investigations of emotions have concentrated on humans, rodents and other non-humans display various cognitive, behavioural, physiological, and neurobiological responses that reflect internal states reminiscent of what in humans are considered emotions (Anderson and Adolphs 2014; Berridge 2018). Through phylogeny animals encounter similar kinds of challenges and opportunities which lead to the elicitation and expression of comparable motivational states and emotional or affective responses. Mice display stereotypic facial expressions in response to

M. Kavaliers (✉)
Department of Psychology, University of Western Ontario, London, ON, Canada

Department of Psychology, University of Guelph, Guelph, ON, Canada
e-mail: kavalier@uwo.ca

K.-P. Ossenkopp
Department of Psychology, University of Western Ontario, London, ON, Canada
e-mail: ossenkop@uwo.ca

E. Choleris
Department of Psychology, University of Guelph, Guelph, ON, Canada
e-mail: echoleri@uoguelph.ca

P. A. Powell, N. S. Consedine (eds.), *The Handbook of Disgust Research*,
https://doi.org/10.1007/978-3-030-84486-8_4

emotionally salient events associated with toxin disgust (Dolensek et al. 2020). These facial expressions encompass the fundamental features, antecedents, and neurobiological properties of human emotions: valence (positive or negative), salience, scalability, flexibility, context dependency, and persistence (Anderson and Adolphs 2014).

Rodents can not only express negatively valenced emotions such as fear, anxiety, and—as addressed here—disgust, but also responses associated with positive emotions (Panksepp and Lahvis 2011). The processing of valence labelled sensory signals provides a foundation for evolutionarily conserved affective and emotional states. The term "affect" as used here refers to valenced states that incorporate approach/withdrawal and reward/punishment associated responses, while "emotion" refers to the whole range of observed aspects of emotional or emotion-like expression and processing. In considering negative emotions in rodents, it is also important to distinguish between fear and disgust. Although both are negatively valenced states, they are qualitatively different in their behavioural and motivational underpinnings, elicited by different stimuli, and have distinct (though overlapping) neural mechanisms (Berridge 2018).

It is also necessary to distinguish between subjective emotional "feelings" and properties of emotional reactions and affective states that can be measured more objectively. The behavioural expression of emotion-like responses by rodents and other non-human animals neither implies nor necessitates a "conscious" affective state as proposed for humans. Rather, emotional behavioural responses in humans and non-human animals are considered to share core commonalities in their expression, functions, origins, and regulation (Case et al. 2020). In particular, there are proposed to be evolutionary continuities in the behavioural consequences and survival value of negative affective states and emotions such as disgust.

Pathogen Disgust

The Landscape of Disgust: A Threat to All Animals

All animals are under constant threat of exposure to and attack by parasites (the terms parasite and pathogen are used interchangeably here). This results in animals facing a so-called "landscape of disgust" (Buck et al. 2018). Parasite detection and avoidance behaviours are proposed to affect animal behaviour and ecology in ways similar to that of predators, leading to both direct consumptive (i.e., killing the host) and indirect non-consumptive effects (non-lethal "risk effects"; Buck et al. 2018). The non-consumptive effects can be considered as the physiological and neurobiological costs (e.g., altered stress, neuromodulatory and immune functions, increased energetic and metabolic costs) associated with the expression of disgust and the detection and avoidance of pathogens (Buck et al. 2018; Kavaliers et al. 2019b). Without infecting their hosts, parasites have a broad range of motivational (e.g., approach-avoidance), social and non-social behavioural, and neurobiological actions

(Kavaliers and Choleris 2018). These indirect effects of pathogens have major impacts on human and non-human animal behaviour, ecology, and community and population dynamics.

Pathogen Disgust Defence Mechanisms

A variety of defence mechanisms that have evolved to reduce the impact of pathogens can be categorised into three broad general strategies: behavioural avoidance and shifts in social interactions and "social distancing" (i.e., disgust-associated responses), immune mediated resistance and tolerance to infection (Kavaliers and Choleris 2018; Stockmaier et al. 2021). Pathogen disgust can be thought of as an affective system that evolved to detect signs of parasites, which includes viruses and bacteria (including shifts in microbiome components), fungi, protozoa, helminth worms, parasitoids, arthropods, and social parasites, and to stimulate behaviours that reduce the risk of their acquisition (Kavaliers and Choleris 2018). The ultimate causes of disgust relate to the evolutionary advantages of the avoidance and aversive behaviours that are expressed. The proximate causes correspond to the immediate environmental/social contexts and concomitant neurobiological mechanisms that trigger disgust. Proximally, disgust stimulates immediate and anticipatory aversive and avoidance behaviours that reduce the risk of the acquisition of, and contact with, the parasites and other contaminants in the environment. These protective responses, which in humans have been termed the "behavioural immune system", are associated with the detection of sickness and actual and potential infection and the elicitation of anticipatory avoidance responses, neurobiological responses, as well as appropriate affective and cognitive responses (Schaller and Park 2011; see also Bradshaw and Gassen, Chap. 3, this volume). Comparable responses to pathogens are evident in non-human animals and may similarly be interpreted as reflecting pathogen disgust and, by extension, potentially the expression of a behavioural immune system (e.g., Hamasato et al. 2017; Kavaliers and Choleris 2018; Poirotte et al. 2017).

Pathogen Disgust Avoidance Behaviours

Pathogen disgust associated behaviours include: (1) reduced interactions with individuals with whom one may share parasites; (2) detection of infective stage(s) of parasites and avoidance of all interactions; (3) detection and avoidance of cues associated with pathogens and infection; and (4) alterations of the behavioural, social, and physical environment occupied by individuals to reduce the likelihood of infection (Kavaliers and Choleris 2018). The efficiency of these avoidance strategies can vary depending on the parasite mode(s) of transmission and context under which different individuals may encounter infectious agents.

Parasite and infection avoidance can be either direct by avoiding or removing parasites themselves, or more indirect through the avoidance of conspecifics with either signs of infection or risky contaminated areas associated with infection (Hart 2011; Kavaliers and Choleris 2018). Susceptible individuals that detect infective and threatening parasites may be able to avoid them directly (e.g., mammals go into cover to avoid biting insects (Hart 2011; Kavaliers et al. 2001). If the infective stages are less detectable infection can be avoided by using cues that are reliably associated with parasites. For example, avoidance of the grooming of parasitized non-kin conspecifics (Poirotte et al. 2017; Sarabian et al. 2017), or selective foraging to avoid contaminated grazing areas (Sarabian and MacIntosh 2015).

Avoidance behaviours may also extend to individuals that are in poorer condition (e.g., old age), which are associated with enhanced likelihood of infection and/or shifts in microbiome components (Sarkar et al. 2020). Avoidance can be displayed either as a direct avoidance of cues associated with infected individuals or a more general augmented vigilance and reduced interactions with other individuals. When infective stages can be directly detected the behavioural responses may be driven in part by fear, but when they are less detectable, the responses are primarily motivated by disgust. It is also important to bear in mind that reduced social interactions can arise from infection and sickness associated changes in affective state and social behaviour (e.g., self-isolation, decreased social contacts; Lopes et al. 2016; Smith and Bilbo 2021; Stockmaier et al. 2020, 2021).

Pathogen Avoidance: Costs and Trade-Offs

The nature and extent of the disgust responses exhibited are further affected by factors such as: immediate and prior environmental and social conditions; social history; individual differences in susceptibility, including that of tolerance to infection; as well as various trade-offs. Depending on the nature of the infection and degree of susceptibility, these factors can differentially influence the behavioural and neurobiological responses elicited.

The expression of disgust can also result in ecological costs, which are a combination of energetic costs (costs of defensive avoidance/aversive behaviours) and missed opportunity costs (reduced foraging or feeding, habitat selection, mate choice). The resultant costs and trade-offs can further influence the degree of behavioural disgust and avoidance responses expressed. Every time a susceptible individual mates with an infectious partner, it exposes itself to pathogens and subsequent loss of fitness. However, the overall benefit of avoidance behaviour might be counterbalanced by the cost of missed mating opportunities. This may be particularly important when considering sexually transmitted infections, which are often minimally detectable (Lockhart et al. 1996). In a similar vein, Japanese macaques adjust the frequency with which they consume food items contaminated with faeces based on the nutritional quality of the food items and degree of their hunger (Sarabian and MacIntosh 2015). These examples highlight that, in decisions related to parasite exposure, energetic, nutritive, and social/sexual benefits may out-weigh the costs associated with infection.

Trade-offs extend to other threats faced. Animals need to balance the risks associated with pathogen avoidance and exposure to other threat. Tadpoles preferentially forage in the presence of trophically transmitted parasites rather than that of predators, likely because infection is less costly to fitness than being eaten (Koprivnikar and Penalva 2015). Rodents are exposed to a variety of potential predators and may allocate relatively more resources and time towards vigilance and predator avoidance than pathogen avoidance and disgust. In humans there are also suggestions that the expression and perception of disgust is sensitive to environmental conditions and threats (e.g., Batres and Perrett 2020).

Pathogen Disgust and Social Information

The expression of disgust involves cognitive mechanisms to anticipate, avoid, and regulate exposure to parasites. Animals can interact directly with the environment and acquire "asocial" information or they can obtain social information from the behaviour or products of other individuals. Social cognition is a process by which animals acquire, process, retain, and act on social information. This includes the acquisition and interpretation of information about others (e.g., social recognition of condition and infection status) as well as information from others (e.g., social learning of parasite detection and avoidance) (Choleris et al. 2009). Social information is used in decision making regarding the presence and extent of pathogen threat and the determination of subsequent approach/avoidance and associated and behavioural responses.

Social factors and social information have a key role in the expression of disgust. Infection by contagious parasites is a cost of sociality, with social interactions between individuals increasing the probability of parasite and pathogen exposure and transmission (Freeland 1976; Moller et al. 1993). This can also lead to trade-offs between the acquisition of information and avoidance of pathogens (Evans et al. 2020; Romano et al. 2020). This includes trade-offs between the expression of positively (non-social and social reward, social and sexual incentives) and negatively (disgust, fear, anxiety) valenced affective states.

In nature, animals experience complex social interactions, contact patterns, and networks, which are not restricted to simple dyadic (e.g., male-female) interactions (Altizier et al. 2003; Ezenwa et al. 2016; Lopes et al. 2016). Individuals that are well connected or highly central in their social networks are more likely to be infected by gastrointestinal parasites and other pathogens (Altizier et al. 2003; Kappeler et al. 2015). However, there is also a co-evolutionary relationship, such that pathogen avoidance increase as sociability increases. In this regard, infection can also increase sensitivity to social information and social interactions with kin (Smith and Bilbo 2021). Even in the absence of infection, mice with less effective physiological defences (e.g., lower immunity) show altered sociability and likely enhanced motivational states akin to disgust that could reduce their likelihood of infection.

How one responds to another individual is further affected by social cues present at an initial appraisal rather than by direction interactions with, and detailed knowledge of, that individual. Social information about potential pathogen threat acquired without any direct behavioural interactions can have significant emotional, motivational, and neurobiological actions affecting the expression of disgust. As parasites can exploit the social and sexual behaviours of the host to increase the likelihood of their own survival and transmission, organisms have evolved diverse strategies and complex aversive and avoidance behaviours. It is, however, important here to distinguish between the risk of pathogen acquisition from social contact (i.e., social transmission) versus shared space use (environmental transmission).

Disgust-related infection avoidance responses that rely on social information and social cognition include: (1) social partner and mate choice and deciding who to interact with and who to avoid (Beltran-Bech and Richard 2014; Kavaliers and Choleris 2018); (2) familiar-unfamiliar discrimination and the recognition and avoidance of strangers to exclude potentially infected conspecifics (Kavaliers et al. 2014); and (3) social learning of disgust utilising the detection of and responses to parasites shown by others (Kavaliers et al. 2017). In addition, there are a number of cognitive and neurobiological levels at which disgust can impact here. These include: (1) sensory input and salience of cues; (2) integration and processing of multimodal sensory inputs; (3) recognising, searching for, and discriminating between various individuals and; (4) turning pathogen cue induced arousal into and deciding who to either interact with or avoid.

Pathogen Disgust and Odours

Animals are exposed to a myriad of cues that give them information about the current state of their social environment. What cues individuals should pay attention to depends on the temporal and spatial availability, costs, timeliness, precision, and redundancy of those cues. Chemical information and odours are especially important for the recognition and assessment of condition and infection status in many species (Kavaliers and Choleris 2018; Olsson et al. 2014; see also Liuzza, Chap. 7, this volume). There is increasing evidence that chemosensory cues associated with parasites, sickness or inflammation, and reduced fitness can be detected and avoided by conspecifics. Hamilton and Zuk (1982) first directly hypothesised that animals should inspect a potential mate's urine and faecal odours to select for parasite free/resistant status. Olfactory information from freshly produced odours can provide an indication of the current condition and infection status of another individual before any actual social interactions occur. Odour-based recognition of infection status is important for determining whether to engage in behavioural interactions (i.e., approach or avoidance) and the nature of the subsequent social behaviours and interactions (e.g. affiliation, aggression, mate choice, and sexual behaviour;

Kavaliers and Choleris 2018). Female and male rodents recognise and display avoidance responses to the odours of individuals infected with a variety of parasites and infectious components (e.g., influenza virus; nematode and protozoan parasites; ectoparasites, such as lice; bacterial endotoxin, including lipopolysaccharide [LPS], the cell wall of Gram-negative bacteria; and specific immune factors; Kavaliers and Choleris 2018).

Odours of infected individuals also affects nociceptive (pain) sensitivity resulting in the induction of a decrease in pain sensitivity (antinociception, analgesia) followed by an increase in pain sensitivity (Kavaliers and Choleris 2018). Humans exposed to disgusting images showed a similar initial analgesic response followed by an increased pain sensitivity (Oaten et al. 2015). These shifts in pain sensitivity and their behavioural and neurobiological correlates can facilitate the expression of disgust responses, shifting the motivational state and enhancing the avoidance of pathogen threat and infected individuals. The initial analgesic response can facilitate the expression of aversive behavioural responses while the later increase in pain sensitivity promotes augmented vigilance to threats (Lister et al. 2020).

Odour provides recognition at a number of levels including; sex; age; reproductive status (e.g., oestrous phase, testosterone levels); social hierarchy; genetic relatedness; familiarity; dietary factors; condition and quality (e.g., infection and immune status, microbiome composition); through to true individual recognition (Choleris et al. 2009). Pathogens can directly influence the expression of disgust-associated behavioural responses in rodents through modifications in the quality and quantity of urine and faecal components, along with chemosignals from other sources (tears, saliva, preputial glands, external and internal microbiome components). In particular, gut microbiota produce a variety of metabolites that are implicated in social odour production and are influenced by infection status.

Two highly polymorphic gene complexes, the major histocompatibility complex (MHC) and the major urinary protein (MUP) cluster are especially important for odour-based recognition and assessment of condition (Hurst 2009; Stowers and Tsuang-Han 2015). Non-volatile MUPs are carriers of volatile ligands associated with individual recognition and the expression of condition (Stowers and Tsuang-Han 2015). Infection-associated changes in the MHC class I gene complex are related to condition through their linkage to immune function and volatile odour composition and production. Volatile components permit individuals to quickly identify infected individuals from a distance while non-volatile components by definition necessitate more intimate contact. In rodents, this odour-mediated recognition involves detection mechanisms and receptors in two complementary, but distinct, olfactory systems—the main olfactory system (MOS) and accessory olfactory system (AOS) and olfactory epithelium in the vomeronasal organ (VNO; Baum and Bakker 2013). There are also a number of other specialised olfactory systems. In mice the Grueneberg ganglion is implicated in the detection and of the odours of predators, stressed conspecifics, and chemical dangers associated with food and alarm (Brechbuhl et al. 2020).

In mice, vomeronasal sensory neurons (VSNrs) detect chemostimuli indicative of infection and health status, immunological fitness, and genetic compatibility (Boillat et al. 2015). In particular, the VSNrs include formyl peptide chemoreceptors that respond to specific bacterial cues (Bufe et al. 2019). In rodents the MOS is also involved in social odour detection and recognition at a distance. In humans the MOS is most probably the sole olfactory system for the detection of social odours associated with infection and pathogens.

Pathogen Disgust and Mate Responses

Disgust-associated mate responses are not static but vary with on-going and prior social conditions. Social experience and cognitive abilities have to be taken into account when considering how the threat of infection may affect the expression of disgust and mate choice. Prior familiarity with infected individuals (including their odours), previous sexual and social experience (e.g., status, stressor exposure), and previous pathogen exposure can affect the level of disgust expressed (Kavaliers and Choleris 2018). Males that are resistant to pathogens may in certain cases be more attractive to females (Adamo and Spiteri 2009; Joye and Kawecki 2019).

There are also indications of bidirectional relationships between disgust and sexual responses. In rodents, bolder and sexually aroused individuals display reduced neophobia and disgust, and enhanced risk taking with reduced responses to predators and pathogens (Kavaliers and Choleris 2018). However, in turn, prior elicitation of pathogen disgust can also inhibit sexual arousal and responses (Kavaliers et al. 2014, 2019a). The immediate social context may "fine tune" vigilance and sensitivity to pathogen threat and the level of disgust expressed.

Transmission risk can also affect the level of disgust and the avoidance of parasite threat. Avoidance behaviours and the level of disgust expressed are sensitive to the infection risk present. Highly susceptible individuals tend to show augmented avoidance behaviours (Stephenson et al. 2018). The degree of infectiveness and the stage of infection is also important here. The odours of coccidia infected male mice at an early non-infective stage, were less aversive to females than the odours from the same males at a later infective stage (i.e., pathogen shedding) (Kavaliers et al. 1997). Infection stage also affected the behaviour of the infected males who displayed maximum sexual interest and motivation when most infective. As such, reciprocal interactions between the effects of parasites on host behaviour and the propensity for infection need to be considered when examining disgust. Since the majority of organisms host multiple parasite species, these interactions likely affect how individuals react to other infected and uninfected individuals and the level of behavioural disgust expressed.

Pathogen Disgust and Unfamiliar Individuals

Social cognition is crucial for recognising and remembering familiar (in-group) and distinguishing them from unfamiliar (out-group) individuals (De Dreu and Kret

2016; Faulkner et al. 2004; Kavaliers and Choleris 2018). The risk(s) of pathogen exposure posed by unfamiliar individuals are important determinants of social interactions in animals. In humans, exposure to pathogen/parasite threat leads to a reduced interest in unfamiliar individuals and a propensity for heightened disgust, though there are also suggestions that disgust is primarily attenuated by in-group relations (De Dreu and Kret 2016). Similar responses are shown by rodents. In outbred mice oestrous females avoided the normally preferred unfamiliar individuals after being exposed to the odours of infected males. In turn exposure to unfamiliar individuals, or their odours, resulted in greater sensitivity to pathogen threat (Kavaliers et al. 2014). This may encompass disgust-related enhancement of neophobia and anxiety seen under high risk and threatening conditions.

Under semi-natural conditions, infection threat also shifted the social connectivity and interactions in wild mice consistent with greater interactions with familiar individuals and avoidance of unfamiliar ones (Lopes et al. 2016). This further indicates that disgust sensitivity and responses influence both individual social behaviour and community dynamics.

Individual and Social Learning of Pathogen Disgust

Learning also affects the elicitation and expression of pathogen disgust. Animals can learn to recognise and respond to negative dangerous and threatening factors through either individual or social learning (Choleris et al. 2009; Debiec and Olsson 2017). There is evidence from mice for social learning of the affective states of others, encompassing emotional contagion of empathy, fear, and disgust (Kavaliers et al. 2017). Individual and social learning influences subsequent decisions and the valence of various sensory cues. For example, uninfected mice that were gauged to be associated with infected individuals subsequently themselves elicited aversive and avoidance responses (Kavaliers et al. 2019a).

Both individual and social learning are involved in the recognition and behavioural avoidance of ectoparasites. Mice that were briefly exposed to attack by biting flies displayed active burying into cover to avoid the flies (Kavaliers et al. 2001, 2005). Fly naïve "observer" mice that witnessed "demonstrator" mice being attacked and bitten by biting flies and engaging in behavioural avoidance subsequently displayed kin and familiarity related socially learned avoidance responses (Kavaliers et al. 2005).

Social learning also influences pathogen modulated mate choice. Individuals utilise the mating choices of others to reduce the risks and uncertainty associated with their own choice (Galef et al. 2008; Kavaliers et al. 2017). Female mate-choice copying is a type of social learning that occurs when a female's likelihood of mating with a male is influenced by the apparent direct or indirect choice of another female. The disgust responses normally displayed to an infected male were markedly reduced when the odours of an oestrous female were associated with that of a parasitized male. Although using the choice of another female and minimising disgust may under certain circumstances be adaptive (e.g., low availability of males; males of low quality) it may also increase the risk of infection. For example, the incidence of socially contagious and sexually transmitted infections in non-human primates is positively associated with social learning and social interactions (McCabe et al. 2015).

Copying of the avoidance of infected males displayed by other females is also evident. Female mice avoid males that are associated with the odour cues of infected individuals (Kavaliers et al. 2019a). This is suggestive of the copying of disgust and is similar to the "stigmatisation by association" and disgust responses proposed in humans (Oaten et al. 2011). As such, social learning influences disgust-related behaviour and parasite and pathogen avoidance in a flexible manner according to the prevailing social context.

Sex Differences in Pathogen Disgust

There is accumulating evidence for male—female differences in infection susceptibility and effects of infection (Klein and Flanagan 2016; Smith and Bilbo 2021). Female-biased disease risk avoidance occurs across species with infection significantly influencing socially related responses (Kavaliers and Choleris 2018). Often males are more susceptible to parasitic infection than females. This can be magnified by differential exposure to pathogens via sex differences in behaviours such as space use, aggression, and sexual behaviour. Males can face higher susceptibility, or exposure, to parasites as a result of the energetic costs of male-male competition for mates, high rates of contact between males and females, and increased levels of potentially immunosuppressive steroids (i.e., testosterone). Females in turn are more susceptible to direct sexually acquired infection, exhibiting stronger avoidance/aversive and hygienic behaviours and greater disgust when confronted with infectious conspecifics (Kavaliers et al. 2019b; Poirotte and Kappeler 2019). Superimposed on this is a lower risk taking by females and greater disgust-related avoidance/aversive behaviours and caution in their non-social and social and sexual interactions. However, the nature of the pathogen threat and specific context of infection exposure needs to be taken into account (Keiser et al. 2019).

Disgust, Toxin Detection, and Avoidance

Disgust responses play a pivotal role in mediating the avoidance of toxins and contaminants. Disgust here is associated with ridding the body of toxins and contaminants that have been either already ingested or to decrease the probability of their ingestion through the expression of anticipatory avoidance and aversive responses (Kavaliers et al. 2019b). In particular, the gut defense system evolved to prevent animals from consuming items that contain toxins of various sorts (Garcia et al. 1974, 1985), by detecting the toxins either before the items are ingested, during the initial oral contact based on taste characteristics, during processing of the substance in the gut, or sensing the toxin in the circulatory system.

Disgust is part of the gut defence system and is associated with a complex of behavioural and physiological defensive reflexes designed to protect the gut and ultimately the rest of the body from the effects of ingested toxins. The powerful visceral sensations (e.g., nausea) motivate additional avoidance behaviour. Toxin elicited disgust responses are speculated to have arisen from a basic system of distaste to bitter substances, which can void the mouth of toxins, to a gut-associated learned preparatory avoidance. Distaste has been considered to reflect innate behavioural responses to eject bitter substances from the mouth.

Disgust involves innate and learned (cognitive) responses that motivate avoidance of and rejection of (e.g., vomiting) suspect ingested food items. This can involve both taste aversion and taste avoidance. Aversion is associated with the active rejection of a tastant at the consummatory level, whereas avoidance occurs at the appetitive levels and involves limiting approach and ingestion (Schier and Spector 2019).

It is perhaps then not surprising that bitter taste receptors have evolved to detect toxins prior to ingestion and that these receptors and responses are evolutionarily conserved. As affective states are considered to involve an integration of interoceptive and extroceptive conditions, disgust may have expanded from an internal toxin and pathogen based food rejection system to an external pathogen/toxin disease avoidance system. These affective reactions to toxins and contaminants can be modified by situational factors and learned preferences or avoidances. In humans, conditioned and unconditioned toxin disgust is characterised by a typical facial expression and a withdrawal response that can be associated with a vomiting (emesis) episode. Rats and other rodents are incapable of vomiting lacking the musculature and brainstem pathway needed to expel harmful substances (Horn et al. 2013). However, they show a characteristic gaping response, which uses the same orofacial musculature and is topographically similar to retching in shrews during an emetic episode (Parker 2014). Furthermore, considerable behavioural evidence shows that only manipulations that produce vomiting in emetic species promotes conditioned gaping in rats (Parker 2014).

Rodents and other mammals, including humans, display taste-elicited affective reactions which reflect hedonic aspects of palatability. The positive hedonic reactions tend to be elicited by sweet tastes and the aversive reactions by bitter tastes (Berridge 2000). In rats the taste reactivity test (Grill and Norgen 1978), which involves infusions of taste solutions directly into the mouth via an intra-oral cannula, shows that a bitter taste, such as a quinine solution, elicits disgust reactions such as gapes, head shakes, forelimb flails and other behaviours indicative of bad taste. These gape reactions have been proposed to reflect affective disgust, rather than a sensory reflex, consummatory behaviours, or just avoidance (Berridge 2000; Parker 2003). In contrast a sweet taste elicits ingestive behaviours, such as rhythmic tongue protrusions and mouth movements. The taste reactivity test has been extremely useful in studying the hedonic aspects of palatability, especially changes in palatability produced by experience with the post-ingestive features of foods (Grill and Berridge 1985). Toxins, such as LiCl, have been shown to drastically alter the palatability of positive tastes, such as sucrose, when ingested with the sucrose solution or paired with the taste by intraperitoneal injection or intragastric infusion following the consumption of sucrose solution (Chambers 2015).

Conditioned Disgust

The association of taste with the effects of toxins can be studied in rats in a very precise fashion using the taste reactivity test (Spector et al. 1988). Various studies have shown that the expression of the initial positive ingestive behaviours to presentation of a novel sucrose solution will shift to the expression of the aversive disgust behaviours of gaping and head shaking, following treatment with LiCl (e.g., Cross-Mellor et al. 2009). Thus, the initially good tasting sucrose solution becomes a bad tasting solution due to the experience with the toxin. In one series of experiments rats were injected with isotonic LiCl or NaCl and then given a series of 1 min intraoral infusions of sucrose every 10 min (Eckel and Ossenkopp 1996; Ossenkopp and Eckel 1995). The rats injected with the control NaCl solution exhibited high levels of ingestive responding over the entire series of infusions. In contrast, the LiCl treated rats exhibited ingestive responding during the initial several intraoral infusions of sucrose, but then exhibited a monotonic decline in ingestive responding, with a concomitant increase in disgust responding.

Anticipatory Disgust

Anticipatory disgust (anticipatory nausea) responses are displayed when an individual is exposed to an environmental or social context that has been previously associated with toxin induced malaise (Cloutier et al. 2017, 2018; Limebeer et al. 2006). Association of a novel taste with the effects of a toxin produces a clear taste aversion (Chambers 2015; Spector et al. 1988). Exposure to a context previously associated with feelings of nausea elicits not only a robust conditioned context place avoidance (Tenk et al. 2005), but also conditioned "gaping" responses in the rat, providing a model for examining conditioned anticipatory nausea in chemotherapy patients (Limebeer et al. 2006, 2008; Parker 2014). Human patients report that with repeated chemotherapy or radiation treatment they experience nausea and sometimes vomiting upon returning to the hospital context were they received treatment, and the evidence suggests that this is a classically conditioned effect (Chambers 2015). Rats have been shown to exhibit conditioned disgust behaviour (gaping responses) when treated with LiCl and other toxins (Parker and Limbeer 2006) and this effect is related to the dose of the toxin. Thus, rats can learn and remember associations, not only between distinctive tastes and experienced nausea, but also between distinctive contexts that were previously paired with nausea. They can subsequently retrieve these associations to show aversion-related disgust behaviours, such as gaping, upon re-exposure to the context in a drug-free state (Cloutier et al. 2018). Learning that cues signal nausea, the rat then reacts to the expected nausea by showing disgust.

Social Factors and Toxin Disgust

Social factors can also serve as cues for the expression of conditioned taste avoidance as well as for conditioned disgust responses and anticipatory nausea in humans and non-humans. In humans, a chemotherapy patient has reported experiencing nausea and vomiting when they saw their oncologist in a mall (Divgi 1989). Similarly, rats showed a decrease in social interactions following re-exposure to a conditioned taste (Guitton et al. 2008). Rats can also associate a social partner with LiCl induced sickness (Boulet et al. 2016). Male rats administered LiCl and conditioned with a social partner displayed significantly more gaping responses when in the presence of that familiar individual but not an unfamiliar rat. As indicated for pathogen disgust, the roles of prior social experience and history need to be examined for toxin and disgust.

Sex Differences in Toxin Disgust

There are also sex differences in the expression of contamination/toxin elicited disgust displayed by rodents. Female rats acquired a higher magnitude of toxin-induced contextual gaping to a novel context than males (Cloutier et al. 2017, 2018). These responses also varied across the oestrous cycle with stronger conditioning occurring during proestrus in females (Cloutier et al. 2018). These responses are consistent with the sex differences in emetic responses and disgust reported from humans (Al-Shawaf et al. 2017; Olatunji et al. 2019).

Neurobiology of Pathogen and Toxin Disgust

The elicitation and expression of pathogen elicited disgust involves a variety of sensory and affective neural mechanisms and central brain networks including the social decision making network (SDMN). This encompasses the social brain network as proposed by Newman (1999), which was merged with mesolimbic reward network to form the social decision making network (O'Connell and Hofmann 2011). This incorporates a number of brain areas including the cortical and medial amygdala and cortical and sub-cortical regions, such as the nucleus accumbens, ventral tegmental area, anterior cingulate cortex, insular cortex, pre-frontal cortex, dorsal hippocampus, thalamus, paraventricular nucleus of the hypothalamus, piriform cortex, and other olfactory regions. These various brain areas are associated with detection and processing of the valence and incentive salience of social and sensory cues (Johnson et al. 2017; Marlin and Froemke 2017; Mitre et al. 2016). In particular, olfactory inputs from the AOS and MOS are conveyed to the medial amygdala where volatile and non-volatile olfactory information is integrated (Baum and Bakker 2013). As a "salience detector" the amygdala is critical to the identification and decoding of social stimuli. This information is relayed to the prefrontal cortex which is essential for social decision making as well as projecting to the mesolimbic system which influences approach and avoidance behaviours. Expression of pathogen disgust associated responses likely involves coordinated activity across these networks and the integration of sensory inputs, salience and reward values, and threat and vigilance, to elicit relevant behavioural responses.

Insular Cortex

The insular cortex in particular is considered to be integral to expression of disgust associated behavioural responses. The insular cortex is involved in the processing of aversive sensory stimuli and bodily states and exerts "top-down" control on avoidance and aversive behaviours having intimate connections with the social brain network and mesolimbic reward areas (Gehrlach et al. 2019; Rogers-Carter and Christianson 2019). Results of imaging studies have revealed that the disgust associated avoidance responses of humans exposed to LPS treated subjects and their odours were associated with increased activity in the orbitofrontal cortex, pyriform cortex, amygdala and especially the insular cortex (Regenbogen et al. 2011). In mice toxin elicited disgust facial expressions were associated with augmented neuronal activity in the anterior insular cortex and could be elicited by optogenetic activation of the anterior insular cortex (Dolensek et al. 2020). These disgust responses involved neural pathways from the anterior insular cortex to the medial amygdala, as well as an additional pathway to the basolateral amygdala that is involved in the expression of conditioned disgust (anticipatory nausea; Kayyal et al. 2019).

The insula is also involved with taste and smell and patients with insula lesions or atrophy of the insula had deficits in identifying or reacting to disgust stimuli (Schier and Spector 2019). In rats, ablation of the insular cortex prevents conditioned gaping (Kiefer and Orr 1992). Neuronal activity in the posterior insular cortex was shown to be involved in, and likely necessary for, the expression of gastrointestinal and disgust responses elicited by LiCl (Aguilar-Rivera et al. 2020). The anterior insula is further associated with the expression of anticipatory nausea (anticipatory disgust) in humans and rats (Tuerke et al. 2012: Wicker et al. 2003). Other brain regions such as the medial parabrachial nucleus and nucleus of the solitary tract have also been implicated in the expression of toxin disgust in rodents (Schier and Spector 2019).

Area Postrema

The area postrema has also been specifically implicated in the expression of toxin disgust. The area postrema is a caudal brainstem, circumventricular organ, located on the floor of the mammalian fourth cerebral ventricle (Borison 1988). This structure is highly vascularised and has a reduced blood-brain barrier providing direct access to blood borne chemicals. As well, the overlying ependyma is differentially weakened providing access to the cerebrospinal fluid (Brizzee and Klara 1984). It has been shown to contain chemosensitive neurons (Adachi and Kobashi 1988) and has been identified as a chemoreceptor trigger zone for vomiting in humans and other species displaying emesis (Borison 1988; Hornby 2001), suggesting that it is part of the gut defence system. Removal of the area postrema in rats results in not only a failure to observe conditioned taste avoidance to a variety of emetic toxins (Ossenkopp and Eckel 1994, 1995; Ossenkopp et al. 1986) but also a failure to exhibit conditioned disgust in the taste reactivity test (Eckel and Ossenkopp 1996; Ossenkopp and Eckel 1995).

The vagus nerve has also been implicated in the regulation of the expression of conditioned taste aversions and potentially the permeability of the blood brain barrier (Schier and Spector 2019). The vagus is part of the gut-brain axis and may be part of the mechanism whereby ingested pathogens may elicit their behavioural and physiological actions.

Neuromodulatory Mechanisms

The expression of disgust associated avoidance and aversive behavioural responses involves various neurotransmitters; sex steroid hormones (in particular, oestrogens and progesterone); other steroid hormones (e.g., corticosteroids, neurosteroids); and nonapeptide systems (oxytocin [OT]), arginine-vasopressin (AVP), and other neuropeptides and their receptors (e.g., thyrotopin releasing hormone [TRH]), as well as immune factors and likely microbiome components (Choleris et al. 2009; Gabor et al. 2012; Goodson 2013; Kwon et al. 2021; Marlin and Froemke 2017; O'Connell and Hofmann 2011). These various systems allow individuals to rapidly evaluate, integrate, and respond to environmental and social information derived from pathogen and toxin threats into adaptive aversive and avoidance behaviours.

Oxytocin and Disgust

The mammalian nonapeptide, oxytocin (OT), plays a major role in modulating the processing of social information and social cognition. OT has been implicated in a number of social domains including: processing of salient social stimuli, social recognition, social interactions, social learning and social memory, social reward; and sexual and parental behaviours (e.g., Choleris et al. 2009; De Dreu and Kret 2016; Kavaliers et al. 2017; Marlin and Froemke 2017; Shamay-Tsoory and Abu-Akel 2016). OT receptors are proposed to modulate the social salience network, the SDMN and related cortical and subcortical structures. As indicated these various brain areas are associated with encoding and processing the positive and negative valence and incentive salience of social and sensory cues such as odours.

The insular cortex in particular is thought to be integral to the expression of affective states, and in particular disgust, and the actions of OT. Inhibition of the insular cortex or blockade of insular OT receptors disrupts social affective behaviour in rats (Rogers-Carter et al. 2018). In mice, an insula-to-central amygdala pathway mediates anxiety-related behaviour, while oxytocinergic projections from the paraventricular nucleus to the central amygdala appear to be crucial for the discrimination of positively and negatively valenced emotional states (Ferretti et al. 2019). OT is also associated with the expression of socially determined anticipatory nausea (anticipatory disgust) in male rats (Boulet et al. 2016) possibly through actions in the anterior cingulate cortex.

OT mediates both approach and avoidance behavioural responses to positive and negative salient social information, respectively. OT acting in the nucleus accumbens and ventral tegmental area facilitates social approach and reward, while aversive contexts which elicit social vigilance and avoidance involve the bed nucleus of the stria terminalis (BNST) (Steinman et al. 2019; Williams et al. 2020). The type of behaviour observed depends on the nature of the social stimulus, social context, and sex of the individual (De Dreu and Kret 2016; Shamay-Tsoory and Abu-Akel 2016). OT enhances social avoidance and aversive responses to threatening social stimuli to a greater extent in female than male rodents (Johnson et al. 2017). OT receptor activation inhibits social approach by increasing vigilance towards unfamiliar and possibly threatening individuals, consistent with the facilitation of disgust responses (Duque-Wilckens et al. 2018). In humans intranasal OT enhanced avoidance of emotional faces indicating disgust, with this effect being greater in females than males (Theodoridou et al. 2013).

Treatment with an OT receptor antagonist attenuated the aversive and avoidance responses of male and female mice and rats to LPS treated conspecifics (Arakawa et al. 2010, 2011; Kavaliers et al. 2019a). The extent of the effects is dependent on the age and sex of the individual providing, as well as receiving, the odours associated with infection. Female mice with either deletions of the OT gene (OT knockout [OTKO] mice), or treated with a selective OT receptor antagonist, were also impaired in their avoidance and aversive responses of the odours of parasitized individuals (Kavaliers et al. 2003, 2019a). Similarly, OTKO female mice and females treated with an OT receptor antagonist were impaired in their use of social information and did not exhibit mate choice copying (Kavaliers et al. 2006).

OT and OT receptors in the medial amygdala are critical for social recognition (Choleris et al. 2003; Ferguson et al. 2001) and as such are an additional target for the altered behavioural responses to infected individuals. Male rats avoided social interactions with LPS treated individuals with this effect dependent on AVP signalling in the medial amygdala (Arakawa As indicated, OT in the insular cortex, pyriform cortex, and anterior olfactory cortex is also involved in the mediation of odour mediated social recognition and the encoding of the saliency of social odours (Marlin and Froemke 2017). This suggests that OT (and possibly AVP) may convey social salience in a variety of sensory modalities leading to a broadly based pathogen perception and expression of disgust and avoidance behaviours.

OT is also associated with the mediation of the in-group bias and out-group avoidance displayed after exposure to pathogen threat (De Dreu and Kret 2016). Treatment with an OT receptor antagonist attenuated the aversive and avoidance responses to unfamiliar males seen in female mice after exposure to infected individuals. Through its modulation of cortico-amygdala circuits, OT also permits the expression of disgust and the display of behavioural avoidance responses to unfamiliar individuals (Mitre et al. 2016). In humans, acute intranasal OT augmented positive responses to members of in-groups and negative responses to out-group members, promoting intergroup discrimination (De Dreu and Kret 2016). OT also amplified intergroup recognition and discrimination leading to enhanced vigilance/anxiety towards out-groups. However, not all studies show a consistent in-group favouritism effect of OT.

Neurotransmitters, Neuropeptides, and Neuroimmune Systems

Opioid systems in particular have been implicated in regulating social behaviours and sexual responses. Differential alterations in opioid activity have been linked to shifts in mate choice and the expression of pathogen avoidance by female mice (Kavaliers et al. 1997). In addition to the opioid system, serotonin and dopamine are implicated in the modulation of pathogen avoidance. Administration of a serotonin (5-HT) antagonist to female mice attenuated the avoidance of the odours of parasitized males while shifts in dopamine (DA) activity were associated with the decreased responses to positively valenced stimuli in contexts associated with disgust (Berridge 2018; Kavaliers and Choleris 2018). OT interacts with these various modulatory neurotransmitter systems and has been shown to influence immune activity, raising additional modes of action (Jurek and Neumann 2018). As well, the roles of the sexually dimorphic immune responses in the expression of pathogen detection and avoidance require further consideration.

Recent studies examining the attenuated mating responses of male mice to LPS treated females also revealed a role for TRH (Kwon et al. 2021). It was found that exposure to LPS treated females (or their odours) lead to the activation of the posteriomedial nucleus of the cortical amygdala which through TRH communicated to the medial amygdala. This TRH dependent input was necessary for the behavioural avoidance of LPS treated females by males.

Manipulations of 5-HT activity have also been shown to influence behaviours associated with toxin disgust. In rats serotonin receptors in the anterior insula are involved in the mediation of anticipatory nausea (Tuerke et al. 2012) with OT modulating 5-HT activity. 5-HT activity in the posterior visceral insular cortex likely produces the nausea-induced reactions of conditioned disgust (Tuerke et al. 2012) and exposing rats to a LiCl paired flavour, but not control flavour, elevates 5-HT release in this region (Limebeer et al. 2018). The activity of the insula to basolateral amygdala neurons is necessary for establishing conditioned taste aversion raising another mode of action (Kayyal et al. 2019). In addition, there is evidence linking the nucleus accumbens to the expression of toxin elicited disgust (Berridge 2018). This may encompass shifts in dopaminergic systems and a reduction in DA mediated reward responses (i.e., reduction of positive valence responses). Glutamatergic, GABAergic and endocabinnoid systems are also either directly or indirectly involved in the mediation of aspects of toxin and pathogen disgust.

Oestrogens and Disgust

Oestrogenic systems are also involved in the central regulation of affective states and responses to social information (Choleris et al. 2009, 2012; Ervin et al. 2015). Oestrogens are involved in the utilisation of social information, particularly in relation to social recognition and social learning. Oestrogen receptors are widespread through the SDMN. Oestrogens exert their actions through a number of receptors: ERα, ERβ, and the G protein coupled ER 1 (GPER1), which can mediate both rapid non-genomic and the more delayed and lasting classical genomic effects (Lymer et al. 2018; Paletta et al. 2019).

ERα and ERβ have been implicated in the expression of disgust and pathogen/toxin detection and avoidance. ERα and ERβ gene deleted mice (ERαKO and ERβKO mice) displayed minimal avoidance of and aversive responses to the odours of infected individuals (Choleris et al. 2012; Ervin et al. 2015). How oestrogenic systems may contribute to the enhanced pathogen disgust seen in females remains to be determined.

There are also sex differences in the conditioned disgust (gaping behaviour) in rats with females showing markedly greater responses than males. Oestrous phase further affects the expression of conditioned disgust with the greatest levels of conditioned disgust seen in proestrous when elevated levels of oestrogens and progesterone are present (Cloutier et al. 2018). These effects may in part be related to oestrogen enhancement of hippocampal memory formation (Paletta et al. 2019). In addition, oestradiol has been shown to produce conditioned taste avoidance when paired with a novel sucrose solution (Ossenkopp et al. 1996) and can enhance the acquisition of LiCl-induced taste aversion in castrated male rats (Lin et al. 2015). It is thus possible that the reduction of food intake elicited by oestradiol and seen after administration of ERα and ERβ receptor agonists is associated with the nausea and disgust elicited by oestrogens.

These effects of oestrogen receptors may further be related to their association with OT. OT has been implicated in the expression of the effects of oestrogen receptors on social recognition and the display of the disgust and avoidance responses to pathogen threat. There is evidence that both ERβ and GPER are involved in the regulation of the synthesis and release of OT at the level of the hypothalamus (Ervin et al. 2015). As such, it is possible that an interactive mechanism between OT and oestrogen is associated with the expression of disgust.

Progesterone and Disgust

Progesterone has been implicated in the enhancement of pathogen and toxin avoidance in women (Compensatory Prophylaxis Hypothesis; Fleischman and Fessler 2011). However, alternative and negative findings are also present. In particular, Jones et al. (2018) found no evidence that pathogen disgust tracked changes in women's progesterone levels. Peripheral administration of progesterone, and the neurosteroid, allopregnalone, enhanced pathogen disgust and the avoidance of the odours of infected males by female mice (Bressan and Kramer 2021; Kavaliers et al. 2021). How this finding may relate toxin disgust and to oestrogen and OT involvement in the modulation of pathogen disgust remains to be determined.

Conclusions and Future Directions

Considerable evidence now exists to suggest that disgust has deep evolutionary and neurobiological underpinnings. From an ecological perspective, disgust permits adaptive conscious or unconscious modifications of contact rates and disease/pathogen transmission from other individuals. In addition, disgust lends itself to the avoidance of toxins and contaminants. The mere presence of parasites/contaminants can, through the elicitation of disgust (or a non-human disgust analogue), alter host behaviour before infection takes place.

The social neuropeptide, oxytocin, as well as oestrogens, progesterone, and other neuromodulators/neurotransmitters/neuropeptides/neuroimmune components are theorised to be involved in the mediation of the social and affective aspects of disgust. In addition, disgust involves various evolutionarily conserved brain circuits. Nevertheless, despite the important behavioural and neurobiological similarities, comparisons of disgust in humans and non-humans remain a scientific and philosophical challenge. Although there is a significant conservation of ultimate and proximate mechanisms across species, disgust in humans likely may possess unique aspects. Understanding further the mechanisms underlying the display of disgust and how pathogen/toxin recognition and avoidance are regulated in both females and males is critical for a fuller understanding of human and non-human animal behaviour.

Acknowledgements The studies described here were supported by Natural Sciences and Engineering Research Council of Canada (NSERC) discovery grants to E.C. and. M.K and K-P.O.

References

Adachi, A., & Kobashi, M. (1988). Electrophysiological analysis of chemosensitive neurons within the area postrema of the rat. *Prog. Brain Res. 74*, 77-83.

Adamo, S. A., & Spiteri, R. J. (2009). He's healthy, but will he survive the plague? Possible constraints on mate choice for disease resistance. *Anim. Behav.*, *77*, 67-78.

Aguilar-Rivera, M., Kin, S., Coleman, T. P., Maldonado, P. E., & Torrealba, F. (2020). Interoceptive insular cortex participates in sensory processing of gastrointestinal malaise and associated behaviors. *Sci. Rep.*, *20*, 21642.

Altizier, S., Nunn, C. L., Thrall, P. H., Gittleman, J. L., Antonovics, J., Cunningham, A. A., Dobson, A. P., Ezenwa, V., Jones, K. E., Pedersen, A. B., Poss, M., & Pulliam, J. R. C. (2003). Social organization and parasite risk in mammals. Integrating theory and empirical studies. *Ann. Rev. Ecol. Sys.*, *34*, 517-547.

Al-Shawaf, L., Lewis, D. M. G., & Buss, D. M. (2017). Sex differences in disgust: Why are women more easily disgusted than men. *Emot. Rev. 10*, 1-12.

Anderson, D. J., & Adolphs, R. (2014). A framework for studying emotions across species. *Cell*, *157*, 187-200.

Arakawa, H., Arakawa, K., & Deak, T. (2010). Oxytocin and vasopressin in the medial amygdala differentially modulate approach and avoidance behaviour towards illness-related social odour. *Neuroscience*, *171*, 1141-1151.

Arakawa, H., Cruz, S., & Deak, T. (2011). From models to mechanisms: odorant communication as a key determinant of social behavior in rodents during illness associated states. *Neurosci. Biobehav. Rev.*, *35*, 1916-1922.

Batres, C., & Perrett, D. I. (2020). Pathogen disgust sensitivity changes according to the perceived harshness of the environment. *Cogn. Emo. 34*, 377-383.

Baum, M. J., & Bakker, J. (2013). Roles of sex and gonadal steroids in mammalian pheromonal communication. *Front. Neuroendocrinol.*, *34*, 268-284.

Beltran-Bech, S. & Richard, F.-J. (2014). Impact of infection on mate choice. *Anim. Behav.*, *90*, 159-170.

Berridge, K. C. (2000). Measuring hedonic impact in animals and infants: microstructure of affective taste reactivity patterns. *Neurosci. Biobehav. Rev.*, *24*, 173-198.

Berridge, K. C. (2018). Evolving concepts of emotion and motivation. *Front. Psychol.*, *9*, 1-20.

Boillat, N., Challet, L., Rossier, D., Kan, C., Carelton, A., & Rodriguez, I. (2015). The vomeronasal system mediates sick conspecific avoidance. *Curr. Biol.*, *25*, 251-255.

Borison, H. L. (1988). Area postrema: Chemoreceptor circumventricular organ of the medulla oblongata. *Prog. Neurobiol.*, *32*, 351-390.

Boulet, N. P., Cloutier, C. J., Ossenkopp, K.-P., & Kavaliers, M. (2016). Oxytocin, social factors and the expression of conditioned disgust (anticipatory nausea) in male rats. *Behav. Pharm.*, *27*, 718-725.

Brechbuhl, J., de Valliere, A., Wood, D., Tosato, M. N., & Broillet, M.-C. (2020). The grueneberg ganglion controls odor-driven food choices in mice under threat. *Comm. Biol*, *3*, 533.

Bressan, P., & Kramer, P. (2021). Progesterone does raise disgust. *Horm. Behav.* Advance online publication. https://doi.org/10.1016/j.yhbeh.2021.104937

Brizzee, K. R., & Klara, P. M. (1984). The structure of the mammalian area postrema. *Fedn. Proc.*, *43*, 2944-2948.

Buck, J. C., Weinstein, S. B., & Young, H. S. (2018). Ecological and evolutionary consequences of parasite avoidance. *Trends Ecol. Evol.*, *33*, 619-632.

Bufe, B., Teuchert, Y., Schmid, A., Pyrski, M., Perez-Gomez, A., Eisenbeis, J., Timm, T., Ishis, T., Lochnit, G., Bischoff, M., Mombaerts, P., Leinders-Zufall, T., & Zufall, F. (2019). Bacterial MgrB peptide activates chemoreceptor Fpr3 in mouse accessory olfactory system and drives avoidance behavior. *Nature Neurosci.*, *10*, 4889.

Case, T. I., Stevenson, R. J., Byrne, R. W., & Hobaiter, C. (2020). The animal origins of disgust: reports of basic disgust in nonhuman great apes. *Evol. Behav. Sci.*, *14*, 231-260.

Chambers, K. C. (2015). Conditioned taste aversions. In: R.L. Doty (Ed.) *Handbook of olfaction and gustation* (3rd ed., pp. 865-886). John Wiley & Sons.

Choleris, E., Clipperton-Allen, A. E., Phan, A., & Kavaliers, M. (2009). Neuroendocrinology of social information processing in rats and mice. *Front. Neuroendocrinol.*, *30*, 442-458.

Choleris, E., Gustafsson, J.-A., Korach, K. S., Mugila, L. J., Pfaff, D. W., & Ogawa, S. (2003). An estrogen dependent 4-gene micronent regulating social recognition: a study with oxytocin and estrogen receptor –α and –β knockout mice. *Proc. Natl. Acad. Sci. USA*, *100*, 6192-6197.

Choleris, E., Phan, A., Clipperton-Allen, A. E., Valsecchi, P., & Kavaliers, M. (2012). Estrogenic involvement in social learning, social recognition and pathogen avoidance. *Front. Neuroendocrinol.*, *33*, 140-159.

Cloutier, C. J., Kavaliers, M., & Ossenkkopp, K.-P. (2017). Sex differences in LiCl-induced context related conditioned disgust behaviors in rats and inhibition following immune system stimulation. *Pharmacol. Biochem. Behav.*, *152*, 4-12.

Cloutier, C. J., Zevy, D., Kavaliers, M., & Ossenkopp, K.-P. (2018). Conditioned disgust in rats (anticipatory nausea) to a context paired with the effects of the toxin LICl: Influences of sex and the estrous cycle. *Pharmacol. Biochem. Behav.*, *173*, 51-57.

Cross-Mellor, S. K., Foley, K. A., Parker, L. A., & Ossenkopp, K.-P. (2009). Lipopolysaccharide dose-dependently impairs the rapid toxin-induced gustatory conditioning in rats consuming a sucrose-lithium chlorides solution. *Brain Behav. Immun.*, *23*, 204-216.

Curtis, V. A. (2014). Infection-avoidance behaviors in humans and other animals. *Trends Immunol.*, *35*, 457-464.

Debiec, J., & Olsson, A. (2017). Social fear learning: from animal models to human function. *Trends Cogn. Sci.*, *21*, 546-555.

De Dreu, C. K. W., & Kret, M. E. (2016). Oxytocin conditions intergroup relations through upregulated in-group empathy, cooperation, conformity, and defense. *Biol. Psych.*, *79*, 165-171.

Divgi, A. B. (1989). Oncologist-induced vomiting: the IGVID syndrome? *New Engl. J. Med.*, *320*, 189-190.

Dolensek, N., Gehrlach, D. A., Klein, A. S., & Gogolla, N. (2020). Facial expressions of emotion states and their neuronal correlates in mice. *Science*, *368*. 89-94.

Duque-Wilckens, N., Steinman, M. Q., Busnelli, M., Chini, B., Yokoyama, S., Pham, M., Laredo, S. A., Hao, R., Perkeybile, A. M., Minie, V. A., Tan, P. R., Bales, K. L., & Trainor, B. C. (2018). Oxytocin receptors in the anteriomedial bed nucleus of the stria terminalis promote stress-induced social avoidance in females. *Biol. Psych.*, *83*, 203-213.

Eckel, L. A., & Ossenkopp, K.-P. (1996). Area postrema mediates the formation of rapid, conditioned palatability shifts in lithium treated rats. *Behav. Neurosci.* *110*, 202-212.

Ervin, K. S. J., Lymer, J. M., Matta, R., Clipperton-Allen, A. E., Kavaliers, M., & Choleris, E. (2015). Estrogen involvement in social behavior in rodents: rapid and long-term actions. *Horm. Behav.*, *74*, 53-76.

Evans, J. C., Silk, M. J., Boogert, M. J., Hodgdon, D. J. (2020). Infected or informed? Social structure and the simultaneous transmission of information and infectious disease? *Oikos*, *129*, 1271-1288.

Ezenwa, V., Ghai, R. R., McKay, A. F., & Williams, A. E. (2016). Group living and pathogen infection revisited. *Curr. Op. Behav. Sci.*, *12*, 66-72.

Faulkner, J., Schaller, M., Park, J. H., & Duncan, L. A. (2004). Evolved disease avoidance mechanisms and contemporary xenophobic attitudes. *Group Proc. Intgr. Rel.*, *7*, 33-53.

Ferguson, J. N., Aldag, J. M., Insel, T. R., & Young, L. J. (2001). Oxytocin in the medial amygdala is essential for social recognition in the mouse. *J. Neurosci.*, *21*, 8278-8285.

Ferretti, V., Maltese, F., Contarini, G., Nigro, M., Bonavia, A., Huang, H., Gigliucci, V., Morelli, G., Scheggia, D., Manago, F., Castellani, G., Lefevre, A., Cancedda, L., Chini, B., Grinevich, V., & Papaleo, F. (2019). Oxytocin signaling in the central amygdala modulates emotion discrimination in mice. *Curr. Biol.*, *29*, 1-16.

Fleischman, D. S., & Fessler, D. M. T. (2011). Progesterone's effects on the psychology of disease avoidance: Support for the compensatory behavioral prophylaxis hypothesis. *Horm. Behav.*, *59*, 271-275.

Freeland, W. J. (1976). Pathogens and the evolution of primate sociality. *Biotropica*, *8*, 12-24.

Gabor, C. S., Phan, A., Clipperton-Allen, A. E., Kavaliers, M., & Choleris, E. (2012). Interplay of oxytocin, vasopressin, and sex hormones in the regulation of social recognition. *Behav. Neurosci.*, *126*, 97-109.

Galef, B. G., Lim, T. C. W., & Gilbert, G. S. (2008). Evidence of mate choice copying in Norway rats, *Rattus norvegicus*. *Anim. Behav.*, *75*, 117-1123.

Garcia, J., Hankins, W. G., & Rusiniak, K. W. (1974). Behavioral regulation of the milieu interne in man and rat. *Science*, *185*, 824-831.

Garcia, J., Laister, P. S., Bermudez-Rattoni, F., & Deems, D. A. (1985). A general theory of aversion learning. *Ann. NY Acad. Sci.*, *443*, 8-21.

Gehrlach, D. A., Dolensek, N., Klein, A. S., Chowdhury, R. R., Matthys, A., Junghanel, M., Gaitanos, T. N., Podgornik, A., Black, T. D., Vaka, R. N., Conzelmann, K.-K., & Gogolla, N. (2019). Aversive state processing in the posterior insular cortex. *Nat. Neurosci.*, *22*, 1424-1437.

Goodson, J. L. (2013). Deconstructing sociality, social evolution and relevant nonapeptide functions. *Psychoneuroendocrinology*, *38*, 465-478.

Grill, H. J., & Berridge, K. C. (1985). Taste reactivity as a measure of the neural control of palatability. *Prog. Psychobiol. Physiol. Psychol.*, *11*, 1-61.

Grill, H. J., & Norgen, R. (1978). The taste reactivity test I. Mimetic responses to gustatory stimuli in neurologically intact rats. *Brain Res. 143*, 263-279.

Guitton, M. J., Klin, Y., & Dudai, Y. (2008). Taste-dependent sociophobia: When food and company do not mix. *Behav. Brain Res.*, *191*, 148-152.

Hamasato, E. K., Lovelock, D., Palermo-Neo, J., & Deak, T. (2017). Assessment of social behaviour directed toward sick partners and its relation to central cytokine expression in rats. *Physiol. Behav.*, *182*, 128-136.

Hamilton, W. D., & Zuk, M. (1982). Heritable true fitness and bright birds: a role for parasites. *Science*, *218*, 384-387.

Hart, B. L. (2011). Behavioural defence in animals against pathogens and parasites: parallels with the pillars of medicine in humans. *Phil. Trans. R. Soc B*, *366*, 3406.

Horn, C. C., Kimball, B. A., Wang, H., Kaus, J., Dienel, S., Nagy, A., Gaithright, G. R., Yates, B. J., & Andrews, P. L. R. (2013). Why can't rodents vomit? A comparative behavioural, anatomical, and physiological study. *PLOS One*, *8*, e60537.

Hornby, P. J. (2001). Central neurocircuitry associated with emesis. *Amer Jour.Med., 111*, 106-112, 2001.

Hurst, J. L. (2009). Female recognition and assessment of males through scent. *Behav. Brain. Res.*, *22*, 295-303.

Johnson, Z. V., Walum, J. H., Xiao, Y., Riefkohl, P. C., & Young, L. J. (2017). Oxytocin receptors modulate a social salience network in male prairie voles. *Horm. Behav.*, *87*, 16-24.

Jones, R. C., Hahn, A. C., Fisher, C. I., Wang, H., Kandrik, M., Lee, A. J., Tybur, J. M., & DeBruine, L. M. (2018). Hormonal correlates of pathogen disgust: testing the compensatory prophylaxis hypothesis. *Evol. Hum. Behav. 39*, 166-169.

Joye, P., & Kawecki, T. J. (2019). Sexual selection favours good or bad genes for pathogen resistance depending on males' pathogen exposure. *Proc. R. Soc. B*, *286*, 20190226.

Jurek, B., & Neumann, I. D. (2018). The oxytocin receptor: from intracellular signaling to behavior. *Physiol. Rev.*, *98*, 1805-1908.

Kappeler, P. M., Cremer, S., & Nunn, C. L. (2015). Sociality and health: impacts of sociality on disease susceptibility and transmission in animal and human societies. *Phil. Trans. Roy. Soc. B*, *370*, 20140116.

Kavaliers, M., Bishnoi, I. R., Ossenkopp, K.-P., & Choleris, E. (2021). Differential effects of progesterone on social recognition and the avoidance of pathogen threat by female mice. *Horm. Behav. 127*, 104873.

Kavaliers, M., & Choleris, E. (2018). The role of social cognition in parasite and pathogen avoidance. *Phil. Trans. Roc. Soc. B*, *373*, 20170206.

Kavaliers, M., Choleris, E., Ågmo, A., Braun, W. J., Colwell, D. D., Muglia, L. J., Ogawa, S., & Pfaff, D.W. (2006). Inadvertent social information and the avoidance of parasitized male mice: a role for oxytocin. *Proc. Natl. Acad. Sci. USA*, *103*, 4293-4298.

Kavaliers, M., Choleris, E., & Colwell, D. D. (2001). Learning from others to cope with biting flies: social learning of fear-induced conditioned analgesia and active avoidance. *Behav. Neurosci.*, *115*, 661-674.

Kavaliers, M., Colwell, D. D., & Choleris, E. (2005). Kinship, familiarity and social status modulate social learning about "micropredators" (biting flies) in deer mice. *Behav. Ecol. Sociobiol.*, *58*, 60-71.

Kavaliers, M., Colwell, D. D., Choleris, E., Agmo, A., Mugila, L. J., Ogawa, S., & Pfaff, D. W. (2003). Impaired discrimination of and aversion to parasitized male odors by female oxytocin knockout mice. *Genes. Brain. Behav.*, *2*, 220-230.

Kavaliers, M., Colwell, D. D., Cloutier, C. J., Ossenkopp, K.-P., & Choleris, E. (2014). Pathogen threat and unfamiliar males rapidly bias the social responses of female mice. *Anim. Behav.*, *97*, 105-111.

Kavaliers, M., Colwell, D. D., Ossenkopp, K.-P., & Perrot-Sinal, T. S. (1997). Altered responses to female odors in parasitized male mice: neuromodulatory mechanisms and relations to female choice. *Behav. Ecol. Sociobiol.*, *40*, 373-384.

Kavaliers, M., Colwell, D. D., Wah, D. T. O., Bishnoi, I. R., Ossenkopp, K.-P., & Choleris, E. (2019a). Conspecific infection threat rapidly biases the social responses of female mice: involvement of oxytocin. *Horm. Behav.*, *113*, 67-75.

Kavaliers, M., Matta, R., & Choleris, E. (2017). Mate-choice copying, social information processing, and the roles of oxytocin. *Neurosci. Biobehav. Rev.*, *72*, 232-242.

Kavaliers, M., Ossenkopp, K. P., & Choleris, E. (2019b). Social neuroscience of disgust. *Genes Brain Behav.*, *18*, e12508.

Kayyal, H., Yiannakas, A., Chandran, S. K., Khamaisy, M., Sharma, V., & Rosenblum, K. (2019). Activity of insula to basolateral amygdala projecting neurons is necessary and sufficient for taste valence representation. *J. Neurosci. 89*, 9369-9382.

Keiser, C. N., Rudolf, V. H. W., Luksik, M. C., & Saltz, J. B. (2019). Sex differences in disease avoidance behaviour vary across modes of pathogen exposure. *Ethology*, *126*, 304-312.

Kiefer, S. W., & Orr, M. R. (1992). Taste avoidance, but not aversion, learning in rats lacking gustatory cortex. *Behav. Neurosci. 106*, 140-146.

Klein, S. L., & Flanagan, K. L. (2016). Sex differences in immune response. *Nat. Rev. Immunol. 16*, 626-638.

Koprivnikar, J., & Penalva, I. (2015). Lesser of two evils? Foraging choices in response to threats of predation and parasitism. *PLOS One*, *19*, 0116589.

Kwon, J.-K., Ryu, C., Lee, H., Sheffield, A., Fan, J., Cho, D. H., Bigler, S., Sullivan, H. A., Choe, H. K., Wickersham, I. R., Heiman, M., & Choi, G. B. (2021). An amygdala circuit that suppresses social engagement. *Nature*. Advance online publication. https://doi.org/10.1038/s41586-021-03413-6

Limebeer, C. L., Hall, G., & Parker, L. A. (2006). Exposure to a lithium-paired context elicits gaping in rats: a model of anticipatory nausea. *Physiol. Behav.*, *88*, 398-403.

Limebeer, C. L., Krohn, J. P., Cross-Mellor, S. K., Litt, D. E., Ossenkopp, K.-P., & Parker, L. A. (2008). Exposure to a context previously associated with nausea elicits conditioned gaping in rats: A model of anticipatory nausea, *Behav. Brain Res. 187*, 33-40.

Limebeer, C. L., Rock, E. M., Sharkey, K. A., & Parker, L. A. (2018). Nausea-induced 5-HT release in the interoceptive insular cortex and regulation by monoacylglycerol lipase (MAGL) inhibition and cannabidiol. eNeuro, *5*, 0256-18

Lin, S. F., Tsai, Y. F., Tai, M. Y., & Yeh, K. Y. (2015). Estradiol enhances the acquisition of lithium chloride-induced conditioned taste aversion in castrated male rats. *Sci. Nat., 102*, 5.

Lister, K. C., Bouchard, S. M., Markova, T., Aternail, A., Denecli, P., Pimentel, S. D., Majeed, M., Austin, J.-S., Williams, A. C., & Mogil, J. S. (2020). Chronic pain produces hypervigilance to predator odor in mice. *Curr. Biol. 15*, R866-R867.

Lockhart, A. B., Thrall, P. H., & Antonovics, J. (1996). Sexually transmitted diseases in animals; Ecological and evolutionary implications. *Biol. Rev., 71*, 415-471.

Lopes, P. C., Block, P., & Konig, B. (2016). Infection-induced behavioural changes reduce connectivity and the potential for disease spread in wild mice contact networks. *Sci. Rep., 6*, 31790

Lymer, J. M., Shepard, P. A. S., Kuun, T., Blackman, A., Jani, N., Mahbub, S., & Choleris E. (2018). Estrogens and their receptors in the medial amygdala rapidly promote social recognition in female mice. *Psychoneuroendocrinology, 89*, 30-38.

Marlin, B. J., & Froemke, R. C. (2017). Oxytocin modulation of neural circuits for social behavior. *Dev. Neurobiol., 77*, 169-182.

McCabe, C. M., Reader, S. M., & Nunn, C. L. (2015). Infectious disease, behavioral flexibility and the evolution of culture in primates. *Proc. R. Soc. B, 282*, 20140862.

Mitre, M., Marlin, B. J., Schiavo, J. K., Morina, E., Norden, S. E., Hackett, T. A., Aoiki, C. J., Chao, M. V., & Froemke, R. C. (2016). A distributed network for social cognition enriched for oxytocin receptors. *J. Neurosci., 36*, 2517-2535.

Moller, A. P., Dufva, R., & Allander, K. (1993). Parasites and the evolution of host social behavior. *Adv. Stud. Behav., 22*, 65-102.

Newman, S. W. (1999). The medial extended amygdala in male reproductive behavior. A node in the mammalian social behavior network. *Ann. N. Y. Acad. Sci., 877*, 242-257.

Oaten, M. J., Stevenson, R. J., & Case, T. I. (2011). Disease avoidance as a functional basis for stigmatization. *Phil. Trans. R. Soc. B, 366*, 3433-3452.

Oaten, M. J., Stevenson, R. J., & Case, T. I. (2015). The effect of disgust on pain sensitivity. *Physiol. Behav., 138*, 107-112.

O'Connell, L. A., & Hofmann, H. A. (2011). The vertebrate mesolimbic reward system and social behavior network: a comparative synthesis. *J. Comp. Neurol., 519*, 3599-3639.

Olatunji, B. O., Taylor, A., & Zald, D. (2019). Sex differences in the etiology of disgust sensitivity: A preliminary behavioral genetic analysis. *J. Anx. Disord., 65*, 41-46.

Olsson, M. J., Lundstrom, J. N., Kimball, B. A., Gorden, A. R., Karshikoff, B., Hosseini, N., Sorjonen, K., Hoglund, O., Solares, C., Soop, A., Axellson, J. B., & Lekander, M. (2014). The scent of disease: human body odor contains an early chemosensory cue of sickness. *Psychol. Sci., 25*, 617-823.

Ossenkopp, K.-P., & Eckel, L. A. (1994) Area postrema as a brainstem chemosensor in defense against ingested toxins. *Adv. Biosci., 93*. 485-490.

Ossenkopp, K.-P., & Eckel, L. A. (1995) Toxin-induced changes in taste reactivity responses and the role of the chemosensitive area postrema. *Neurosci. Biobehav. Rev. 19*, 99-108.

Ossenkopp, K.-P., Rabi, Y. J., & Eckel, L. A. (1996). Oestradiol-induced taste avoidance is the result of a conditioned palatability shift. *NeuroReport, 7*, 2777-27.

Ossenkopp, K.-P., Sutherland, C., & Ladowsky, R. L. (1986). Motor activity changes and conditioned taste aversions induced by administration of scopolamine in rats: Role of the area postrema. *Pharmacol. Biochem. Behav. 25*, 269-276.

Paletta, P., Sheppard, P. A. S., Matta, R., Ervin, K. S. J., & Choleris, E. (2019). Rapid effects of estrogens on short term memeory: possible mechanisms. *Horm. Behav., 104*, 88-99.

Panksepp, J. B., & Lahvis, G. P. (2011). Rodent empathy and affective neuroscience. *Neurosci. Biobehav. Rev., 35*, 1864-1875.

Parker, L. A. (2003). Taste avoidance and taste aversion: evidence for two different processes. *Anim. Learn. Behav.*, *31*, 165-172.
Parker, L. A. (2014). Conditioned flavour avoidance and conditioned gaping: rat models of conditioned nausea. *Eur. J. Pharmacol. 722*, 122-133.
Parker, L. A., & Limbeer, C. L. (2006). Conditioned gaping in rats: a selective measure of nausea. *Auton. Neurosci. Basic Clin.*, *129*, 36-41.
Poirotte, C., & Kappeler, P. M. (2019). Hygenic personalities in wild grey mouse lemurs vary adaptively with sex. *Proc. R. Soc. B*, *286*, 20190863.
Poirotte, C., Massol, F., Herbert, A., Willaume, E., Borno, P. M., Kappeler, P. M., & Charpentier, G. (2017). Mandrills use olfaction to socially avoid parasitized conspecifics. *Sci. Adv.*, *3*, e160172.
Regenbogen, C., Axelsson, J., Lasselin, J., Porada, D. K., Sundelin, T., Peter, M. G., Lekander, M., Lundstrom, J. N., & Olsson, M. J. (2011). Behavioral and neural correlates to multisensory detection of sick humans. *Proc. Natl. Acad. Sci. USA*, *114*, 6499-6405.
Rogers-Carter, M. M., & Christianson, J. P. (2019). An insular view of the social decision-making network. *Neurosci. Biobehav. Rev.*, *103*, 119-132.
Rogers-Carter, M. M., Varela, J. A., Gribbons, K.-B., Pierce, A. F., McGoey, M. T., Ritchey, M., & Christianson, J. B. (2018). Insular cortex mediates approach and avoidance responses to social affective stimuli. *Nat. Neurosci.*, *21*, 404-414.
Romano, V., MacIntosh, A. J. J., & Sueur, C. (2020). Stemming the flow: information, infection and social evolution. *Trends Ecol. Evol.*, *39*, 849-845.
Sarabian, C., & MacIntosh, A. J. J. (2015). Hygenic tendencies correlate with low geohelminth infections in free-ranging macaques. *Biol. Lett.*, *11*, 20150757
Sarabian, C., Ngoubangoye, B,, & MacIntosh, A. J. J. (2017). Avoidance of biological contaminants through sight, smell and touch in chimpanzees. *R. Soc. Open Sci.*, *4*, 170968.
Sarkar, A., Harty, S., Johnson, K. V.-A., Moeller, A. H., Archie, E.A., Schell, L. D., Carmody, R. N., Clutton-Brock, T. H., Dungar, R. I. M., & Burnet, P. W. J. (2020). Microbial transmission in animal social networks and the social microbiome. *Nat Ecol. Evol.,*, *4*, 1020-1035.
Schaller, M., & Park, J. H. (2011). The behavioral immune system (and why it matters). *Curr. Dir. Psychol. Sci.*, *20*, 99-103.
Schier, L. A., & Spector, A. C. (2019). The functional and neurobiological properties of bad taste. *Physiol. Rev.*, *99*, 605-663.
Shamay-Tsoory, S. G., & Abu-Akel, A. (2016). The social salience hypothesis of oxytocin. *Biol. Psych.*, *79*, 194-202.
Smith, C., & Bilbo, S. D. (2021). Sickness and the social brain: Love in the time of COVID. *Front. Psycht.*, *12*, 633664.
Spector, A. C., Breslin, P., & Grill, H. J., (1988). Taste reactivity as a dependent measure of the rapid formation of a conditioned taste aversion: A tool for the analysis of taste-visceral associations. *Behav. Neurosci.*, *102*, 942-952.
Steinman, M. Q., Duque-Wilckens, N., & Trainor, B. C. (2019). Complementary neural circuits for divergent effects of oxytocin: social approach versus social anxiety. *Biol. Psych.*, *85*, 792-801.
Stephenson, J. F., Perkins, S. E., & Cable, J. (2018). Transmission risk predicts avoidance of infected conspecifics in Trinidadian guppies. *J. Anim. Ecol.*, *87*, 1525-1533.
Stevenson, R. J., Case, T. I., Oaten, M. J., Stafford, L., & Saluja, S. (2019). A proximal perspective on disgust. *Emot. Rev.*, *11*, 209-225.
Stockmaier, S., Bolnick, D. L., Page, R. A., & Carter, G. G. (2020). Sickness effects depend on the type of behaviour and relationship. *J. Anim. Ecol.*, *6*, 1387-1394.
Stockmaier, S., Stroeymeyt, N., Shattuck, E. C., Hawley, D. M., Meyers, L.A., & Bolnick, D. I. (2021) Infectious diseases and social distancing in nature. *Science*, *371*, eabc8881.
Stowers, L., & Tsuang-Han, K. (2015). Mammalian pheromones: emerging properties and mechanisms of detection. *Curr. Op. Neurobiol.*, *34*, 103-109.
Tenk, C. M., Kavaliers, M., & Ossenkopp, K.-P. (2005). Dose response effects of lithium chloride on conditioned place aversions and locomotor activity. *Eur. J. Pharmacol.*, *515*, 117-127.

Theodoridou, A., Penton-Voak, I. S., & Rowe, A. C. (2013). A direct examination of the effect of intranasal administration of oxytocin on approach-avoidance motor responses to emotional stimuli. *PLOS One*, *8*, e58113.

Tuerke, K. J., Limebeer, C. L., Fletcher, P. J., & Parker, L. A. (2012). Double dissociation between the regulation of conditioned disgust and taste avoidance by serotonin availability at the 5-HT3 receptor in the posterior and anterior insular cortex. *J. Neurosci.*, *32*, 13709-13717.

Wicker, B., Keysers, C., Plailly, J., Royet, J. P., Gallese, V., & Rizzolatti, G. (2003). Both of us disgusted in my insula: the common neural basis of seeing and feeling disgust. *Neuron*, *40*, 655-664.

Williams, A. V., Duque-Wilckens, N., Ramos-Maciel, S., Campi, K.L., Bhela, S., Xu, K., Jackson, K., Chini, B., Pesavento, P. A., & Trainor, B. V. (2020). Social approach and social vigilance are differentially regulated by oxytocin receptors in the nucleus accumbens. *Neuropsychopharmacology*, *45*, 1423-1430.

Chapter 5
Direct and Indirect Learning Pathways to Disgust

Gemma Reynolds and Chris Askew

Disgust, as a *food-related emotion*, or distaste (thought to be a precursor to disgust) has often been studied within an associative learning framework (e.g., Garcia et al. 1955; Rozin and Fallon 1987). However, it is only within the last few decades that disgust, as a *basic emotion*, has been considered from a conditioning perspective (e.g., Armstrong et al. 2014; de Jong 2013; Mason and Richardson 2010; Olatunji et al. 2007). This chapter aims to explore the theoretical perspectives that consider the development and acquisition of disgust as resulting from direct and indirect associative learning mechanisms. It will also briefly explore the contribution of indirect disgust-related learning experiences to the promotion of fear.

Direct Learning

Associative learning mechanisms account for the direct association between a stimulus and the threat of disease or contamination. The most well-known form of this type of learning is classical or "Pavlovian" conditioning, which involves the formation of learnt associations between two stimuli: a neutral conditioned stimulus (CS; e.g., a tone) and an unconditioned stimulus (US; e.g., a shock) that elicits an unconditioned response (UR; e.g., pain). Following CS-US pairing trials, conditioning is deemed to have occurred when CS-alone presentations elicit a conditioned response (CR; e.g., fear) related to the UR.

G. Reynolds (✉)
Department of Psychology, Middlesex University, London, UK
e-mail: G.Reynolds@mdx.ac.uk

C. Askew
School of Psychology, University of Surrey, Guildford, UK
e-mail: c.askew@surrey.ac.uk

P. A. Powell, N. S. Consedine (eds.), *The Handbook of Disgust Research*,
https://doi.org/10.1007/978-3-030-84486-8_5

It has been argued that two types of learning can occur in classical conditioning: expectancy learning and referential learning (e.g., Baeyens and De Houwer 1995). *Expectancy learning* refers to the acquisition of a CR as a result of learning that the CS predicts the occurrence of the US. For example, following pairing trials a tone (CS) may predict, and therefore activate an expectation of, the occurrence of subsequent shock (US). On the other hand, *merely referential learning* is said to occur when the CS activates a representation of the US without eliciting expectations of the US occurrence (Hermans et al. 2002b). For example, specific sights or sounds might trigger an individual's memory of traumatic childhood abuse without expectation that it is about to happen in the present. One form of referential learning is argued to be evaluative conditioning, in which the CS itself is argued to take on the positive or negative valence of the US it has been associated with (Baeyens et al. 1992). Thus, evaluative conditioning and expectancy learning are theorised to be underpinned by two different learning mechanisms, with the CS acting as a predictor of the US in expectancy learning, and the CS serving as merely a reference to the US in evaluative learning.

Expectancy Learning

In the context of disgust acquisition, expectancy learning most likely occurs when an individual experiences a neutral CS together with a disgust-related negative stimulus or event (US) and subsequently shows increased expectation that the CS will be followed by a disgust-related outcome. For example, learned avoidance (CR) for a particular food (CS) can be acquired when an individual experiences illness (US) after ingestion of that food. This is known as conditioned taste aversion or the "Garcia effect" after John Garcia who studied the effect extensively in lab experiments with rats (e.g., Garcia and Koelling 1966; see also Kavaliers et al., Chap. 4, this volume). The illness must not actually be caused by the food (indeed, Garcia frequently used radiation to induce nausea in rats after they had consumed sweetened water) but nausea will more readily become associated with taste cues than with say audiovisual cues in the environment. Thus, learned aversion in this scenario is typically only to the taste of the food and not other features of the learning event (Garcia et al. 1974). Outside of the laboratory, a survey by Garb and Stunkard (1974) found that 87% of participants reported having acquired taste aversions after associating food with gastrointestinal illness. In a study of university students, Midkiff and Bernstein (1985) discovered that 14% of aversions were aversions to alcohol (see also Logue et al. 1981). Arguably, students begin to consume greater levels of alcohol during their university years (e.g., drinking games involving spirits), which is more likely to produce symptoms (e.g., nausea) that play an important role in the conditioning of taste aversions.

Contrary to early understanding of associative learning, taste aversion can occur even after a relatively long-time delay between food consumption (Garcia et al. 1966) and subsequent illness. Unfamilar foods are more likely than familiar foods to become aversive during conditioning procedures (Revusky and Bedarf 1967). Critically, just one pairing of food with illness is sufficient to elicit long-lasting taste aversions (Garb and Stunkard 1974; Garcia et al. 1974). This ability to rapidly learn potent taste aversions following an aversive experience is likely to be an evolutionarily adaptive defence mechanism. From an evolutionary perspective, disgust may have evolved as a disease-avoidance mechanism to protect us from fatal harm from potentially life-threatening stimuli such as pathogens (Davey and Marzillier 2009; Oaten et al. 2009). As such, the rapid acquisition of disgust via direct learning pathways, and the subsequent avoidance of the disgust-related stimulus, has obvious survival importance.

Evaluative Conditioning

Another commonly studied form of disgust learning is the *law of contagion*, which refers to the ability of a stimulus to elicit disgust after contact with another stimulus considered disgusting (Rozin and Fallon 1987). That is, disgust properties can be transmitted from an object perceived as disgusting to another previously neutral object by very brief, direct contact. For example, if food comes into contact with something we consider disgusting (e.g., a bug), the food may be considered "contaminated". Given the assertion that one stimulus may change the evaluative properties of another stimulus, the law of contagion is conceptually similar to evaluative conditioning (Baeyens et al. 1988). While expectancy learning may explain some examples of disgust learning, some authors have proposed that evaluative conditioning provides a better understanding of the learning process (e.g., Olatunji et al. 2007; Schienle et al. 2001; Woody and Teachman 2000). As discussed, evaluative conditioning (see de Houwer 2007; de Houwer et al. 2001) refers to the change in valence of an initially neutral stimulus (CS) as a result of its contingent pairing with another, emotionally-valenced, stimulus (US). Thus, in the context of disgust acquisition, evaluative learning occurs when a neutral stimulus (CS) is experienced together with a stimulus that evokes disgust (US), leading to an aversive disgust response for the CS when it is presented alone (e.g., Olatunji et al. 2007). This argument can explain how children develop disgust for a variety of stimuli during childhood. Activation of representations of the disgust-evoking US are often argued to underpin this learned response for the CS. However, it is worth noting that alternative explanations for evaluative conditioning also exist. For example, Davey (1994) argued that during evaluative conditioning features of the CS that are emotionally congruent with the US become more salient, while incongruent features become less salient. So, for example, the disgusting features of a CS become more prominent after it has been seen together with a disgusting US, causing a re-categorisation of the CS as disgusting.

A wealth of research supports the involvement of evaluative conditioning in the acquisition of disgust. For example, disgust has been shown to increase for CSs paired with aversive USs in evaluative conditioning procedures. Olatunji et al.

(2013) presented participants with 12 reinforced CS+ pairings that consisted of either the word "part" or "some" together with 12 aversive US images (e.g., bodily mutilation). Participants also saw 12 pairings of a CS- (the word "cylinder") with 12 neutral images (e.g., a pencil). Self-reported ratings of the CS on an evaluative dimension revealed that participants reported a significant increase in disgust towards the CS+ following evaluative conditioning, and a significant increase in happiness towards the CS-. Similarly, Rozin et al. (1986) found that participants had reduced preference for a glass of juice (CS) after seeing a dead cockroach (US) dropped into a similar glass of juice. However, it is worth noting that wide variability in the strength and reliability of evaluative conditioning is reported, with some authors not finding effects at all (e.g., Field et al. 2008a). This is important because if the evaluative conditioning process is relatively weak and unreliable, it would not be a good evolutionary mechanism for avoiding pathogens.

One criticism of disgust evaluative conditioning studies is the reliance on self-report measures to assess conditioned disgust responses (for more on disgust measurement see Consedine, Chap. 2, this volume). To overcome this, studies have utilised more objective measures such as facial electromyogram (EMG) recordings of the levator labii muscle, which retracts upper lip skin to produce the typical disgust expression (e.g., Bosman et al. 2016). Following the presentation of neutral pictures (CS) with disgust-relevant pictures depicting a dirty toilet, a decaying dog, and faeces (US), Schienle et al. (2001) found that participants with blood-injury fear showed a stronger increase in facial EMG activity of the levator muscles towards the CS. However, this evaluative conditioning effect was not reflected in global affective ratings and specific disgust ratings, potentially due to insufficient intensity of the US disgust stimuli or habituation to the disgust pictures meaning that they no longer act as an effective US. The findings highlight the importance of selecting appropriate measures of evaluative conditioning.

Research has also found evaluative conditioning effects using measures of skin conductance, heart rate (e.g., Olatunji et al. 2007), and visual avoidance (Mason and Richardson 2010). The use of implicit measures that do not require conscious awareness is important in that they not only overcome the issues of demand characteristics and negativity biases, but they also allow a differentiation to be made between automatic and deliberate reflexive responses (Borg et al. 2016). However, given that participants may also report greater levels of anxiety, anger and sadness towards the CS, questions remain about whether some reported evaluative conditioning effects are actually the result of learned changes in global affective evaluations of the CS, rather than specific disgust-learning effects (Olatunji et al. 2013).

Disgust acquired via expectancy learning can be reversed in extinction procedures in which the CS is presented without the US; extinction trials result in the CS no longer activating the US expectancy (e.g., Hermans et al. 2002a). In contrast, disgust responses acquired via evaluative conditioning have frequently been shown to be resistant to extinction (e.g., Armstrong et al. 2014; Bosman et al. 2016; Engelhard et al. 2014; Hofmann et al. 2010; Mason and Richardson 2010; Olatunji et al. 2007). This can be explained if evaluative conditioning is a form of referential learning, because unreinforced presentations of the CS during extinction would not modify the intrinsic valence of the US referenced by the CS. Counterconditioning, on the other hand, does appear to reduce disgust acquired via evaluative conditioning; possibly because while extinction addresses the predictive value of the CS, counterconditioning addresses the valence of the CS. Counterconditioning may impart a new positive representation that is able to compete with the original disgust representation (Engelhard et al. 2014). It is worth noting though, that counterconditioning procedures following evaluative conditioning have not always been successful (e.g., de Jong et al. 2000; Olatunji et al. 2009), possibly due to the counterconditioning duration being too short or the positively valenced stimuli lacking sufficient intensity.

While some common disgust-associated animals (e.g., rats) may be carriers of potentially lethal diseases that are best avoided, many other disgust-associated creatures (e.g., slugs) do not present any danger (McNally 2002). Evaluative conditioning can account for the acquisition of disgust responses for stimuli for which avoidance does not have adaptive value. However, an alternative explanation for the disgust relevance of harmless slimy invertebrates, such as slugs and worms, is that they resemble stimuli such as mucus and faeces, which are already disgust-evoking due to their role in contamination, illness, and disease (Davey 2011). This is important if it is the disgust-relevant stimulus characteristics of the CS that come to be associated with the US.

Co-occurrence of Disgust and Fear

Disgust expectancy learning is underpinned by the same learning mechanisms as fear expectancy learning, in which the contiguous occurrence of a stimulus (CS) and an aversive event (US) can lead to learned fear (CR) of the stimulus. Cognitive responses to disgust and fear are often considered distinct, with fear responding typified by instinctive and largely automatic responses, and disgust responding more dependent on focused attention and characterised by slower development (e.g., Anderson et al. 2003; Woody and Teachman 2000). That is, fear heightens environmental input to expedite a fast response to threatening stimuli, whereas disgust diminishes attention to non-threatening environmental input in order to avoid contamination (e.g., Krusemark and Li 2011). Other studies have indicated fear and disgust also have a different effect on memory, with greater recall and recognition seen for disgust-related images, suggesting enhanced salience in memory compared to fear (Chapman et al. 2013). Fear and disgust also diverge in terms of their biological functions in that fear is likely a survival response to immediate threat (Gray 1987) and disgust a response to potential life-threatening contamination (e.g., Tybur et al. 2013).

Nonetheless, fear and disgust do share many similar characteristics and are thought to be closely related (e.g., Woody and Teachman 2000). For example, both conditioned fear and disgust responses depend on a common neural network, which includes the cingulate cortex, the nucleus accumbens, the orbitofrontal cortex, and the occipital cortex (Klucken et al. 2012). Fear and disgust frequently co-occur and are both considered to be defensive emotions causing arousal in order to protect from threat or harm (Davey 2011). Moreover, research suggests that disgust is involved in some phobias in adulthood, such as animal phobias (e.g., Cisler et al. 2009; Davey 1992; Matchett and Davey 1991) and blood-injection-injury phobia (e.g., Page 1994; Tolin et al. 1997), and similar involvement is seen in childhood phobias (Muris et al. 2008b). Disgust has also been found to increase susceptibility to anxiety pathology (see Merckelbach et al. 1996; Muris and Merckelbach 2001) and is implicated in post-traumatic stress disorder (Engelhard et al. 2002) and contamination-based obsessive-compulsive disorder (Cisler et al. 2009).

Given the frequent co-occurrence of disgust and fear, learning mechanisms involved in fear acquisition may also contribute to the acquisition of disgust (Olatunji and Sawchuk 2005; Rozin and Fallon 1987). But fear is not only acquired via direct learning: fear can also be learned indirectly via the verbal and nonverbal expressions of others (Rachman 1977). Likewise, there is now good evidence that disgust can also be acquired via these indirect learning pathways.

Indirect Learning

Curtis et al. (2011) refer to humans as "informavores" to describe the constant information-seeking and observation of others that is used to make decisions about how to behave. Stevenson et al. (2010) give the example of a child beginning to make contact with a dirty sweet on the ground and an adult responding with a facial expression showing disgust, vocally stating "don't touch that", and physically moving the child away from the sweet (behavioural avoidance). They explored parent-child transmission by measuring children's disgust responding and behavioural avoidance to core disgust-eliciting stimuli (e.g., dirty socks), animal elicitors (e.g., mealworms), and sociomoral elicitors (e.g., a park covered in litter). Children were presented with the disgust elicitors when they were either with the experimenter only, or when they also had a parent present. They discovered greater avoidance of core and animal disgust elicitors in parents of younger children compared to older children, suggesting that parents play an important role in disgust acquisition in young children. When children's animal disgust responding was measured when they were alone, their disgust responses mirrored those of their parents. The frequency of disgust responding in parents also predicted whether children took a sweet from the bottom of a potty. Similarly, correlational research has demonstrated a moderate relationship between parents' self-reported measures of disgust and those of their adult offspring (Rozin et al. 1984; see Rottman et al. 2019, for a detailed discussion on the enculturation of disgust).

Recent experimental research has allowed specific pathways in disgust transmission to be identified. There is a long history of studies showing that fear can be acquired indirectly via the transmission of verbal threat-related information or vicarious learning (also referred to as “observational learning” or “modelling”). More recently, this work has been extended to demonstrate that disgust can also be acquired indirectly via disgust-related verbal information and vicarious learning. Indirect learning pathways make adaptive sense from an evolutionary perspective, because they do not involve potentially harmful contact with a stimulus before learning occurs and are therefore far less risky than direct learning (Rachman 1977). Consequently, they provide a survival advantage over potentially fatal direct learning events.

Verbal Information Pathway

A plethora of studies have demonstrated that fear can be acquired via verbal information using a paradigm originally developed by Field et al. (2001). Children were presented with threat-related or positive information about novel stimuli. Field et al. (2001) found that, following threat-related information about monster dolls, children’s self-reported fear beliefs significantly increased. Field and Lawson (2003) extended this to demonstrate increases in fear responses towards Australian marsupials (expected to be unfamiliar to UK children) following negative information compared to positive or no information. Children reported greater fear beliefs for the marsupials, a reluctance to approach them in a behavioural approach task, and a tendency to link them with an “unpleasant” category during an implicit association task. Subsequent studies have demonstrated that the verbal information pathway can lead to a range of fear-related changes, including changes in children’s fear cognitions, avoidance behaviour, physiological responding, and attentional bias (e.g., Field 2006a, b; Field et al. 2008b; Field and Schorah 2007; Muris et al. 2003).

This paradigm has also been used to explore disgust acquisition. Muris et al. (2008a) found significantly increased disgust for novel animals (Australian marsupials) after children (aged 9–13 years) were given disgust-related information about them. Furthermore, children’s fear responses for the animals also increased following the disgust-related information. Cleanliness-related information, on the other hand, decreased levels of disgust and fear for animals. The authors concluded that disgust-related information can not only lead to the acquisition of disgust but can also contribute to the development of fear in children. However, it is important to note that the study was unable to determine whether it was disgust-related information per se that produced high levels of disgust and fear, or whether disgust increased due to the increases in fear produced by the information. Indeed, inducing anxious mood in participants has been shown to increase reported disgust (Marzillier and Davey 2005), though similar findings were not found for self-reported anxiety when disgust was induced. Induced disgust then does not necessarily cause increases in anxiety. Typically, disgust emotions are focused on specific disgust-eliciting stimuli; thus one possible explanation for this finding is that disgust does not affect anxiety if it is induced without reference to a stimulus to direct fear or anxiety towards. Other experimental studies have found that inducing disgust via disgust-related visual and auditory stimuli increases fear responses (e.g., Webb and Davey 1992). As such, it may be too simplistic to conclude that disgust merely facilities fear, but rather the role of disgust in fear development may be more complex.

In two studies, Muris et al. (2009) attempted to explore the relation between fear and disgust further. The first study used a within-subjects design in which participants were presented with disgust-related, cleanliness-related, and threat-related information about three novel animals. Higher levels of disgust and fear beliefs followed disgust-related information, supporting Muris et al. (2008a) earlier findings. They also found greater avoidance of the disgust-paired animal in a paper and pencil "Jungle Task" in which children indicated where they would position themselves in relation to the animals in a jungle. Similarly, enhanced fear beliefs, avoidance tendencies and disgust feelings towards the animals were also found following threat-related information. Thus, the findings showed again that disgust and fear are closely related emotions, with disgust being involved in the production of fear, and threat-related information activating disgust. In their second study, Muris et al. (2009) used a between-group design in which children received *either* disgust-related, cleanliness-related, or threat-related information. Rather than relying only on self-report or subjective measures, the researchers also used a behavioural approach task in which children were instructed to approach a box they believed contained the animal. The study replicated the results of study one, with the additional finding that children were less inclined to approach the animal following disgust-related or threat-related information. Crucially, they also found that increases in fear following disgust-related information were explained by changes in disgust, and increases in disgust following threat-related information were explained by changes in fear levels.

Muris et al. (2012) followed-up the second study with a nonverbal disgust manipulation. Children were given a set of glass jars containing disgust-eliciting or neutral specimen characteristics related to novel animals and asked to form an impression of the animals. For example, for an animal's "sleeping place" children received either a specimen displaying a nest of mud and slush (disgust-eliciting specimen) or a nest of leaves, petals, and flowers (neutral specimen). The researchers found that children who were presented with the disgust-related characteristics showed increased feelings of disgust and fear towards the animals. Further, the disgust-exposed children were more likely to interpret ambiguous situations involving the animal negatively. These findings offer further evidence that disgust can be involved in the development of some fears in children.

Muris et al. (2012) suggested several possible explanations for why disgust-related information leads not only to increases in disgust feelings, but also increases in fear beliefs. From a disease-avoidance perspective, disgust elicited by a disgust-related animal may promote children's fear because of the belief they could contract disease from the dirty animal (see Davey and Marzillier 2009). However, Muris et al. (2009) did not find any evidence that children made illness-related interpretations for ambiguous situations involving the animals. Alternatively, it may be that the disgust-related information activates a negative interpretation bias that makes a child more susceptible to a range of negative emotions, including fear and anxiety (e.g., Davey et al. 2006).

Vicarious Learning

Rachman's (1977) second indirect learning pathway to fear acquisition is vicarious learning. This refers to situations in which individuals learn to fear a stimulus after observing someone else acting fearfully towards it (e.g., a family member acting scared of a dog), or after observing someone experiencing a traumatic event involving the stimulus (e.g., a family member being bitten by a dog). Studies with adults have found that fear can be acquired via observing someone else acting fearfully (e.g., Golkar et al. 2013; Olsson et al. 2016; Olsson and Phelps 2004) and similar findings have been found with young infants (e.g., Dubi et al. 2008; Gerull and Rapee 2002). Likewise, a series of experiments with 7- to 10-year-olds has indicated that vicarious fear learning can increase fear-related cognitions (Askew and Field 2007; Askew et al. 2008, 2015, 2016; Dunne and Askew 2013, 2018; Dunne et al. 2017; Reynolds et al. 2015), behavioural avoidance (Askew and Field 2007; Askew et al. 2013; Reynolds et al. 2017a), physiological responses (heart rate), and attentional bias (Reynolds and Askew 2019; Reynolds et al. 2014, 2016), during middle childhood.

Given that fear and disgust often co-occur and can develop in similar ways, it seems likely that disgust might also develop via vicarious learning. That is, disgust responses for a stimulus might be transmitted from, for example, parent to child by the child appraising the stimulus as disgusting after observing a parent responding to it with disgust. Research has suggested that the crucial mechanism of person-to-person transmission of disgust is the universally recognised disgust facial expression (e.g., Curtis et al. 2011; Izard 1994; Oaten et al. 2009; Rozin and Fallon 1987; Rozin et al. 2008; Stevenson et al. 2010). Children have the ability to acquire disgust-relevant information from the facial responses of significant others such as their parents, particularly about sources of contamination and disease (Rottman 2014). In support of this, studies with infants have found that pairing novel stimuli with disgusted facial expressions leads infants to avoid them (e.g., Moses et al. 2001). Similarly, Hertenstein and Campos (2004) demonstrated that infants observing verbal and visual responses of disgust towards a novel object were less likely to make contact with the object at a later date. Significantly, this effect persisted for at least 1 h in 14-month-olds, suggesting some evidence of persistence.

Using an adaptation of Askew and Field's (2007) fear learning paradigm, Askew et al. (2014), conducted a pair of experiments exploring the acquisition of disgust via vicarious learning. They presented children (aged 7–10 years) with pictures of novel animals (CS) together with adult faces expressing disgust (US). Following this disgust vicarious learning procedure, children rated animals as more disgusting and fear-evoking. Notably, animals were also rated with increased disgust following fear vicarious learning, with children reporting similar levels of increased disgust (relative to a control animal) regardless of whether animals were seen with disgust or fear faces. When asked how close to the animals they would like to stand in a make-believe nature reserve, children also placed themselves further away from animals seen with disgust faces, avoiding them compared to control animals. These avoidance preferences were comparable to the significant avoidance preferences seen for animals following fear vicarious learning. This avoidance was found to be predicted by increases in fear beliefs and mediated by increases in disgust beliefs. Thus, the study provided convincing evidence that disgust responses can be acquired vicariously in the same way as fear responses.

In a more recent study, Reynolds and Askew (2019) extended the findings further, showing that disgust vicarious learning significantly increases fear cognitions, disgust cognitions, and avoidance preferences for animals, as well as how scary children feel animals are. Similar to the earlier study, no evidence was found that fear and disgust responses following disgust vicarious learning differed in magnitude to those found following fear vicarious learning. Reynolds and Askew (2019) also extended the original findings by investigating the effects of disgust vicarious learning on the development of attentional biases. Research with similar-aged children has established that fear vicarious learning can create attentional bias towards animals seen with fearful faces (Reynolds et al. 2014, 2016). Using a dot probe task, Reynolds and Askew (2019) found that disgust-related vicarious learning also creates attentional bias towards animals as well as an increased preference to avoid them. This is interesting, not only because attentional biases are a central feature of many fears (see Koster et al. 2004), but also because responses in computerised response time tasks like the dot probe are under less conscious control than self-report measures; thus, the findings provide particularly compelling evidence of disgust vicarious learning. The findings also support Knowles et al.'s (2019) proposition that disgust-prone individuals exhibit an initial enhanced orienting response and attentional bias towards disgusting cues, which is then followed by a significant avoidance of these stimuli. From an evolutionary perspective, the finding that disgust-associated stimuli can capture attention at early stages of processing makes adaptive sense because it allows rapid detection of potentially harmful stimuli that can then be swiftly avoided.

It is widely accepted that vicarious fear learning is a form of associative learning (e.g., Askew and Field 2007, 2008; Bandura 1969; Berger 1962; Mineka and Cook 1993; Olsson and Phelps 2004) and is likely underpinned by stimulus-stimulus (CS-US) associative learning. In vicarious learning, the animal acts as the CS and the fear-related responses of the model act as the US, so that during a learning episode a CS-US association is formed. This form of expectancy learning is supported by the fact that devaluation of the US (i.e., model's response) following vicarious learning reduces the learned fear response to the CS, although the CS has not been seen again either alone or in pairing trials (Reynolds et al. 2015). This finding indicates that responses to the CS must be dependent on associations between the CS and US. It seems likely that vicarious *disgust* acquisition is underpinned by similar CS-US associative learning, with the disgust-related responses of the models acting as the US. However, the mechanisms underpinning disgust vicarious learning are not yet established experimentally and may not be the same as for fear.

The finding that vicarious learning with models expressing disgust can lead to the acquisition of both disgust and fear is in line with models that assert that disgust may contribute to fear acquisition by enhancing avoidance of disease-related stimuli to prevent contamination (Matchett and Davey 1991). From this perspective, a stimulus is first associated with disgust-evoking characteristics either by being associated with the spread of contamination or via disgust-evoking perceptual qualities. This is followed by cultural or familial learning processes that aim to prevent transmission of contamination, which result in avoidance of the stimulus and subsequently fear acquisition.

Not all studies have found evidence supporting the observational learning of disgust. For example, Muris et al. (2013) explored both the verbal information and vicarious learning pathways to disgust acquisition and found little evidence for the latter. Children in the study observed their mothers being shown boxes by the experimenter, each of which conveyed information about attributes of the eating, sleeping, and grooming habits of novel animals unfamiliar to the participant. These materials were either neutral (e.g., fruit as food) or contained disgust-evoking properties (e.g., worms as food). Results indicated that children showed higher levels of disgust if they had received disgust-related verbal information from their mother, but similar results were not found after they observed mothers' nonverbal facial expressions and gestures of disgust.

One explanation for the absence of vicarious learning in Muris et al. (2013) might be that mothers' nonverbal expressions of disgust, which were not under the control of the researchers, may have been inconsistent and not potent enough to produce learning. The researchers also acknowledged several other limitations. First, the order of communication from the mothers was fixed in that mothers first communicated nonverbally followed by the verbal communication. Therefore, it is possible that there was a systematic order effect. Additionally, only mothers were used as models. Research has demonstrated that the correlation between fathers' and children's disgust sensitivity is greater than that between mothers and children. Correlation between fathers' disgust sensitivity and children's fear of disgust-related animals is also greater than for mothers and children (Davey et al. 1993). While the precise reason for greater correlations for fathers than for mothers is not clear, Bögels and Phares (2008) provide a review of fathers' roles in child anxiety suggesting that father roles are characterised by play, challenge, risk taking, encouraging independence, and in guiding the transition to the outside world in later childhood, all of which may be important in the context of understanding their role in the development of child anxiety. Future research is required to determine the impact these fatherly roles may have on disgust sensitivity and fear of disgust-related animals in offspring.

Individual Differences in Disgust Learning

A final area of research worth discussing is the contribution of individual differences to disgust learning (see also Tybur, Chap. 6, this volume). Beyond sociodemographic differences, two factors likely to influence disgust learning are *disgust sensitivity* - the tendency to appraise the elicitation or experience of disgust as particularly negative - and *disgust propensity* - the tendency to respond with disgust more frequently and intensely (van Overveld et al. 2006). Previous research has demonstrated that disgust sensitivity predicts some phobias such as spider phobia, while both disgust sensitivity and disgust propensity predict other phobias such as a fear of blood (van Overveld et al. 2006; see also Olatunji and Sawchuk 2005).

One possible explanation is that having higher levels of disgust propensity causes individuals to be more emotionally reactive during aversive events, which then increases their tendency to respond with disgust. Using an evaluative conditioning paradigm, Olatunji et al. (2013) demonstrated that individuals with greater disgust propensity showed larger increases in disgust, anger, and anxiety when presented with initially neutral words (CS) that had been paired with aversive images (US). The researchers argued that disgust propensity may serve to potentiate or facilitate learned aversions that contribute to behavioural avoidance and increase risk for anxiety disorders. They also suggested that disgust propensity may influence the underlying learning processes, consequently changing the quality of the learned response. That is, individuals high in disgust propensity may display enhanced memory for disgust-related events, and greater difficulty in disengaging attention from disgust related stimuli. These memory and attentional processes are likely to facilitate more intense learned aversive responses to the CS.

Disgust propensity and/or disgust sensitivity might also increase susceptibility to disgust vicarious learning in children. Muris et al. (2013) found an additive effect of disgust sensitivity on disgust responses, in addition to the effects of modelled disgust-related information about animals. Disgust sensitivity was also significantly related to feelings of fear about the animal. Research has also demonstrated that disgust conditioning to blood-injection-injury-relevant stimuli can be influenced by levels of disgust sensitivity (Olatunji et al. 2009). Reynolds and Askew (2019) study did not show any relationship between disgust propensity or sensitivity and any of the vicariously learned disgust and fear responses measured, including attentional bias. On the face of it, this seems a surprising finding given that other studies show attentional bias towards disgust is only found in disgust prone individuals (e.g., Cisler et al. 2009). However, it may be that higher-order traits like disgust propensity and sensitivity are too interwoven with other traits like trait anxiety to be dissociated in this type of procedure and trait anxiety has not been found to have a mediating effect on vicarious fear learning (Reynolds et al. 2014).

Other research has suggested that disgust sensitivity may itself be shaped by parental modelling of disgust reactions (e.g., Kim et al. 2013; Rozin et al. 2008; Widen and Olatunji 2016). Support comes from studies demonstrating that children do not appear to experience disgust until early childhood; Rottman (2014) has argued that the age of disgust recognition is around 5 years and Rozin and Fallon (1987) argued that children under the age of 8 years lack the necessary cognitive prerequisites to experience disgust. It is worth noting, however, that although it is thought infants cannot experience disgust per se, there is evidence that infants as young as 12 months old have the ability to acquire *disgust-relevant information* and subsequently avoid disgust-related stimuli (e.g., Moses et al. 2001). Furthermore, it does not necessarily follow that the later development of a trait initially absent at birth is the product of socialisation (Pinker 2002). Tybur et al. (2018) point out that traits such as object permanence are not present at birth but reliably develop without the need of socialisation. Evidence suggesting that disgust sensitivity may indeed be a product of socialisation comes from research showing that children demonstrate similar levels of disgust sensitivity to their parents and their siblings (e.g., Davey et al. 1993; Stevenson et al. 2010). However, dissociating whether familial similarities are due to environmental or genetic transmission is not straightforward (Tybur et al. 2018; see also Tybur, Chap. 6, this volume). Indeed, twin research has demonstrated that 50% of the variance in disgust sensitivity is genetic and 50% is due to non-shared environmental influences (Sherlock et al. 2016), while other researchers have found that both disgust sensitivity and disgust responding seem to be only minimally heritable (Rozin and Millman 1987).

Summary

There is a growing body of research demonstrating the involvement of both direct and indirect conditioning pathways in the acquisition of disgust. There is compelling evidence for conditioned aversion and a role for evaluative conditioning in disgust development, as well as the indirect learning pathways of verbal information and vicarious learning. These two learning processes may be involved at two different levels. First, they may be involved in children developing an initial disgust response. This could be via interactions with their parents. Second, they may be involved in the acquisition of individual childhood fears and phobias.

A good understanding of the learning mechanisms underpinning disgust development is required, not only because it improves our theoretical understanding of the emotion, but because it may also inform effective preventions, interventions, and treatments of disgust-based and, to an extent, fear-based disorders. For example, research has already demonstrated that positive modelling can prevent the acquisition of fear responses if presented to children immediately following a fear-related vicarious learning episode (e.g., Reynolds et al. 2016, 2017b). Similarities between fear and disgust learning mechanisms suggest results from vicarious fear learning may be mirrored in vicarious disgust learning, although it is yet to be seen whether this is the case. The finding that disgust and fear responses can be acquired following verbal information transmission and vicarious disgust learning in childhood has particular implications for clinicians, caregivers, and those working directly with children. It also suggests that the role of media-transmitted (e.g., via television, computer games, and the internet) negative verbal and visual information in disgust formation should not be underestimated.

In conclusion then, it is important that we continue to study disgust-related direct and indirect learning and unlearning mechanisms, as well as individual differences in potential risk factors for the development of pathology, such as high disgust propensity and sensitivity.

References

Anderson, A. K., Christoff, K., Panitz, D., De Rosa, E., & Gabrieli, J. D. E. (2003). Neural correlates of the automatic processing of threat facial signals. *Journal of Neuroscience, 23,* 5627-5633. https://doi.org/10.1523/JNEUROSCI.23-13-05627.2003

Armstrong, T., McClenahan, L., Kittle, J., & Olatunji, B. O. (2014). Don't look now! Oculomotor avoidance as a conditioned disgust response. *Emotion, 14,* 95-104. https://doi.org/10.1037/a0034558

Askew, C., Cakir, K., Poldsam, L., & Reynolds, G. (2014). The effect of disgust and fear modeling on children's disgust and fear for animals. *Journal of Abnormal Psychology, 123,* 566-577. https://doi.org/10.1037/a0037228

Askew, C., Dunne, G., Özdil, Z., Reynolds, G., & Field, A. P. (2013). Stimulus fear-relevance and the vicarious learning pathway to fear. *Emotion, 13,* 915–925. https://doi.org/10.1037/a0032714

Askew, C., & Field, A. P. (2007). Vicarious learning and the development of fears during childhood. *Behaviour, Research and Therapy, 45,* 2616–2627. https://doi.org/10.1016/j.brat.2007.06.008

Askew, C., & Field, A. P. (2008). The vicarious learning pathway to fear 40 years on. *Clinical Psychology Review, 28,* 1249-1265. https://doi.org/10.1016/j.cpr.2008.05.003

Askew, C., Hagel, A., & Morgan, J. (2015). Vicarious learning of children's social anxiety-related fear beliefs and emotional Stroop bias. *Emotion, 15,* 501-510. https://doi.org/10.1037/emo0000083

Askew, C., Kessock-Philip, H., & Field, A. P. (2008). Interactions between the indirect pathways to fear in children: what happens when verbal threat information and vicarious learning combine? *Behavioural and Cognitive Psychotherapy, 36,* 491-505. https://doi.org/10.1017/S1352465808004402

Askew, C., Reynolds, G., Fielding-Smith, S., & Field, A. P. (2016). Inhibition of vicariously learned fear in children using positive modeling and prior exposure. *Journal of Abnormal Psychology, 125,* 279-291. https://doi.org/10.1037/abn0000131

Baeyens, F., Crombez, G., Van den Bergh, O., & Eelen, P. (1988). Once in contact always in contact: evaluative conditioning is resistant to extinction. *Advances in Behaviour Research and Therapy, 10,* 179-199. https://doi.org/10.1016/0146-6402(88)90014-8

Baeyens, F., & De Houwer, J. (1995). Evaluative conditioning is a qualitatively distinct form of classical conditioning: A reply to Davey (1994). *Behaviour Research and Therapy, 33,* 825-831. https://doi.org/10.1016/0005-7967(95)00021-O

Baeyens, F., Eelen, P., Crombez, G., & Van den Bergh, O. (1992). Human evaluative conditioning: Acquisition trials, presentation schedule, evaluative style, and contingency awareness. *Behaviour Research and Therapy, 30,* 133–142. https://doi.org/10.1016/0005-7967(92)90136-5

Bandura, A. (1969). *Principles of Behavior Modification.* Holt, Rinehart & Winston.

Berger, S. M. (1962). Conditioning through vicarious instigation. *Psychological Review, 69,* 450-466. https://doi.org/10.1037/h0046466

Bögels, S., & Phares, V. (2008). Fathers' role in the etiology, prevention and treatment of childhood anxiety: A review and new model. *Clinical Psychology Review, 28,* 539-558. https://doi.org/10.1016/j.cpr.2007.07.011

Borg, C., Bosman, R. C., Engelhard, I., Olatunji, B. O., & de Jong, P. J. (2016). Is disgust sensitive to classical conditioning as indexed by facial electromyography and behavioural responses? *Cognition and Emotion, 30,* 669-686. https://doi.org/10.1080/02699931.2015.1022512

Bosman, R. C., Borg, C., & de Jong, P. J. (2016). Optimising extinction of conditioned disgust. *Plos One, 11,* e0148626. https://doi.org/10.1371/journal.pone.0148626

Chapman, H. A., Johannes, K., Poppenk, J. L., Moscovitch, M., & Anderson, A. K. (2013). Evidence for the differential salience of disgust and fear in episodic memory. *Journal of Experimental Psychology General, 142,* 1100–1112. https://doi.org/10.1037/a0030503

Cisler, J. M., Olatunji, B. O., Lohr, J. M., & Williams, N. L. (2009). Attentional bias differences between fear and disgust: Implications for the role of disgust in disgust-related anxiety disorders. *Cognition and Emotion, 23,* 675-687. https://doi.org/10.1080/02699930802051599

Curtis, V., de Barra, M., & Aunger, R. (2011). Disgust as an adaptive system for disease avoidance behaviour. *Philosophical Transactions of The Royal Society, 366,* 389-401. https://doi.org/10.1098/rstb.2010.0117

Davey, G. C. L. (1992). Characteristics of individuals with fear of spiders. *Anxiety Research, 4,* 299-314. https://doi.org/10.1080/08917779208248798

Davey, G. C. L. (1994). Defining the Important Theoretical Questions to Ask About Evaluative Conditioning - a reply to Martin and Levey (1994). *Behaviour Research and Therapy, 32*(3), 307-310. https://doi.org/10.1016/0005-7967(94)90126-0

Davey, G. C. L. (2011). Disgust: The disease-avoidance emotion and its dysfunctions. *Philosophical Transactions of The Royal Society, 366,* 3453-3465. https://doi.org/10.1098/rstb.2011.0039

Davey, G. C. L., Bickerstaffe, S., & MacDonald, B. A. (2006). Experienced disgust causes a negative interpretation bias: A causal role for disgust in anxious psychopathology. *Behaviour Research and Therapy, 44,* 1375-1384. https://doi.org/10.1016/j.brat.2005.10.006

Davey, G. C. L., Forster, L., & Mayhew, G. (1993). Familal resemblances in disgust sensitivity and animal phobias. *Behaviour Research and Therapy, 31,* 41-50. https://doi.org/10.1016/0005-7967(93)90041-R

Davey, G. C. L., & Marzillier, S. (2009). *Disgust and animal phobias.* In B. O. Olatunji & D. McKay (Eds.), *Disgust and its disorders: Theory, assessment, and treatment implications* (pp. 169–190). American Psychological Association. https://doi.org/10.1037/11856-008

de Houwer, J. (2007). A conceptual and theoretical analysis of evaluative conditioning. *Spanish Journal of Psychology, 10,* 230-241. https://doi.org/10.1017/S1138741600006491

de Houwer, J., Thomas, S., & Baeyens, F. (2001). Association learning of likes and dislikes: A review of 25 years of research on human evaluative conditioning. *Psychological Bulletin, 127,* 853. https://doi.org/10.1037/0033-2909.127.6.853

de Jong, P. (2013). Learning mechanisms in the acquisition of disgust. In D. Hermans, B. Rimé, & B. Mesquita, (Eds.). *Changing Emotions* (pp. 74-80). Psychology Press.

de Jong, P. J., Vorage, I., & van den Hout, M. A. (2000). Counterconditioning in the treatment of spider phobia: effects on disgust, fear and valence. *Behaviour Research and Therapy, 38,* 1055-1069. https://doi.org/10.1016/s0005-7967(99)00135-7

Dubi, K., Rapee, R. M., Emerton, J., & Schniering, C. (2008). Maternal modeling and the acquisition of fear and avoidance in toddlers: Influence of stimulus preparedness and child temperament. *Journal of Abnormal Child Psychology, 36,* 499–512. https://doi.org/10.1007/s10802-007-9195-3

Dunne, G., & Askew, C. (2013). Vicarious learning and un-learning of fear in children via mother and stranger models. *Emotion, 13,* 974-980. https://doi.org/10.1037/a0032994

Dunne, G., & Askew, C. (2018). Vicarious Learning and Reduction of Fear in Children via Adult and Child Models. *Emotion, 18,* 528-534. https://doi.org/10.1037/emo0000341

Dunne, G., Reynolds, G, & Askew, C. (2017). Stimulus fear-relevance and the speed, magnitude, and robustness of vicariously learned fear. *Behaviour Research and Therapy, 95,* 1-18. https://doi.org/10.1016/j.brat.2017.05.002

Engelhard, I. M., Leer, A., Lange, E., & Olatunji, B. O. (2014). Shaking that icky feeling: Effects of extinction and counterconditioning on disgust-related evaluative learning. *Behavior Therapy, 45,* 708-719. https://doi.org/10.1016/j.beth.2014.04.003

Engelhard, I. M., van den Hout, M. A., Arntz, A., & McNally, R. J. (2002). A longitudinal study of 'intrusion-based reasoning' and PTSD after a train disaster. *Behaviour Research and Therapy, 40,* 1415-1424. https://doi.org/10.1016/s0005-7967(02)00018-9

Field, A. P. (2006a). The behavioural inhibition system and the verbal information pathway to children's fears. *Journal of Abnormal Psychology, 115,* 742-752. https://doi.org/10.1037/0021-843x.115.4.742

Field, A. P. (2006b). Watch out for the beast: Fear information and attentional bias in children. *Journal of Clinical Child and Adolescent Psychology, 35,* 431-439. https://doi.org/10.1207/s15374424jccp3503_8

Field, A. P., Argyris, N. G., & Knowles, K. A. (2001). Who's afraid of the big bad wolf: a prospective paradigm to test Rachman's indirect pathways in children. *Behaviour Research and Therapy, 39,* 1259-1276. https://doi.org/10.1016/s0005-7967(00)00080-2

Field, A. P., Lascelles, K. R. R., Lester, K. J., Askew, C., & Davey, G. C. L. (2008a). Evaluative conditioning: Missing, presumed dead. *Netherlands Journal of Psychology, 64*, 46-64. https://doi.org/10.1007/BF03076407

Field, A. P., & Lawson, J. (2003). Fear information and the development of fears during childhood: Effects on implicit fear responses and behavioural avoidance. *Behaviour Research and Therapy, 41,* 1277–1293. https://doi.org/10.1016/S0005-7967(03)00034-2

Field, A. P., Lawson, J., & Banerjee, R. (2008b). The verbal threat information pathway to fear in children: The longitudinal effects on fear cognitions and the immediate effects on avoidance behavior. *Journal of Abnormal Psychology, 117,* 214–224. https://doi.org/10.1037/0021-843x.117.1.214

Field, A. P., & Schorah, H. (2007). The verbal information pathway to fear and heart rate changes in children. *Journal of Child Psychology and Psychiatry, 48,* 1088–1093. https://doi.org/10.1111/j.1469-7610.2007.01772.x

Garb, J. L., & Stunkard, A. J. (1974). Taste aversions in man. *The American Journal of Psychiatry, 131,* 1204-1207. https://doi.org/10.1176/ajp.131.11.1204

Garcia, J., Ervin, R. R., & Koelling, R. A. (1966). Learning with prolonged delay of reinforcement. *Psychonomic Science, 5,* 121-122. https://doi.org/10.3758/BF03328311

Garcia, J., Hankins, W. G., & Rusiniak, K. W. (1974). Behavioral regulation of the milieu interne in man and rat. Food preferences set by delayed visceral effects facilitate memory research and predator control. *Science 185,* 824–831. https://doi.org/10.1126/science.185.4154.824

Garcia, J., Kimeldorf, D. J., & Koelling, R. A. (1955). Conditioned aversion to saccharin resulting from exposure to gamma radiation. *Science, 122,* 157-158.

Garcia, J., & Koelling, R. A. (1966). Relation of cue to consequence in avoidance learning. *Psychonomic Science, 4*(3), 123-124. https://doi.org/10.3758/BF03342209

Gerull, F. C., & Rapee, R. M. (2002). Mother knows best: Effects of maternal modeling on the acquisition of fear and avoidance behaviour in toddlers. *Behaviour Research and Therapy, 40,* 279–287. https://doi.org/10.1016/s0005-7967(01)00013-4

Golkar, A., Selbing, I., Flygare, O., Öhman, A., & Olsson, A. (2013). Other People as Means to a Safe End: Vicarious Extinction Blocks the Return of Learned Fear. *Psychological Science, 24,* 2182-2190. https://doi.org/10.1177/0956797613489890

Gray, J. A. (1987). *The Psychology of Fear and Stress. CUP Archive.*

Hermans, D., Crombez, G., Vansteenwegen, D., Baeyens, F., & Eelen, P. (2002a). Expectancy-learning and evaluative learning in human classical conditioning: Differential effects of extinction. In S. P. Shohov (Ed.), *Advances in psychology research,* Vol. 12, pp. 17-40). Nova Science Publishers. https://doi.org/10.1016/s0005-7967(01)00006-7

Hermans, D., Vansteenwegen, D., Crombez, G., Baeyens, F., & Eelen, P. (2002b). Expectancy-learning and evaluative learning in human classical conditioning: affective priming as an indirect and unobstrusive measure of conditioned stimulus valence. *Behaviour, Research and Therapy, 40,* 217-234. https://doi.org/10.1016/s0005-7967(01)00006-7

Hertenstein, M. J., & Campos, J. J. (2004). The retention effects of an adult's emotional displays on infant behavior. *Child Development, 75,* 595-613. https://doi.org/10.1111/j.1467-8624.2004.00695.x

Hofmann, W., de Houwer, J., Perugini, M., Baeyens, F., & Crombez, G. (2010). Evaluative conditioning in humans: A meta-analysis. *Psychological Bulletin, 136,* 39-421. https://doi.org/10.1037/a0018916

Izard, C. E. (1994). Innate and universal facial expressions: Evidence from developmental and cross-cultural research. *Psychological Bulletin, 115,* 288-299. https://doi.org/10.1037/0033-2909.115.2.288

Kim, E. H., Ebesutani, C., Young, J., & Olatunji, B. O. (2013). Factor structure of the Disgust Scale–Revised in an adolescent sample. *Assessment, 20,* 620–631. https://doi.org/10.1177/1073191111434200

Klucken, T., Schweckendiek, J., Koppe, G., Merz, C. J., Kagerer, S., Walter, B., Sammer, G., Vaitl, D., & Stark, R. (2012). Neural correlates of disgust- and fear-conditioned responses. *Neuroscience, 201,* 209-218. https://doi.org/10.1016/j.neuroscience.2011.11.007

Knowles, K. A., Cox, R. C., Armstrong, T., & Olatunji, B. O. (2019). Cognitive mechanisms of disgust in the development and maintenance of psychopathology: A qualitative review and synthesis. *Clinical Psychology Review, 69,* 30-50. https://doi.org/10.1016/j.cpr.2018.06.002

Koster, E. H. W., Crombez, G., Van Damme, S., Verschuere, B., & De Houwer, J. (2004). Does imminent threat capture and hold attention? *Emotion, 4,* 312-317. https://doi.org/10.1037/1528-3542.4.3.312

Krusemark, E. A., & Li, W. (2011). Do all threats work the same way? Divergent effects of fear and disgust on sensory perception and attention. *The Journal of Neuroscience 31*, 3429–3434. https://doi.org/10.1523/JNEUROSCI.4394-10.2011

Logue, A. W., Ophir, I., & Strauss, K. E. (1981). The acquisition of taste aversions in humans. *Behaviour, Research and Therapy, 19,* 319-333. https://doi.org/10.1016/0005-7967(81)90053-x

Marzillier, S. J., & Davey, G. C. L. (2005). Anxiety and disgust: Evidence for a unidirectional relationship. *Cognition and Emotion, 19*, 729–750.https://psycnet.apa.org/doi/10.1080/02699930441000436

Mason, E. C., & Richardson, R. (2010). Looking beyond fear: The extinction of other emotions implicated in anxiety disorders. *Journal of Anxiety Disorders, 24,* 63-70. https://doi.org/10.1016/j.janxdis.2009.08.007

Matchett, G., & Davey, G. C. L. (1991). A test of a disease-avoidance model of animal phobias. *Behaviour Research and Therapy, 29,* 91-94. https://doi.org/10.1016/s0005-7967(09)80011-9

McNally, R. J. (2002). Disgust has arrived. *Journal of Anxiety Disorders, 16,* 561-566. https://doi.org/10.1016/s0887-6185(02)00174-3

Merckelbach, H., De Jong, P. J., Muris, P., & Van den Hout, M. A. (1996). The etiology of specific phobias: A review. *Clinical Psychology Review, 16*, 337-361. https://doi.org/10.1016/0272-7358(96)00014-1

Midkiff, E. E., & Bernstein, I. L. (1985). Targets of learned food aversions in humans. Physiology and Behavior, 34, 839-841. https://doi.org/10.1016/0031-9384(85)90387-7

Mineka, S., & Cook, M. (1993). Mechanisms involved in the observational conditioning of fear. *Journal of Experimental Psychology: General, 122,* 23-38. https://doi.org/10.1037//0096-3445.122.1.23

Moses, L. J., Baldwin, D. A., Rosicky, J. G., & Tidball, G. (2001). Evidence for referential understanding in the emotions domain at twelve and eighteen months. *Child Development, 72,* 718-735. https://doi.org/10.1111/1467-8624.00311

Muris, P., Bodden, D., Merckelbach, H., Ollendick, T. H., & King, N. J. (2003). Fear of the beast: A prospective study on the effects of negative information on childhood fear. *Behaviour Research and Therapy, 41,* 195-208. https://doi.org/10.1016/s0005-7967(01)00137-1

Muris, P., Huijding, J., Mayer, B., & de Vries, H. (2012). Does "yuck" mean "eek"? Fear responses in children after a disgust manipulation. *Journal of Behavior Therapy and Experimental Psychiatry, 43,* 765-769. https://doi.org/10.1016/j.jbtep.2011.10.002

Muris, P., Huijding, J., Mayer, B., Leemreis, W., Passchier, S., & Bouwmeester, S. (2009). The effects of verbal disgust- and threat-related information about novel animals on disgust and fear beliefs and avoidance in children. *Journal of Clinical Child and Adolescent Psychology, 38,* 551–563. https://doi.org/10.1080/15374410902976379

Muris, P., Mayer, B., Borth, M., & Vos, M. (2013). Nonverbal and verbal transmission of disgust from mothers to offspring: Effects on children's evaluation of a novel animal. *Behavior Therapy, 44,* 293-301. https://doi.org/10.1016/j.beth.2012.10.002

Muris, P., Mayer, B., Huijding, J., & Konings, T. (2008a). A dirty animal is a scary animal! Effects of disgust-related information on fear beliefs in children. *Behaviour Research and Therapy, 46,* 137-144. https://doi.org/10.1016/j.brat.2007.09.005

Muris, P., & Merckelbach, H. (2001). The etiology of childhood specific phobias: A multifactorial model. In M. W. Vasey, M. R. Dadds (Eds.), *The Developmental Psychopathology of Anxiety* (pp. 355-385). Oxford University Press. https://psycnet.apa.org/doi/10.1093/med:psych/9780195123630.003.0016

Muris, P., van der Heiden, S., & Rassin, E. (2008b). Disgust sensitivity and psychopathological symptoms in non-clinical children. *Journal of Behavior Therapy and Experimental Psychiatry, 39,* 133-146. https://doi.org/10.1016/j.jbtep.2007.02.001

Oaten, M., Stevenson, R., & Case, T. (2009). Disgust as a disease-avoidance mechanism. *Psychology Bulletin, 135,* 303-321. https://doi.org/10.1037/a0014823

Olatunji, B. O., Forsyth, J. P., & Cherian, A. (2007). Evaluative differential conditioning of disgust: a sticky form of relational learning that is resistant to extinction. *Journal of Anxiety Disorders, 21,* 820–834. https://doi.org/10.1016/j.janxdis.2006.11.004

Olatunji, B. O., Lohr, J. M., Smits, J. A. J., Sawchuk, C. N., & Patten, K. (2009). Evaluative conditioning of fear and disgust in blood-injection-injury phobia: Specificity and impact of individual differences in disgust sensitivity. *Journal of Anxiety Disorders, 23,* 153-159. https://doi.org/10.1016/j.janxdis.2008.06.002

Olatunji, B. O., & Sawchuk, C. N. (2005). Disgust: characteristic features, social manifestations, and clinical implications. *Journal of Social and Clinical Psychology, 27,* 932-962. https://doi.org/10.1521/jscp.2005.24.7.932

Olatunji, B. O., Tomarken, A., & Puncochar, B. D. (2013). Disgust propensity potentiates evaluative learning of aversion. *Emotion, 13,* 881-890. https://doi.org/10.1037/a0032743

Olsson, A., McMahon, K., Papenberg, G., Zaki, J., Bolger, N., & Ochsner, K. N. (2016). Vicarious fear learning depends on empathic appraisals and trait empathy. *Psychological Science, 27,* 25-33. https://doi.org/10.1177/0956797615604124

Olsson, A., & Phelps, E. A. (2004). Learned fear of "unseen" faces after Pavlovian, observational, and instructed fear. *Psychological Science, 15,* 822–828. https://doi.org/10.1111/j.0956-7976.2004.00762.x

Page, A. P. (1994). Blood-injury phobia. *Clinical Psychology Review, 14,* 443-461. https://doi.org/10.1016/0272-7358(94)90036-1

Pinker, S. (2002). *The Blank Slate: The Modern Denial of Human Nature.* Penguin Putnam.

Rachman, S. (1977). The conditioning theory of fear acquisition: A critical examination. *Behaviour Research and Therapy, 15,* 375–387. https://doi.org/10.1016/0005-7967(77)90041-9

Revusky, S. H., & Bedarf, E. W. (1967). Association of illness with prior ingestion of novel foods. *Science, 155,* 219-220. https://doi.org/10.1126/science.155.3759.219

Reynolds, G., & Askew, C. (2019). Effects of Vicarious Disgust Learning on the Development of Fear, Disgust and Attentional Biases in Children. *Emotion*, 19(7), 1268-1283. https://doi.org/10.1037/emo0000511

Reynolds, G., Field, A. P., & Askew, C. (2014). The effect of vicarious fear learning on children's heart rate responses and attentional bias for novel animals. *Emotion, 14,* 995–1006. https://doi.org/10.1037/a0037225

Reynolds, G., Field, A. P., & Askew, C. (2015). Preventing the development of observationally learnt fears in children by devaluing the model's negative response. *Journal of Abnormal Child Psychology, 43,* 1355–1367. https://doi.org/10.1007/s10802-015-0004-0

Reynolds, G., Field, A. P., & Askew, C. (2016). Reductions in Children's Vicariously Learnt Avoidance and Heart Rate Responses Using Positive Modeling. *Journal of Clinical Child & Adolescent Psychology, 23,* 1-14. https://doi.org/10.1080/15374416.2016.1138410

Reynolds, G., Field, A. P., & Askew, C. (2017a). Learning to fear a second-order stimulus following vicarious learning. *Cognition and Emotion, 31,* 572-579. https://doi.org/10.1080/02699931.2015.1116978

Reynolds, G., Wasely, D., Dunne, G., & Askew, C. (2017b). A comparison of positive vicarious learning and verbal information for reducing vicariously learned fear. *Cognition and Emotion, 19,* 1-12. https://doi.org/10.1080/02699931.2017.1389695

Rottman, J. (2014). Evolution, development, and the emergence of disgust. *Evolutionary Psychology, 12,* 417–433. https://doi.org/10.1177/147470491401200209

Rottman J., DeJesus J., & Greenebaum H. (2019). Developing Disgust: Theory, Measurement, and Application. In V. LoBue, K. Pérez-Edgar, K. Buss (Eds.), *Handbook of Emotional Development* (pp. 283-209). Springer, Cham. https://doi.org/10.1007/978-3-030-17332-6_12

Rozin, P., & Fallon, A. E. (1987). A perspective on disgust. *Psychological Review, 94,* 23-41. https://doi.org/10.1037/0033-295X.94.1.23

Rozin, P., Fallon, A., & Mandell, R. (1984). Family resemblance in attitudes to foods. *Developmental Psychology, 20,* 309–314. https://doi.org/10.1037/0012-1649.20.2.309

Rozin, P., Haidt, J., & McCauley, C. R. (2008). Disgust. In M. Lewis, J. M. Haviland-Jones, L. F. Barrett (Eds.), *Handbook of emotions* (3rd ed., pp. 757-776). Guilford Press.

Rozin, P., & Millman, L. (1987). Family environment, not heredity, accounts for family resemblances in food preferences and attitudes: A twin study. Appetite, *8*, 125-134. https://doi.org/10.1016/s0195-6663(87)80005-3

Rozin, P., Millman, L., & Nemeroff, C. (1986). Operation of the laws of sympathetic magic in disgust and other domains. *Journal of Personality and Social Psychology, 50,* 703-712. https://doi.org/10.1037/0022-3514.50.4.703

Schienle, A., Stark, R., & Vaitl, D. (2001). Evaluative conditioning: A possible explanation for the acquisition of disgust responses? *Learning and Motivation, 32,* 65-83. https://psycnet.apa.org/doi/10.1006/lmot.2000.1067

Sherlock, J. M., Zietsch, B. P., Tybur, J. M., & Jern, P. (2016). The quantitative genetics of disgust sensitivity. *Emotion, 16,* 43–51. https://doi.org/10.1037/emo0000101

Stevenson, R. J., Oaten, M. J., Case, T. I., Repacholi, B. M., & Wagland, P. (2010). Children's response to adult disgust elicitors: development and acquisition. *Developmental Psychology, 46,* 165–177. https://doi.org/10.1037/a0016692

Tolin, D. F., Lohr, J. M., Sawchuk, C. N., & Lee, T. C. (1997). Disgust and disgust sensitivity in blood-injection-injury and spider phobia. *Behaviour, Research and Therapy, 35,* 949-953. https://doi.org/10.1016/s0005-7967(97)00048-x

Tybur, J. M., Çınar, Ç., Karinen, A. K., & Perone, P. (2018). Why do people vary in disgust? *Philosophical Transactions of the Royal Society,* 373:20170204. https://doi.org/10.1098/rstb.2017.0204

Tybur, J. M., Lieberman, D., Kurzban, R., & DeScioli, P. (2013). Disgust: evolved function and structure. *Psychological Review, 120*, 65–84. https://doi.org/10.1037/a0030778

van Overveld, W. J. M., de Jong, P. J., Peters, M. L., Cavanagh, K., & Davey, G. C. L. (2006). Disgust propensity and disgust sensitivity: Separate constructs that are differentially related to specific fears. *Personality and Individual Differences, 41,* 1241-1252. https://doi.org/10.1016/j.paid.2006.04.021

Webb, K., & Davey, G. C. L. (1992). Disgust sensitivity and fear of animals: effect of exposure to violent or revulsive material. *Anxiety, Stress & Coping*, *5,* 329-335. https://doi.org/10.1080/10615809208248369

Widen, S. C., & Olatunji, B.O. (2016). A developmental perspective on disgust: implications for obsessive-compulsive disorder. *Current Behavioral Neuroscience Reports, 3,* 204–210. https://doi.org/10.1007/s40473-016-0087-0

Woody, S. R., & Teachman, B. A. (2000). Intersection of disgust and fear: Normative and pathological views. *Clinical Psychology: Science and Practice, 7,* 291-311. https://doi.org/10.1093/clipsy.7.3.291

Chapter 6
Individual Differences in Disgust

Joshua M. Tybur

Imagine entering a public restroom and seeing a film of vomit coating the surface of a toilet's water. Your face contorts into a grimace, with muscles acting to narrow your eyelids and tighten your jaw. Your breath catches, and you inhale only sparingly. All other thoughts leave your mind, and you are focused solely on avoiding physical contact with that toilet.

These responses serve a common function: to avoid becoming a host for infectious microbes (Tybur et al. 2013). The coordinated facial actions reduce the likelihood of pathogens entering the body via the eyes, nose, and mouth (Susskind et al. 2008). The strong compulsion to avoid contact reduces the likelihood of those same pathogens resting on the hands, where they could then be transported to the mouth while eating or eyes while mindlessly touching the face. Because these events unfold in response to things like vomit, faeces, and decaying tissue—the types of substances that have reliably housed microbes infectious to humans and their recent ancestors for millions of years—and because they almost seem engineered solutions to the problems posed by pathogens, researchers largely agree that disgust is an anti-pathogen adaptation, forged through millions of years of natural selection (Curtis and Biran 2001; Tybur and Lieberman 2016). Consistent with this proposition—and the idea that adaptations (and their by-products) are the building blocks of human universals (Tooby and Cosmides 1992)—researchers have observed remarkable cross-cultural consistency in: (1) what elicits disgust (e.g., Curtis and Biran 2001; Curtis et al. 2004); (2) the capacity for disgust-elicitors to "contaminate" objects that they come into contact with (Apicella et al. 2018); and (3) people's ability to recognise disgust vocalisations, even from voices recorded from cultures on the other side of the globe (Sauter et al. 2010).

J. M. Tybur (✉)
Department of Experimental and Applied Psychology, Vrije Universiteit Amsterdam, Amsterdam, Netherlands
e-mail: j.m.tybur@vu.nl

P. A. Powell, N. S. Consedine (eds.), *The Handbook of Disgust Research*,
https://doi.org/10.1007/978-3-030-84486-8_6

The fact that other experiences of disgust (e.g., considering sex with a parent or sibling, or hearing a loathed politician's latest example of dishonesty and self-interest) do not follow from cues to pathogens and correspond with different behavioural tendencies suggests that the folk category "disgust" is actually comprised of distinct adaptations (e.g., pathogen disgust, sexual disgust, moral disgust). These distinct adaptations serve different functions (Tybur et al. 2013; cf. Scarantino 2012) and are undergirded by distinct psychological architectures (e.g., de Smet et al. 2014; Fessler et al. 2005; Kupfer and Giner-Sorolla 2017; Lieberman 2009; Molho et al. 2017).

This brief summary describes the core of contemporary theory on disgust (see also Chapman and Anderson 2013; Curtis et al. 2011; Oaten et al. 2009; Rozin et al. 2008; Tybur et al. 2013; Tybur and Lieberman 2016). But it alone might not have precipitated the massive upswing in empirical work on the topic seen over the past decade. The recent surge in interest has instead been largely driven by the fact that people vary in their tendencies to experience disgust, and this variation apparently has many consequences. Returning to the example above, one person might shrug off the smell and sight of vomit and simply flush the toilet before using it, and another might not even be able to stomach staying in the bathroom. Researchers have (perhaps surprisingly) found that individuals reporting more disgust towards these types of situations have more negative attitudes towards foreigners (Aarøe et al. 2017; Karinen et al. 2019b; Zakrzewska et al. 2019; see also Vartanian et al., Chap. 10, this volume); more broadly conservative political stances (e.g., Inbar et al. 2012; Liuzza et al. 2018; Tybur et al. 2015; see also Shook et al., Chap. 14, this volume); tend to suffer more from anxiety disorders (e.g., Olatunji et al. 2017; see also Davey, Chap. 11, this volume); and have more positive or negative attitudes towards health screenings, depending on their invasiveness (Gruijters et al. 2016; Reynolds et al. 2014; see also Reynolds and Dev, Chap. 13, this volume). By better understanding variation in disgust, researchers will hopefully better be able to understand phenomena ranging from political sentiments to psychopathology to health maintenance.

But how should we understand this variability, which is typically estimated via self-report instruments in which respondents are asked to report how disgusting they would find a hypothetical situation, such as stepping in dog poop? Consider the trope occasionally derisively directed towards intelligence tests—that they only measure how good someone is at taking the test. A wealth of findings suggests that this comment is inaccurate (Ritchie 2015). But is a similar critique valid for disgust sensitivity instruments, that is, do resulting scores have any meaning outside of a testing context? And, if they do, do such scores simply reflect a broader disposition to experience negative emotions, rather than a disposition to experience disgust specifically (see also Consedine, Chap. 2, this volume)? Where does such variation come from—are some people more easily disgusted because their parents were especially pernickety about cleaning or often expressed disgust? And, at a more proximate level, where does such variation arise in the complex cognitive and perceptual processes that underlie disgust? This chapter aims to briefly summarise the progress that has been made in answering these and related questions. It does so by first describing the validity of disgust sensitivity instruments and then summarising the state of the field's knowledge on the developmental roots of variation in disgust. After covering what we *do* know about disgust sensitivity, it outlines some of the major questions that disgust sensitivity researchers might tackle in the coming years.

What Is Disgust Sensitivity?

Put simply, disgust sensitivity refers to the degree to which individuals experience disgust towards objects or situations that tend to elicit some degree of disgust in most people. The field owes the current use of the term to Haidt et al. (1994), who referred to their widely used Disgust Scale as a disgust sensitivity instrument. Other researchers have suggested that the variation tapped by such measures is better described as *propensity* to experience disgust (van Overveld et al. 2006). We will stick with disgust sensitivity here, with the caveat that some literatures refer to this same trait as disgust propensity, while using the term disgust sensitivity to refer to variation in how negative people find the experience of disgust.

In completing a typical disgust sensitivity instrument, an individual imagines a variety of disgust-eliciting scenarios, and he or she clicks a button (or circles a number, if we're going back to paper-pencil assessment) to indicate the degree to which they find those scenarios disgusting. Several critical points could be raised regarding this procedure's validity. We will focus on four here. First, given the domain differentiation detailed above (e.g., pathogen, sexual, moral), references to disgust sensitivity might gloss over important differentiation across categories of disgust elicitors. Second, responses to disgust instruments might largely reflect random noise, fluctuating across time of assessments and in response to transient mood or interpretations of the situations described in the items. Third, even if responses are systematic, they might more broadly assess emotionality or neuroticism (i.e., tendencies to experience negative affect) rather than specifically disgust. Fourth, even if responses are consistent across time and specific to disgust, they might simply reflect variation in vividness of imagination. Each of these possibilities could cut the validity of inferences based on disgust sensitivity instruments off at the knees. We'll briefly review the evidence supporting or contradicting these potential problems with disgust sensitivity instruments.

The Domains of Disgust Sensitivity

In developing the Disgust Scale, Haidt et al. (1994) solicited descriptions of disgust elicitors from 20 respondents. After reviewing the 221 descriptions, they proposed that disgust sensitivity might vary along eight dimensions: food, sex, body products, body envelope violations, animals, hygiene, death, and moral violations. Statistical analyses of a preliminary version of the Disgust Scale designed to cover these putative dimensions indicated that responses to the moral violation items did not covary sufficiently to be included in a global measure of disgust sensitivity. The subsequently developed final version of the instrument was labelled as having eight domains—those listed above, without moral violations, but with a replacement group of items labelled as magical thinking. Although this taxonomy would go on to strongly influence the disgust literature, no quantitative evidence supported the validity of these domains. Indeed, a later study developing the Disgust Scale-Revised concluded that the instrument has three factors if seven of the original items were removed (Olatunji et al. 2007). Nevertheless, correlations between these three dimensions—labelled "core," "animal reminder," and "contamination"—ranged from $r = .66$ to $r = .84$, suggesting limited distinctiveness across these types of domains. A recent alternative approach examined disgust endorsed towards items involving each of six pathogen-transmission pathways: (1) direct interpersonal contact; (2) indirect interpersonal contact with aerosolised droplets; (3) sexual contact; (4) contact with a secondary host (e.g., an insect); (5) ingestion of contaminated food or water; and (6) contact with a fomite (a pathogen-contaminated object; Curtis and de Barra 2018). Factor analyses of endorsed disgust towards items across these categories did not favour this taxonomy. Importantly, when items were aggregated across five putatively distinct factors implied by a factor analysis, the average correlation between each of the five factors was $r = .58$. Hence, we again have evidence for a general tendency to be disgusted by pathogen cues, or a common factor.

One approach that has identified greater distinctiveness between domains of disgust sensitivity has involved broadening items to include sexual and moral content. The Three-Domain Disgust Scale (Tybur et al. 2009) contains three factors—pathogen, sexual, and moral—which are more modestly correlated, with *r*'s ranging from .20 to .40 in the sample in which the instrument was developed. Importantly, the pathogen domain of the Three-Domain Disgust Scale correlates strongly with all three subscales of the Disgust Scale-Revised (*r*'s ranging from .61 to .92), suggesting that it captures the same higher order factor captured by the Disgust Scale-Revised. This instrument's sexual and moral domains correlate less strongly with these same Disgust Scale-Revised subscales (*r*'s between −.01 and .49), and they relate differently to a number of variables (e.g., personality, respondent sex, political sentiments; see Tybur and Karinen 2018, for a review).

The Three-Domain Disgust Scale and the Disgust Scale (and its revised version) are the two most widely-used measures in the literature. Given that the Disgust Scale largely corresponds with the pathogen domain of the Three-Domain Disgust Scale—and given that much of the knowledge we have about the Three-Domain Disgust Scale concerns the pathogen domain—references to disgust sensitivity should be interpreted as reflecting *pathogen* disgust sensitivity throughout most of this chapter.

How Should We Interpret Disgust Sensitivity Scores?

A variety of studies have aimed to inform the reliability and validity of disgust sensitivity scores. For example, one study found that intraclass correlation coefficients for Three-Domain Disgust Scale scores are around .80 across 5 weekly administrations (Jones et al. 2018), suggesting that disgust sensitivity scores are consistent across assessments. Scores on the Disgust Scale also predict engagement in disgust-related behaviours requested by an experimenter (e.g., touching someone else's used hair comb; touching the inside of a toilet bowl), with an estimated correlation of $r = .58$ (Rozin et al. 1999). Of course, such a relationship does not rule out the possibility that disgust sensitivity simply reflects emotionality (in the HEXACO personality framework) or neuroticism (in the Big Five personality framework). And, indeed, some findings suggest that disgust sensitivity is moderately related to neuroticism ($r = .45$ and .46; Druschel and Sherman 1999; Olatunji et al. 2008), though others have reported weaker relationships (e.g., $r = .10$; Tybur et al. 2011).

One recent study (Karinen et al. 2019b) has examined self-other agreement in disgust sensitivity to assess the degree to which disgust sensitivity instruments measure systematic trait variance rather than response bias. The logic behind this approach is simple and elegant. If disgust sensitivity scores relate to behaviours that are stable across time and visible to others, then there should be a correspondence between how one rates themselves on the scale and how an observer (e.g., a romantic partner) rates the same person on the same scale. In this study, pairs of participants completed the Three-Domain Disgust Scale and the HEXACO-100 (Lee and Ashton 2018), a measure of the six HEXACO personality dimensions. Each member of the pair rated themselves and their partner on the same items; hence, all participants had both a self-reported disgust sensitivity score and other-reported disgust sensitivity score. Self-other agreement in disgust sensitivity was $r = .46$—similar to self-other agreement in personality. Controlling for personality (both self-rated and other-rated) only modestly reduced this agreement (to $r = .41$), indicating that self-other agreement in disgust sensitivity is not a by-product of self-other agreement in broader personality domains.

With some comfort in the assumption that we're measuring something that is: (1) relatively stable across time, (2) corresponds with a willingness to engage in disgust-eliciting acts in lab settings, (3) is detectable by others over long periods of time, and (4) is mostly independent from emotionality (and other lexically-derived personality dimensions), we can move on to ask why people vary in disgust in the first place.

Why Do People Vary in Disgust Sensitivity?

This "why" question can be answered in many ways. At a proximate level, it could refer to any neurobiological differences that underlie variation in disgust. At a phylogenetic level, it could refer to the evolutionary forces that have allowed for this variation. And at a developmental level, it could refer to the environmental inputs that give way to this variation. While the former two have scarcely been addressed in the disgust sensitivity literature (though see Case et al. 2020, and Sarabian et al. 2017, for recent exceptions), the developmental origins have received substantial attention (Rottman 2014; Rozin et al. 2008; Stevenson et al. 2010; Widen and Olatunji 2016; Widen and Russell 2013; see also Reynolds and Askew, Chap. 5, this volume). And, based on recent work, we're now able to evaluate some developmental accounts of disgust sensitivity.

Various lines of evidence suggest that young children do not experience disgust towards pathogen cues in the same way that adults do (Rottman 2014). This apparent absence has been interpreted as suggesting that disgust elicitors are transmitted from parents to offspring—that is, that children learn what to feel disgusted by after observing the targets of parents' facial and vocal expressions of disgust (Rozin and Fallon 1987). Similarly, researchers have interpreted multiple lines of evidence as indicating that children's disgust sensitivity is shaped by observations of their parents' disgust expressions (Olatunji et al. 2017; Widen and Olatunji 2016). For example, one study taken as support for this hypothesis reports that parents in a lab context avoid disgust-eliciting stimuli around their children more if those children are young (around 2.5 years old) than if they are old (e.g., age 10; Stevenson et al. 2010). Such a behaviour might reflect strategic modelling during a critical learning period. Other studies report that parents and offspring score similarly on measures of contamination sensitivity (Davey et al. 1993; Rozin et al. 1984). And another reports that twins score similarly on measures of contamination sensitivity, but that monozygotic twins score no more similarly than dizygotic twins, and hence the similarities between twins stem from shared environment (which could include observations of parents) rather than shared genes (Rozin and Millman 1987; see also Olatunji et al. 2019).

Even if initially compelling, each of these pieces of evidence provides only limited support for the hypothesis that disgust sensitivity emerges from parental modelling. The study reporting that parents of younger children avoid disgust-eliciting stimuli more than parents of older children also found no difference in reported or emoted disgust across parents of older and younger children. Other studies of older children (age 7–10) report that a conditioning task in which animals are paired with images of strangers posing disgusted facial expressions are rated more negatively than animals paired with images of strangers posing happy facial expressions, but not more negatively than strangers posing fearful facial expressions (Askew et al. 2014). These results suggest that expressions of disgust from any source—not just parents—can influence assessments of targets of those expressions. If disgust sensitivity is in fact influenced by observations of others' disgust responses, then parental influence might be diluted by observations of the thousands of other individuals a child witnesses during development.

Further, because within-family similarities can arise via shared genes rather than the type of parental modelling described above, studies reporting similarities between parents and offspring do not adjudicate between environmental and genetic transmission accounts. Twin studies are uniquely positioned to do so, since researchers can compare similarities between monozygotic twins (who share nearly all of their segregating genes, as well as a common environment, which includes exposure to the same parents) and dizygotic twins (who share, on average, 50% of their genes, as well as exposure to the same parents). If correlations between dizygotic twins are more than half as large as correlations between dizygotic twins, then some shared environment effects (e.g., being exposed to a parent who expresses some level of disgust) are present; if the correlations are the same, then environmental features that twins share entirely underlie within-family similarities. If the monozygotic twin correlations are at least double the dizygotic twin correlations, then similarities within families result predominantly from shared genes. These interpretations rest on the assumption that monozygotic twin pairs share environments to the same extent as dizygotic twin pairs—an assumption that is robust to myriad challenges (e.g., Barnes et al. 2014).

The one study that reported no difference between monozygotic twins and dizygotic twins had a small sample (38 monozygotic twin pairs and 34 dizygotic twin pairs; Rozin and Millman 1987) that was underpowered to detect genetic effects. Another study reporting genetic effects for women, but not men, was similarly underpowered, with 41 monozygotic twin pairs and 19 dizygotic twin pairs (Olatunji et al. 2019). Two larger studies of Finnish twins, with 123 and 553 monozygotic twin pairs and 127 and 490 dizygotic twin pairs, respectively, both found that correlations between monozygotic twins (r's = .49 and .42) were more than double those between dizygotic twins (r's = .23 and .11; Sherlock et al. 2016; Tybur et al. 2020b). Results from these twin studies—like many before them—can conflict with folk intuitions that variation in psychological traits arises from parental behaviours (Pinker 2003). They also provide critical new information on disgust sensitivity by suggesting that about half the trait variation in population studies is accounted for by genetic differences between people in that population, with the other half being accounted for by environmental influences that children do not share.

Naturally, it is important to keep straight what these results mean and what they do not. Rather than suggesting that environment inputs have no influence on disgust sensitivity, they indicate that such inputs are either typically unshared by twins, or shared by unrelated individuals within a population. Again, though, parental modelling is an unlikely candidate for such environmental inputs. If disgust sensitivity does not arise based on parental modelling, then how and why does it arise? This question can be used generate directions for future research. The next section covers four of such questions.

Four Questions for Future Disgust Sensitivity Research

Question 1: What Maintains Genetic Variance in Disgust?

Intuitively, the existence of heritable variation in traits with survival and reproductive consequences should approximate zero, since natural selection should shave away those genetic variants that lead to suboptimal outcomes (Tooby and Cosmides 1990). Multiple evolutionary mechanisms can maintain variance in fitness-relevant traits, though (Penke et al. 2007). If variation is blind to selection (i.e., has no consequence for survival and reproduction), then it can merely reflect genetic noise, which itself might be beneficial (e.g., to neutralize pathogens). Alternatively, mutation-selection balance can maintain variance in genetically complex traits, with deleterious mutations with small negative effects accumulating across generations, and selection purging those mutations that most reduce fitness. Given that most heritable psychological traits are undergirded not by a handful of genes but by thousands (Chabris et al. 2015), mutation-selection balance is a plausible mechanism for underlying genetic variants in some traits (Penke et al. 2007). However, not all traits have optimum levels. Personality offers a good example. Although high levels of extraversion are associated with some outcomes that are positive for reproductive fitness (e.g., lifetime sexual partner number), they are also associated with negative outcomes (e.g., the likelihood of being hospitalized due to accident or illness; Nettle 2005). Similar trade-offs likely exist for other dimensions of personality (de Vries et al. 2016). The fact that high and low levels of personality traits each have benefits—and the fact that the costs and benefits of different levels of personality traits vary across ecologies and situations—suggests that genetic variation in personality is maintained through a different mechanism: balancing selection (Penke and Jokela 2016).

The logic underlying the balancing selection view of personality applies more cleanly to disgust sensitivity. Indeed, whereas proposed costs and benefits of high and low levels of personality traits have been admittedly speculative (e.g., Nettle 2005), they are clear for disgust sensitivity. Higher levels correspond with less contact with substances harbouring pathogens, and hence mitigate the consequences of infectious disease. Simultaneously, though, they shrink the pool of individuals one is willing to share the contact necessary for cooperative interactions (Tybur et al. 2020a). Conversely, lower levels are associated with less avoidance of potentially-beneficial social contact, but an increased probability of infection. No optimal level of disgust sensitivity exists; high and low levels correspond with different classes of costs and benefits, and hence selection might not eliminate genes that lead to higher or low levels of the trait.

Balancing selection is not the only account that can accommodate these considerations of costs and benefits. An alternative, reactive heritability, proposes that behavioural traits emerge contingent on other aspects of the phenotype, which themselves are heritable (Tooby and Cosmides 1990). Again, take extraversion as an example. Not all individuals are equally able to convert the highly extraverted behaviours into reproductive benefits, and not all individuals are equally vulnerable to the costs associated with high extraversion. In other words, the optimal level of extraversion presumably varies across individuals depending on moderating characteristics, such as physical strength (in men) and attractiveness (in both sexes) (Lukaszewski and Roney 2011). According to this perspective, behavioural traits are not directly heritable, but instead reflect facultative calibration to other heritable aspects of the phenotype. The few genetically-informed studies testing reactive heritability (Haysom et al. 2015; von Rueden et al. 2015) have important limitations and conflicting results, and whether reactive heritability can explain psychological variation is an open question. Regardless, it provides another avenue through which to explain genetic variance in disgust sensitivity. Along the lines of proposals from several authors (e.g., Curtis et al. 2011; Fessler et al. 2003; Oaten et al. 2009; Tybur et al. 2013), higher disgust sensitivity might result from lower abilities to resist pathogens, which itself has a heritable component (de Craen et al. 2005). But is the literature consistent with this proposal—that is, are those less able to resist infection more disgust sensitive?

Question 2: What Is the Connection Between Disgust Sensitivity and Pathogen Resistance?

Multiple studies have been interpreted as suggesting that pathogen resistance (i.e., ability to resist infection) is negatively associated with pathogen avoidance (e.g., disgust sensitivity) (Ersche et al. 2014; Fessler et al. 2005; Fleischman and Fessler 2011; Oaten et al. 2017). Conclusions from these studies should be treated with caution, though, as they use cocaine dependency, pregnancy trimester, progesterone, and rheumatoid arthritis, respectively, as proxies for immune function. Other work suggests little if any relationship between immune function and disgust sensitivity. One study found that disgust sensitivity is unrelated to self-reports of infection recency and frequency in a sample of 616 Australians (Stevenson et al. 2009). Another study of 284 rural Bangladeshi's found that disgust sensitivity is unrelated to both illness recency and illness experienced during childhood (de Barra et al. 2014). Of course, self-reports of ill health are noisy measures, prone to recall error and unable to determine the cause of the illness. Perhaps more importantly, illness can arise based on both exposure to pathogens and ability to combat them. Imagine that individuals with a low ability to resist infections indeed compensate by developing higher disgust sensitivity. Under such circumstances, the relationship between illness frequency and disgust sensitivity might be negative (if higher disgust sensitivity results in low-resistance individuals avoiding pathogens enough so as to not fall ill in the first place), zero (if higher disgust sensitivity results in low-resistance individuals falling ill as often as their higher-resistance peers), or positive (if higher disgust sensitivity is unable to fully compensate for lower resistance).

Similar issues complicate the perhaps intuitive hypothesis that individuals living in areas with more infectious disease should have higher disgust sensitivity to defend against pathogens (see, e.g., Rozin and Fallon 1987, and Fincher and Thornhill 2012, for such arguments). Speaking against this hypothesis, a recent study of over 11,000 individuals across 30 nations found that national parasite stress was unrelated to disgust sensitivity—that is, that individuals in parasite-rich countries such as India are no more disgust sensitive than individuals in low-parasite countries such as Canada (Tybur et al. 2016). Limitations apply here, too. Parasite stress is not homogenous within nations, and national averages obscure potentially important differences within countries. Further, the aforementioned study was not nationally representative, and we cannot rule out the possibility that individuals from the lower parasite-stress areas of high parasite-stress countries were sampled, and vice versa.

The literature on pathogen resistance and disgust sensitivity is small and consists of studies with sizable limitations. Nevertheless, existing evidence suggests that there is little-to-no relationship between disgust sensitivity and history of infection or the concentration of pathogens within the ecology. At first blush, such findings might seem damning to the proposal that pathogen disgust functions to motivate pathogen avoidance. A closer look at the computational processes underlying disgust suggests that such scepticism would be premature.

Question 3: Where in Information Processing Does Variation in Pathogen Disgust Arise?

Proposals of function are ubiquitous in the disgust literature, with most researchers endorsing a pathogen-avoidance account (e.g., Curtis and Biran 2001; Haidt et al. 1994; Oaten et al. 2009; Rozin and Fallon 1987; Tybur et al. 2009). While useful, exclusive attention to function can render multiple aspects of disgust puzzling. After all, if disgust functions to neutralise pathogen transmission, why are people not disgusted by many pathogen-harbouring substances (e.g., cholera-contaminated water; salmonella-contaminated chicken)? Why do many people seem to experience little disgust towards so many pathogen-risky behaviours, like sex with strangers (see also Borg and de Jong, Chap. 9, this volume)? And, relevant to the section above, why wouldn't people at a greater risk for infection be more prone to experiencing disgust? Considerations of the mechanisms (e.g., information processing) that might execute pathogen avoidance can illuminate these and other puzzles (Cosmides and Tooby 1994, 2000).

All five senses can detect cues to pathogens. We can see the redness of a feverish face or the yellow-white of puss; we can smell a rotting corpse; we can feel the warmth and moisture of another person's sweaty palms; we can hear the coughs, sneezes, and retching that result from influenza; and we can taste the sour flavours corresponding with elevated bacterial concentrations. Apparently, disgust (perhaps uniquely among emotions) can be elicited through each of the senses. But is it triggered automatically upon detection of any one of myriad cues to pathogens? Consider how a lover experiences no disgust at his or her partner's mouth during a kiss (but would experience intense disgust towards the same act with a stranger), a parent does not recoil in revulsion when his or her infant soils a diaper (but would be grossed out by another baby's diaper), and many people think nothing of taking a sip from a friend's water bottle (but would have serious reservations about drinking from the bottle of someone they dislike). These observations suggest that the computational systems underlying disgust process information apart from infection likelihood in generating disgust. Broadly speaking, such information should inform the benefits of contact (or, perhaps, costs of avoidance), which might or might not outweigh the corresponding pathogen costs of contact. This is precisely how a pathogen-avoidance system should have evolved—to integrate multiple pieces of information rather than reflexively generate revulsions when pathogen cues are detected (Tybur and Lieberman 2016). Evidence suggests such trade-offs between costs and benefits reflect variation in how much one invests in pathogen avoidance across individual targets, with one study reporting that people are more comfortable with infection-risky contact with close friends than with people they dislike, and more comfortable with infection-risky contact with honest and kind strangers than with dishonest and unkind ones (Tybur et al. 2020a; similar discrimination appears between kin and non-kin in mandrills; Poirotte and Charpentier 2020).

Viewed through this lens, what are the sources of variation in disgust sensitivity? Does high disgust sensitivity arise from sensory systems that are better able to detect cues to pathogens? Does it arise later, with the minds of more disgust sensitive individuals more heavily invested in avoiding infection? Or does it arise because higher disgust sensitive individuals are less averse to paying the costs of avoidance? Asking and answering these types of questions might prove instrumental in moving forward our understanding of disgust sensitivity.

Take as an example a recent study (Perone et al. 2020) that used the emotional attentional blink paradigm to better understand where in cognitive processing variation in disgust sensitivity is apparent. In this task, participants see 18 images appear on a screen one after another, each for only 100 ms. One of these images—in this study, a building rotated 90 °—is the target, which participants must identify. The other 17 images per trial are irrelevant to the task that participants are told to complete, and one of them is emotionally evocative. As in other work (Ciesielski et al. 2010), participants were worse at identifying the target when it was preceded by a disgust-eliciting image 200 ms (i.e., two images) earlier. With temporal attention occupied after detecting cues to pathogens, it is as if attention "blinks" for a moment. This attentional blink was unrelated to disgust sensitivity, though. That is, any differences in the early-level processing of visual cues to pathogens across participants did not correspond with differences in disgust sensitivity. However, when participants were later asked to rate those same disgust-eliciting images on valence (i.e., how positive versus negative they were), differences across disgust sensitivity did emerge, with more disgust sensitive participants rating the images more negatively.

Other work using different methods has not detected evidence that disgust sensitivity is apparent at early levels of sensory detection. For example, Hunt et al.

(2017) used scarcely-detectable nylon fibre monofilaments to apply progressively greater pressure to the hands of participants, who indicated when they could feel the fibres. Tactile sensitivity was unrelated to disgust sensitivity, though it was higher after participants viewed disgust-eliciting images. In another study on olfaction, Chan et al. (2016) used a three-alternative forced-choice staircase paradigm in which blindfolded participants smelled the contents of two sticks that were empty and one stick that contained a dilute concentration of n-Butanol (an odour evocative of white-board marker ink), with dilutions increasing or decreasing until a threshold could be estimated. The researchers found that disgust sensitivity was unrelated to olfactory thresholds. In sum, these findings suggest that more-disgust-sensitive individuals are no better at detecting tactile and olfactory information than are less-disgust-sensitivity individuals (see also Liuzza, Chap. 7, this volume). Such findings might lead us to conclude that variation in disgust sensitivity emerges from how information about pathogen cues is processed. That said, neither Hunt et al. (2017) nor Chan et al. (2016) assessed thresholds specific to pathogen cues, with the former not using tactile cues such as moisture, and the latter not using olfactory cues associated with bodily wastes or decay. Future work can assess whether disgust sensitivity reflects variation in abilities to detect such cues, or only the motivational tendencies to avoid such cues once detected.

Question 4: What Are Sexual and Moral Disgust Sensitivity?

As noted earlier, people are disgusted by stimuli that presumably elicit disgust for reasons apart from pathogen avoidance—notably those with sexual or moral content. Of course, sexual fluids transmit pathogens, and may elicit pathogen disgust for this reason (Curtis and Biran 2001). But much of the disgust associated with sexuality concerns pathogen-irrelevant aspects of a potential partner (e.g., whether they are relatives, as in the case of incest). In addition to assessing pathogen disgust, the widely used Three-Domain Disgust Scale also includes sexual and moral domains, which researchers have used to test hypotheses across a number of research areas. Examples include work indicating that the widely-cited relation between political ideology and pathogen disgust is entirely accounted for by sexual disgust (Billingsley et al. 2018; Tybur et al. 2015), that food neophobia relates to sexual disgust rather than pathogen disgust (Al-Shawaf et al. 2015a), and that women's preferences for facial masculinity relate to moral disgust rather than pathogen disgust (McIntosh et al. 2017). What are we to make of such findings? Answers to this question improve with our understanding of the functions of sexual and moral disgust. Such understandings of function lag behind our understanding of the function of pathogen disgust, though. Sexual disgust could, among other things, function to signal a lack of interest to other parties, to motivate avoidance after interest is detected, or to inhibit sexual approach to otherwise appetitive stimuli (e.g., a sibling who is otherwise physically attractive), and moral disgust could, among other things, motivate social avoidance, signal moral stances to observers, or motivate indirect aggression. Comparatively limited understanding of sexual disgust and moral disgust likely results from the ethical perils of putting research participants in situations intended to arouse such disgust and from long-standing disagreements in the literature regarding whether moral disgust is "really" disgust (compare, e.g., Chapman and Anderson 2011, 2013 with Royzman and Kurzban 2011 and Royzman and Sabini 2001). Research that circumvents these challenges can make large strides towards understanding the functions of sexual and moral disgust (see also Borg and de Jong, Chap. 9, this volume and Giner-Sorolla, Chap. 8, this volume).

Progress in understanding the functions of sexual and moral disgust would not necessarily inform how we understand the sexual and moral domains of the Three-Domain Disgust Scale (or other disgust sensitivity instruments), though. Consider the revised version of the Disgust Scale (Olatunji et al. 2007), which labels a cluster of items as "animal reminder" disgust, named after Rozin et al.' (2008) proposal that some instances of disgust function to protect people from being reminded that they are animals—a recognition that would putatively lead to crippling anxiety. Myriad potential reminders that humans are animals (e.g., the fact that we run, jump, sleep, eat, and care for young, just as many other non-human animals do) elicit no disgust, and no evidence indicates that disgust reduces anxiety (indeed, people who experience disgust more intensely are at higher risk for anxiety disorders; see Olatunji et al. 2017). Put straightforwardly, the proposed functions of animal reminder disgust appear to be wrong (see Kollareth and Russell 2017; Royzman and Sabini 2001; and Tybur et al. 2013 for more thorough discussions). But we can still make progress in understanding what the scale labelled as "animal reminder" on the Disgust Scale-Revised measures. As noted above, latent variable estimates indicate that this subscale correlates around $r = .80$ with other subscales of the same measure, and it has virtually identical relationships with personality (Olatunji et al. 2008). Hence, we can understand this subscale as reflecting pathogen disgust, regardless of the scale's label.

How does the literature suggest that we should interpret the sexual and moral domains of the Three-Domain Disgust Scale, independent of the veracity of any theoretical accounts of sexual or moral disgust? As with pathogen disgust sensitivity, self-other agreement is about as high for sexual disgust sensitivity ($r = .66$) as it is for personality dimensions like extraversion and openness to experience, and self-other agreement for moral disgust sensitivity ($r = .36$) is a bit lower (Karinen et al. 2019a). This same study found that sexual disgust sensitivity correlates only modestly with the six HEXACO personality dimensions (all r's $< .30$), though moral disgust sensitivity correlates sizably with honesty-humility, $r = .45$, and also moderately with conscientiousness, $r = .29$. The general picture indicates the sexual and moral domains of the Three-Domain Disgust Scale correspond with characteristics that are stable and visible enough to be assessed by perceivers, and that much of the variability in moral disgust sensitivity reflects tendencies to be hard working, modest, humble, and straightforward. Other work suggests that sexual disgust sensitivity, while only modestly related to lexically-based personality dimensions, is strongly related to socio-sexual orientation (i.e., openness to sex outside of a committed relationship), with studies reporting correlations of around $r = -.55$ between the two traits (Al-Shawaf et al. 2015b; Tybur et al. 2015). Further, while women score modestly higher on the pathogen and moral domains of the Three-Domain Disgust Scale (*d* around .30), they score much higher on the sexual domain (*d* around 1.40; Tybur et al. 2011). For some context, a *d* of .30 would mean that about 62% of women score higher on the pathogen and moral domains than the average man; a *d* of 1.40 would means that 92% of women score higher on the sexual domain than the average man.

Interpretations of pathogen disgust sensitivity as reflecting motivations to avoid pathogen cues are coloured by our understanding of the function of pathogen disgust. Given ambiguities in the functions of moral and sexual disgust, we can interpret scale item content and the findings reviewed above as suggesting that moral disgust sensitivity (at least as measured by the Three-Domain Disgust Scale) reflects objections to non-cooperative behaviours, and sexual disgust sensitivity reflects aversions to intercourse outside of a monogamous relationship. Future work can address the degree to which moral disgust sensitivity reflects general versus disgust-specific outrage. Other work can test whether sexual disgust sensitivity has a different meaning than not only socio-sexual orientation, but also other measures of sexual attitudes and behaviour (e.g., sex drive). And, based on such findings, work can further aim to develop better instruments, perhaps those inspired by progress in functional understandings of moral and sexual disgust.

Conclusions

Investments in the science of disgust can help us better understand our relationship with health, food, politics, sex, and prejudice, among other things. And illuminating individual differences play a key role in this science. To realise such benefits, disgust researchers are best served by delving both deeply and broadly into the toolkits available to psychologists. Methods employed by developmental psychologists, behavioural geneticists, personality psychologists, and cognitive psychologists can converge to help us understand why people vary in disgust, which can in turn help us understand the myriad features of human nature that disgust relates to. This chapter has provided a summary of some of the latest developments in this enterprise, and it hopefully offers some useful directions for the next wave of disgust sensitivity research.

References

Aarøe, L., Petersen, M. B., & Arceneaux, K. (2017). The behavioral immune system shapes political intuitions: Why and how individual differences in disgust sensitivity underlie opposition to immigration. *American Political Science Review, 111*, 277-294.

Al-Shawaf, L., Lewis, D. M., Alley, T. R., & Buss, D. M. (2015a). Mating strategy, disgust, and food neophobia. *Appetite, 85,* 30-35.

Al-Shawaf, L., Lewis, D. M., & Buss, D. M. (2015b). Disgust and mating strategy. *Evolution and Human Behavior, 36,* 199-205.

Apicella, C. L., Rozin, P., Busch, J. T., Watson-Jones, R. E., & Legare, C. H. (2018). Evidence from hunter-gatherer and subsistence agricultural populations for the universality of contagion sensitivity. *Evolution and Human Behavior, 39,* 355-363.

Askew, C., Çakır, K., Põldsam, L., & Reynolds, G. (2014). The effect of disgust and fear modeling on children's disgust and fear for animals. *Journal of Abnormal Psychology, 123,* 566-577.

Barnes, J. C., Wright, J. P., Boutwell, B. B., Schwartz, J. A., Connolly, E. J., Nedelec, J. L., & Beaver, K. M. (2014). Demonstrating the validity of twin research in criminology. *Criminology, 52*(4), 588-626.

Billingsley, J., Lieberman, D., & Tybur, J. M. (2018). Sexual disgust trumps pathogen disgust in predicting voter behavior during the 2016 U.S. presidential election. *Evolutionary Psychology, 16,* 1474704918764170.

Case, T. I., Stevenson, R. J., Byrne, R. W., & Hobaiter, C. (2020). The animal origins of disgust: Reports of basic disgust in nonhuman great apes. *Evolutionary Behavioral Sciences, 14,* 231-260.

Chabris, C. F., Lee, J. J., Cesarini, D., Benjamin, D. J., & Laibson, D. I. (2015). The fourth law of behavior genetics. *Current Directions in Psychological Science, 24,* 304-312.

Chan, K. Q., Holland, R. W., van Loon, R., Arts, R., & van Knippenberg, A. (2016). Disgust and fear lower olfactory threshold. *Emotion, 16,* 740-749.

Chapman, H. A., & Anderson, A. K. (2011). Response to Royzman and Kurzban. *Emotion Review, 3,* 272-273.

Chapman, H. A., & Anderson, A. K. (2013). Things rank and gross in nature: a review and synthesis of moral disgust. *Psychological Bulletin, 139,* 300-327.

Ciesielski, B. G., Armstrong, T., Zald, D. H., & Olatunji, B. O. (2010). Emotion modulation of visual attention: Categorical and temporal characteristics. *PLoS ONE, 5,* e13860.

Cosmides, L., & Tooby, J. (1994). Origins of domain specificity: The evolution of functional organization. In L. Hirschfeld & S. German (Eds.), *Mapping the mind: Domain specificity in cognition and culture* (pp. 85-116). Cambridge University Press.

Cosmides, L., & Tooby, J. (2000). Evolutionary psychology and the emotions. In M. Lewis & J. M. Haviland-Jones (Eds.), *Handbook of Emotions* (pp. 91–115). Guilford Press.

Curtis, V., & Biran, A. (2001). Dirt, disgust, and disease: Isis hygiene in our genes? *Perspectives in Biology and Medicine, 44,* 17–31.

Curtis, V., & de Barra, M. (2018). The structure and function of pathogen disgust. *Philosophical Transactions of the Royal Society B: Biological Sciences, 373*(1751), 20170208.

Curtis, V., Aunger, R., & Rabie, T. (2004). Evidence that disgust evolved to protect from risk of disease. *Proceedings of the Royal Society of London. Series B: Biological Sciences, 271 (suppl_4),* S131-S133.

Curtis, V., de Barra, M., & Aunger, R. (2011). Disgust as an adaptive system for disease avoidance behaviour. *Philosophical Transactions of the Royal Society B: Biological Sciences, 366*(1563), 389-401.

Davey, G. C., Forster, L., & Mayhew, G. (1993). Familial resemblances in disgust sensitivity and animal phobias. *Behaviour Research and Therapy, 31,* 41-50.

de Barra, M., Islam, M. S., & Curtis, V. (2014). Disgust sensitivity is not associated with health in a rural Bangladeshi sample. *PloS one, 9,* e100444.

de Craen, A. J. M., Posthuma, D., Remarque, E. J., van Den Biggelaar, A. H. J., Westendorp, R. G. J., & Boomsma, D. I. (2005). Heritability estimates of innate immunity: An extended twin study. *Genes and immunity, 6,* 167-170.

de Smet, D., van Speybroeck, L., & Verplaetse, J. (2014). The Westermarck effect revisited: A psychophysiological study of sibling incest aversion in young female adults. *Evolution and Human Behavior, 35,* 34-42.

de Vries, R. E., Tybur, J. M., Pollet, T. V., & van Vugt, M. (2016). Evolution, situational affordances, and the HEXACO model of personality. *Evolution and Human Behavior, 37,* 407-421.

Druschel, B. A., & Sherman, M. F. (1999). Disgust sensitivity as a function of the Big Five and gender. *Personality and Individual Differences, 26,* 739-748.

Ersche, K. D., Hagan, C. C., Smith, D. G., Abbott, S., Jones, P. S., Apergis-Schoute, A. M., & Döffinger, R. (2014). Aberrant disgust responses and immune reactivity in cocaine dependent men. *Biological Psychiatry, 75,* 140–147.

Fessler, D. M., Arguello, A. P., Mekdara, J. M., & Macias, R. (2003). Disgust sensitivity and meat consumption: A test of an emotivist account of moral vegetarianism. *Appetite, 41,* 31-41.

Fessler, D. M., Eng, S. J., & Navarrete, C. D. (2005). Elevated disgust sensitivity in the first trimester of pregnancy: Evidence supporting the compensatory prophylaxis hypothesis. *Evolution and Human Behavior, 26,* 344-351.

Fincher, C. L., & Thornhill, R. (2012). Parasite-stress promotes in-group assortative sociality: The cases of strong family ties and heightened religiosity. *Behavioral and Brain Sciences, 35,* 61-79.

Fleischman, D. S., & Fessler, D. M. (2011). Progesterone's effects on the psychology of disease avoidance: Support for the compensatory behavioral prophylaxis hypothesis. *Hormones and Behavior, 59,* 271–275.

Gruijters, S. L., Tybur, J. M., Ruiter, R. A., & Massar, K. (2016). Sex, germs, and health: Pathogen-avoidance motives and health-protective behaviour. *Psychology & Health, 31,* 959-975.

Haidt, J., McCauley, C., & Rozin, P. (1994). Individual differences in sensitivity to disgust: A scale sampling seven domains of disgust elicitors. *Personality and Individual differences, 16,* 701-713.

Haysom, H. J., Mitchem, D. G., Lee, A. J., Wright, M. J., Martin, N. G., Keller, M. C., & Zietsch, B. P. (2015). A test of the facultative calibration/reactive heritability model of extraversion. *Evolution and Human Behavior, 36,* 414-419.

Hunt, D. F., Cannell, G., Davenhill, N. A., Horsford, S. A., Fleischman, D. S., & Park, J. H. (2017). Making your skin crawl: The role of tactile sensitivity in disease avoidance. *Biological Psychology, 127,* 40-45.

Inbar, Y., Pizarro, D., Iyer, R., & Haidt, J. (2012). Disgust sensitivity, political conservatism, and voting. *Social Psychological and Personality Science, 3,* 537-544.

Jones, B. C., Hahn, A. C., Fisher, C. I., Wang, H., Kandrik, M., Lee, A. J., Tybur, J.M., & DeBruine, L. M. (2018). Hormonal correlates of pathogen disgust: testing the compensatory prophylaxis hypothesis. *Evolution and Human Behavior, 39,* 464-467.

Karinen, A.K., Molho, C., Kupfer, T.R., & Tybur, J.M. (2019a). Disgust sensitivity and opposition to immigration: Does contact avoidance or resistance to foreign norms explain the relationship? *Journal of Experimental Social Psychology, 84,* 103817.

Karinen, A., Tybur, J. M., & de Vries, R. E. (2019b). The disgust trait: Self-other agreement in disgust sensitivity and its independence from HEXACO personality. *PsyArXiv.* https://doi.org/10.1037/emo0000795

Kollareth, D., & Russell, J. A. (2017). Is it disgusting to be reminded that you are an animal? *Cognition and Emotion, 31,* 1318-1332.

Kupfer, T. R., & Giner-Sorolla, R. (2017). Communicating moral motives: The social signaling function of disgust. *Social Psychological and Personality Science, 8,* 632-640.

Lee, K., & Ashton, M. C. (2018). Psychometric Properties of the HEXACO-100. *Assessment, 25,* 543-556.

Lieberman, D. (2009). Rethinking the Taiwanese minor marriage data: evidence the mind uses multiple kinship cues to regulate inbreeding avoidance. *Evolution and Human Behavior, 30,* 153-160.

Liuzza, M. T., Lindholm, T., Hawley, C. B., Gustafsson Sendén, M., Ekström, I., Olsson, M. J., & Olofsson, J. K. (2018). Body odour disgust sensitivity predicts authoritarian attitudes. *Royal Society Open Science, 5*(2), 171091.

Lukaszewski, A. W., & Roney, J. R. (2011). The origins of extraversion: Joint effects of facultative calibration and genetic polymorphism. *Personality and Social Psychology Bulletin, 37,* 409-421.

McIntosh, T. L., Lee, A. J., Sidari, M. J., Stower, R. E., Sherlock, J. M., & Dixson, B. J. (2017). Microbes and masculinity: Does exposure to pathogenic cues alter women's preferences for male facial masculinity and beardedness? *PloS one, 12*(6), e0178206.

Molho, C., Tybur, J. M., Güler, E., Balliet, D., & Hofmann, W. (2017). Disgust and anger relate to different aggressive responses to moral violations. *Psychological Science, 28,* 609-619.

Nettle, D. (2005). An evolutionary approach to the extraversion continuum. *Evolution and Human Behavior, 26,* 363-373.

Oaten, M., Stevenson, R. J., & Case, T. I. (2009). Disgust as a disease-avoidance mechanism. *Psychological Bulletin, 135,* 303–321.

Oaten, M. J., Stevenson, R. J., & Case, T. I. (2017). Compensatory up-regulation of behavioral disease avoidance in immuno-compromised people with rheumatoid arthritis. *Evolution and Human Behavior, 38,* 350–356.

Olatunji, B. O., Williams, N. L., Tolin, D. F., Sawchuk, C. N., Abramowitz, J. S., Lohr, J. M., & Elwood, L. (2007). The Disgust Scale: Item analysis, factor structure, and suggestions for refinement. *Psychological Assessment, 19,* 281-297.

Olatunji, B. O., Haidt, J., McKay, D., & David, B. (2008). Core, animal reminder, and contamination disgust: three kinds of disgust with distinct personality, behavioral, physiological, and clinical correlates. *Journal of Research in Personality, 42,* 1243–1259.

Olatunji, B. O., Armstrong, T., & Elwood, L. (2017). Is disgust proneness associated with anxiety and related disorders? A qualitative review and meta-analysis of group comparison and correlational studies. *Perspectives on Psychological Science, 12,* 613-648.

Olatunji, B. O., Taylor, S., & Zald, D. (2019). Sex differences in the etiology of disgust sensitivity: A preliminary behavioral genetic analysis. *Journal of Anxiety Disorders, 65,* 41-46.

Penke, L., & Jokela, M. (2016). The evolutionary genetics of personality revisited. *Current Opinion in Psychology, 7,* 104-109.

Penke, L., Denissen, J. J., & Miller, G. F. (2007). The evolutionary genetics of personality. *European Journal of Personality*, 21, 549-587.

Perone, P., Becker, D. V., & Tybur, J. M. (2020). Visual disgust-elicitors produce an attentional blink independent of contextual and trait-level pathogen avoidance. *Emotion.* Advance online publication. https://doi.org/10.1037/emo0000751

Pinker, S. (2003). *The blank slate: The modern denial of human nature.* Penguin.

Poirotte, C., & Charpentier, M. J. (2020). Unconditional care from close maternal kin in the face of parasites. *Biology Letters, 16*(2), 20190869.

Reynolds, L. M., McCambridge, S. A., Bissett, I. P., & Consedine, N. S. (2014). Trait and state disgust: An experimental investigation of disgust and avoidance in colorectal cancer decision scenarios. *Health Psychology, 33,* 1495-1506.

Ritchie, S. (2015). *Intelligence: All that matters.* Hodder & Stoughton.

Rottman, J. (2014). Evolution, development, and the emergence of disgust. *Evolutionary Psychology, 12,* 147470491401200209.

Royzman, E., & Kurzban, R. (2011). Minding the metaphor: The elusive character of moral disgust. *Emotion Review, 3,* 269-271.

Royzman, E. B., & Sabini, J. (2001). Something it takes to be an emotion: The interesting case of disgust. *Journal for the Theory of Social Behaviour, 31,* 29-59.

Rozin, P., & Fallon, A. E. (1987). A perspective on disgust. *Psychological Review, 94,* 23-41

Rozin, P., & Millman, L. (1987). Family environment, not heredity, accounts for family resemblances in food preferences and attitudes: A twin study. *Appetite, 8,* 125-134.

Rozin P., Fallon A. E., & Mandell, R. (1984). Family resemblance in attitudes to food. *Developmental Psychology, 20,* 309–314.

Rozin, P., Haidt, J., McCauley, C., Dunlop, L., & Ashmore, M. (1999). Individual differences in disgust sensitivity: Comparisons and evaluations of paper-and-pencil versus behavioral measures. *Journal of Research in Personality, 33,* 330-351.

Rozin, P., Haidt, J., & McCauley, C. R. (2008). Disgust. In M. Lewis, J. M. Haviland-Jones, & L. F. Barrett (Eds.), *Handbook of Emotions* (3rd ed., pp. 757–776). Guilford Press.

Sarabian, C., Ngoubangoye, B., & MacIntosh, A. J. J. (2017). Avoidance of biological contaminants through sight, smell and touch in chimpanzees. *Royal Society Open Science, 4,* 170968

Sauter, D. A., Eisner, F., Ekman, P., & Scott, S. K. (2010). Cross-cultural recognition of basic emotions through nonverbal emotional vocalizations. *Proceedings of the National Academy of Sciences, 107,* 2408-2412.

Scarantino, A. (2012). How to define emotions scientifically. *Emotion Review, 4,* 358-368.

Sherlock, J. M., Zietsch, B.P., Tybur, J. M., & Jern, P. (2016). The quantitative genetics of disgust sensitivity. *Emotion, 16,* 43-51.

Stevenson, R. J., Case, T. I., & Oaten, M. J. (2009). Frequency and recency of infection and their relationship with disgust and contamination sensitivity. *Evolution and Human Behavior, 30,* 363-368.

Stevenson, R. J., Oaten, M. J., Case, T. I., Repacholi, B. M., & Wagland, P. (2010). Children's response to adult disgust elicitors: Development and acquisition. *Developmental Psychology, 46,* 165-177.

Susskind, J. M., Lee, D. H., Cusi, A., Feinman, R., Grabski, W., & Anderson, A. K. (2008). Expressing fear enhances sensory acquisition. *Nature Neuroscience, 11,* 843–850

Tooby, J., & Cosmides, L. (1990). On the universality of human nature and the uniqueness of the individual: The role of genetics and adaptation. *Journal of Personality, 58,* 17–67.

Tooby, J., & Cosmides, L. (1992). The psychological foundations of culture. In J. Barkow, L. Cosmides, & J. Tooby (Eds.), *The adapted mind: Evolutionary psychology and the generation of culture* (pp. 19-136). Oxford University Press.

Tybur, J. M., & Karinen, A. K. (2018). Measurement and theory in disgust sensitivity. In V. Zeigler-Hill & T. K. Shackelford (Eds.), *The SAGE handbook of personality and individual differences: Applications of personality and individual differences* (pp. 159-179). Sage Reference.

Tybur, J. M., & Lieberman, D. (2016). Human pathogen avoidance adaptations. *Current Opinion in Psychology, 7,* 6–11.

Tybur, J. M., Lieberman, D., & Griskevicius, V. (2009). Microbes, mating, and morality: Individual differences in three functional domains of disgust. *Journal of Personality and Social Psychology, 97,* 103-122.

Tybur, J. M., Bryan, A. D., Lieberman, D., Hooper, A. E., & Merriman, L. A. (2011). Sex differences and sex similarities in disgust sensitivity. *Personality and Individual Differences, 51,* 343–348.

Tybur, J. M., Lieberman, D., Kurzban, R., & DeScioli, P. (2013). Disgust: evolved function and structure. *Psychological Review, 120,* 65–84.

Tybur, J. M., Inbar, Y., Güler, E., & Molho, C. (2015). Is the relationship between pathogen avoidance and ideological conservatism explained by sexual strategies? *Evolution and Human Behavior, 36,* 489-497.

Tybur, J. M., Inbar, Y., Aarøe, L., Barclay, P., Barlow, F. K., de Barra, M., Becker, D. V., Borovoi, L., Choi, I., Choi, J. A., Consedine, N. S., Conway, A., Conway, R. J., Conway, P., Cubela Adoric, V., Demirci, D. E., Fernández, A. M., Ferreira, D. C. S., Ishii, K.,. . . Žeželj, I. (2016). Parasite stress and pathogen avoidance relate to distinct dimensions of political ideology across 30 nations. *Proceedings of the National Academy of Sciences, 113,* 12408-12413.

Tybur, J. M., Lieberman, D., Fan, L., Kupfer, T. R., & de Vries, R. E. (2020a). Behavioral immune trade-offs: Interpersonal value relaxes social pathogen avoidance. *Psychological Science, 31,* 1211-1221.

Tybur, J. M., Wesseldijk, L. W., & Jern, P. (2020b). Genetic and environmental influences on disgust proneness, contamination sensitivity, and their covariance. *Clinical Psychological Science, 8,* 1054-1061.

van Overveld, W. J. M., de Jong, P. D., Peters, M. L., Cavanagh, K., & Davey, G. C. L. (2006). Disgust propensity and disgust sensitivity: Separate constructs that are differentially related to specific fears. *Personality and Individual Differences, 41,* 1241–1252.

von Rueden, C. R., Lukaszewski, A. W., & Gurven, M. (2015). Adaptive personality calibration in a human society: effects of embodied capital on prosocial traits. *Behavioral Ecology, 26,* 1071-1082.

Widen, S. C., & Olatunji, B. O. (2016). A developmental perspective on disgust: Implications for obsessive-compulsive disorder. *Current Behavioral Neuroscience Reports, 3,* 204-210.

Widen, S. C., & Russell, J. A. (2013). Children's recognition of disgust in others. *Psychological Bulletin, 139,* 271-299.

Zakrzewska, M., Olofsson, J. K., Lindholm, T., Blomkvist, A., & Liuzza, M. T. (2019). Body odor disgust sensitivity is associated with prejudice towards a fictive group of immigrants. *Physiology & Behavior, 201,* 221-227.

Chapter 7
Olfaction and Disgust: Sensory and Affective Processes to Avoid Disease

Marco Tullio Liuzza

Disgust and the Chemical Senses

Charles Darwin defined disgust as "something revolting, primarily in relation to the *sense of taste* [emphasis added], as actually perceived or vividly imagined; and secondarily to anything which causes a similar feeling, through the *sense of smell* [emphasis added], touch, and even of eyesight" (Darwin 1872, p. 254). Similarly, the facial expression of disgust seems to serve the function of expelling distasteful food (Darwin 1872) and minimise air inflow (Susskind et al. 2008). This observation attests to the primary involvement of the *chemical senses*, taste and olfaction, in an emotion that probably evolved to help people avoid poisonous substances (Rozin and Fallon 1987) and/or contaminants (Oaten et al. 2009).

The view of disgust as a disease-avoidance, survival mechanism predicts that the disgust reaction should emerge early in childhood (Rottman et al. 2019), especially in light of the fact that pathogen-borne diseases are most deadly during the first 5 years of life (Bryce et al. 2005). In fact, using their chemical senses, newborns react with disgust to bitter tastes, to the odour of rotten food (Steiner 1979), and other malodorous or bitter stimuli (Soussignan et al. 1997). Stevenson et al. (2010) showed that children as young as 2.5 years of age started to exhibit avoidance behaviours, such as unwillingness to sniff body products (e.g., dirty socks) and unwillingness to eat a candy that was placed in a potty. Toddlers ranging from 16 to 29 months of age, seem to be willing to put disgusting items in their mouths, such as faeces-shaped edible items (Rozin et al. 1986). A more marked preference for uncontaminated food seems to emerge after 5 years of age (Blacker and LoBue 2016; DeJesus et al. 2015). However, it should be observed that, at a younger age,

M. T. Liuzza (✉)
Department of Medical and Surgical Sciences, "Magna Graecia" University of Catanzaro, Catanzaro, Italy
e-mail: liuzza@unicz.it

P. A. Powell, N. S. Consedine (eds.), *The Handbook of Disgust Research*,
https://doi.org/10.1007/978-3-030-84486-8_7

children are fed by parents, or their eating behaviour is heavily parentally modelled (Stevenson et al. 2010). Moreover, it should be noted that when items produce chemosensory cues of poisoning or contamination (i.e., malodour and/or bitter taste), then the avoidance behaviour is automatically activated.

This brief review on disgust through the lenses of its ontogeny suggests a prominent role played by chemical senses in avoiding disease and poison. In particular, we will now focus on olfaction, which might be considered as a guardian of the mouth able to trigger the appropriate emotion (disgust) and behaviour (avoidance) before the contaminant makes contact with and enters our body.

Olfaction

Olfaction is intimately related to motivations. In bacteria, the ability to redirect movements as a function of chemical gradients (chemotaxis) serves the purpose of maximising contact with behaviourally beneficial (i.e., rewarding) chemical stimuli (Gottfried and Wilson 2011). Hence, odour processing is crucial to evaluate the edibility of food (Walliczek-Dworschak and Hummel 2017). Our sense of smell helps us to prevent many kinds of potential harms (Santos et al. 2004), mainly microbial threats (Stevenson 2010), as testified by the observation that patients with olfactory disorders are very likely to eat spoiled food, at least once in their life (Temmel et al. 2002). Moreover, these patients often show personal hygiene issues, due to their inability to perceive their own smell (Temmel et al. 2002). In fact, it has been shown that disgust is the primary emotional response to unpleasant odours (Alaoui-Ismaïli et al. 1997a, b), and unpleasant odours are the easiest to memorise. This negative bias is often encountered, and it is probably due to the evolutionary advantage provided by avoiding potentially life-threatening stimuli (Nesse 2005), thus abiding to a "better-safe-than-sorry" strategy. Furthermore, odour-evoked disgust is less permeable to top-down influences (Ferdenzi et al. 2013), and harder to suppress than disgust evoked by visual stimuli (Adolph and Pause 2012), testifying to the power of olfaction.

Besides its obvious connection with the search and selection of food, olfaction is linked to mating behaviour in many species (Doty 1976; Pfeiffer and Johnston 1994) and, to some extent, in humans (e.g., Havlicek et al. 2005; Thornhill and Gangestad 1999). In particular, olfactory attractiveness (odours that provide cues to the physical quality of a potential sexual partner concerning health and reproductive status, see Ferdenzi et al. 2020) in humans seems to promote mating with people having a dissimilar major histocompatibility complex (MHC). Such an olfactory-guided disassortative mating behaviour, in order to maximise MHC-heterozygosity, is associated with an advantage under immune challenge and therefore leads to an increase in pathogen resistance in the progeny (Havlicek and Roberts 2009, but see Lobmaier et al. 2018).

Disgust, Olfaction, and Disease Avoidance

Research has shown that unpleasant odours are perceived faster (Bensafi et al. 2002), and are processed earlier in the brain (Croy et al. 2013b) than pleasant odours. As compared to other sensory modalities, olfaction seems to be more closely connected to areas of the brain that are traditionally involved in emotional reactivity, as it has direct connections with the limbic system (Soudry et al. 2011; and see Walliczek-Dworschak and Hummel 2017, Figure 2). The olfactory system, differently from other senses, projects ipsilaterally, and most fibres bypass the thalamus and project directly into the amygdala, piriform cortex, and entorhinal cortex (Gottfried 2006). Odours reliably induce disgust (Alaoui-Ismaïli et al. 1997a; Bensafi et al. 2002; Croy et al. 2011), and, as compared with disgusting visual stimuli, disgusting odours provoke a stronger decrease of systolic blood pressure, which has been interpreted as a sign of preparation to vomiting (Croy et al. 2013b).

The Direct Relationship Between Olfaction and Disgust

A few studies have directly investigated the relationship between disgust and olfaction; some of them contributed to a 2017 special issue of the journal *Chemical Senses* (Stafford 2017). Individuals differ in their olfactory threshold, which is the lowest concentration of an odour command that can be reliably detected by their sense of smell. In other words, individuals with a lower olfactory threshold are more sensitive to odours. Ilona Croy et al. (2017) investigated whether the olfactory threshold (or sensitivity) directly related to individuals' disgust sensitivity (i.e., how easily one is disgusted). They found a specific relation between the level of disgust and olfactory sensitivity in men, but not in women, an interaction that might be due to ceiling effects (i.e., women were exhibiting already high levels of disgust even when low in olfactory sensitivity). In a study from Rotraut Ille and colleagues, published in the same issue, they tested a similar hypothesis on a clinical population of patients with an olfactory impairment (Ille et al. 2017). The authors found that hyposmic patients were higher on self-disgust than healthy controls; reported elevated sensitivity to poor hygiene; and expressed greater concerns with regards to personal cleanliness, a result that was more marked among men (thus replicating a previous finding from Ille et al. 2016). Even though it could have been hypothesised that those with less smell may find themselves less (physically) disgusting, this finding is consistent with the idea that disgust sensitivity may compensate for a deficit in a sensory modality that plays such an important role in detecting pathogen cues.

Focusing on state disgust, Kai Chan and colleagues found that induction of disgust lowers the sensory threshold (increases sensitivity) to both mildly negative (i.e., *n-butanol*; Chan et al. 2016), and mildly positive (*phenylethanol*; Chan et al. 2019) odours. In an elegant study, Lorenzo Stafford et al. (2018) explored the relationship between trait disgust and sense of smell: they induced either moral or pathogen disgust (vs. a control condition) experimentally, and tested whether this manipulation affected the rating of three affectively distinct odours (disgust-evoking, pleasant, and neutral). However, their study led to mixed results, with a paradoxical decrease in reported disgust in the pathogen disgust condition (regardless of the odour rated) in experiment one and an increase to the disgusting odour in experiment two. Although the authors attribute the differences between to possible habituation or to time-dependent effects that could be at play in study one, future better-powered replication studies are warranted. Taken together, these results stack up in favour of a functional connection between olfaction and disgust (Soudry et al. 2011; Stevenson 2010).

One of the domains where olfactorily mediated disease avoidance mechanisms play a delicate role is sexual behaviour. Although the study on the relationship between disgust and sexual behaviour is deepened elsewhere in this book (see Borg and De Jong, Chap. 9, this volume), here, I am focusing on its under-investigated relationship with the sense of olfaction, and, in particular, on two domains where these differences can be crucial: gender and aging.

Differences by Gender

Gender Differences in Disgust Sensitivity

Gender differences in dispositions towards experiencing disgust are observed consistently across many studies (Al-Shawaf and Lewis 2013; Curtis et al. 2004; Haidt et al. 1994; Liuzza et al. 2016; Liuzza et al. 2019; Mancini et al. 2001; Oaten et al. 2009; Olatunji et al. 2007b; see also Tybur, Chap. 6, this volume). However, it should be noted that gender differences in self-reported disgust sensitivity are much more marked in the domain of sexual disgust (Al-Shawaf et al. 2014; Tybur et al. 2009) than in the domain of pathogen disgust (Liuzza et al. 2016; Tybur et al. 2009). In a recent review, Laith Al-Shawaf et al. (2018) provide a list of theoretical accounts for gender difference in disgust sensitivity, some of which call directly into question some form of disease avoidance mechanism. In fact, women are more likely to transmit infections to offspring, as well as having a greater role in keeping children away from external pathogens. However, it is not always clear whether this difference is grounded in a genuine physiological or neural difference (see Schienle et al. 2005 and Smith et al. 2011, for a dissociation between self-report and physiological measures), which begs the question of whether gender differences reflect response biases—possibly driven by cultural norms—rather than real gut reactions.

Laith Al-Shawaf et al. (2018) have proposed that these gender-based response biases may have, at their turn, an evolutionary grounded account. In fact, men may down-regulate their expression of disgust to signal their immune strength (Al-Shawaf et al. 2015; Fessler et al. 2004). Such an explanation relies heavily on the so-called "compensatory prophylaxis hypothesis" (Fessler et al. 2005;

Fleischman and Fessler 2011; see also Bradshaw and Gassen, Chap. 3, this volume) that contends that the behavioural immune system should compensate for higher vulnerability to disease due to suppression of the physiological immune system (e.g., due to high progesterone luteal phase of the menstrual cycle). However, a well-powered longitudinal study failed to replicate one of the most critical findings originating by this hypothesis (Jones et al. 2018).

Although there is overwhelming evidence that women tend to report higher levels of disgust, as outlined above it is not clear whether this difference is rooted in biology or is a measurement artefact. Accordingly, more research is warranted in the domain of disgust measurement in order to better test if the items used to measure individual differences in disgust exhibit satisfying psychometric properties and a similar functioning across genders and cultures. A similar remark can be raised in the context of measuring olfactory function, as we will see in the next paragraph. Another issue that arises from the literature reviewed so far is whether the compensatory prophylaxis is non-tenable, or whether it should be tested in a more rigorous way, for instance by focusing on vulnerable populations (e.g., the elderly, or patients with conditions that suppress the immune system). Alternatively, as suggested in the review by Tybur et al. (2018), experimental paradigms could be leveraged to induce inflammatory processes, such as the ones used by Mats Olsson and colleagues, who induced inflammatory processes via lipopolysaccharide administration (Axelsson et al. 2018; Olsson et al. 2014; Regenbogen et al. 2017; Sarolidou et al. 2019).

Gender Differences in Olfaction

It is common sense that women typically outperform men in olfactory function, namely it is typically assumed that women are more sensitive to odours. Many findings do suggest that women perform better in tasks that require olfaction, as confirmed by recent meta-analytical evidence (Sorokowski et al. 2019). However, as the same authors emphasise, olfactory functionality can be assessed by different tests and subtests that may tap into different sensory and cognitive processes, and, when investigating gender differences, it is important to look at more elementary components. Odour identification, namely the ability to recognise and name a smell, is an ability that heavily depends on semantic memory (Economou 2003; Larsson et al. 2004) and speed of processing (Larsson et al. 2004, 2005). On the other hand, odour threshold, or the lowest concentration at which the presence of an odorant is reliably detected, taps more into sensory processes, and seems to rely less on cognitive abilities (Hedner et al. 2010) or the influence of cultural factors (Hoshika et al. 1994). Olfactory discrimination ability is typically based on a task where subjects are confronted with three smells and have to decide which of the three odours is different from the other two (Hummel et al. 1997), and does require cognitive abilities such as semantic memory (Sabiniewicz 2021).

Sorokowski et al. (2019) showed reliable, but not large, gender differences in the three olfactory abilities but, puzzlingly, a considerable variation is observed in odour identification depending on the test used: the Sniffin' Sticks Test (SST, Hummel et al. 1997; Hummel et al. 2007, Hedge's $g = 0.08$) or the University of Pennsylvania Smell Identification Test (UPSIT, Doty et al. 1984a; Doty et al. 1984b, Hedge's $g = 0.3$). As discussed by Piotr Sorokowski et al. (2019), such a large difference in the effect size for gender in the two tests may be due to the differences in the items (odours) used in the two tests, a concern that is related to the concept of differential item functioning (DIF), as I will further discuss below. In other words, before drawing inferences on group differences in raw scores, one should make sure that the items work in the same way in both of the groups. In this context, it means that men and women with the same level of latent ability (odour identification) should have the same probability to correctly respond to that item, otherwise gender differences (or lack of a difference) on this item are due to measurement artefacts rather than to real differences in odour identification.

Sexual Behaviour

Disgust and Sex

Disgust, and down-regulation of disgust, plays a crucial role in sexual behaviour. Sexual behaviour is fundamental for the survival of the species but, at the same time, is a behaviour that comes with high risks in terms of pathogen threats (see Borg and de Jong, Chap. 9, this volume, for a more detailed account on the relationship between sex and disgust). Mark Schaller and Damian Murray, from the University of British Columbia, compared the prevalence of disease from 71 countries from which personality data and sociosexual orientation inventory (SOI, Simpson and Gangestad 1991) data were available. They found a significant negative correlation between their index of disease prevalence and preference for sexual promiscuity (e.g., having or desiring more sexual encounters outside from a long-term relationship). However, this correlation was much more marked among women ($r = -.62$) than men ($r = -.27$; Schaller and Murray 2008). In terms of individual differences, it has been found that people with higher levels of germ aversion (Murray et al. 2013; Tybur et al. 2015) and disgust (Tybur et al. 2015) sensitivity are less inclined towards sexual promiscuity to minimise contacts with pathogens. However, some authors found that this association was confined to the sexual domain of disgust (Tybur et al. 2009), and failed to find a statistically significant association with pathogen disgust (Al-Shawaf et al. 2014), a finding recently replicated on a larger sample (O'Shea et al. 2019).

Evidence for an effect of pathogen threat on sexual behaviour also comes from experimental paradigms, which found that the induction of a disease threat interacts with individual differences in germ aversion in decreasing the willingness to engage in promiscuous sexual behaviours (Murray et al. 2013). Overall, these studies show that people higher in disgust sensitivity or, more generally, in disease avoidance, tend to avoid sexual promiscuity. In a recent study, Laith Al-Shawaf et al. (2019)

found that an experimental induction of disgust reduces the willingness to engage in short-term strategies. Moreover, Diana Fleischman et al. (2015) found that women exposed to disgusting stimuli displayed decreased physiological sexual arousal. Interestingly, the association between sexuality and disgust has also been shown to go in the opposite direction. Induction of sexual arousal decreases disgust reactivity in women (Borg and De Jong 2012; although Fleischman et al. 2015 did not replicate this result), and in men (Oaten et al. 2018; Stevenson et al. 2011).

To conclude, we have seen that disgust, being an emotion that primarily serves the function of avoiding diseases, naturally plays a crucial role in sexual behaviour, where the risks of contracting diseases can outnumber the reproduction benefits. Olfaction also plays a critical role in this relationship, as a sensory modality involved in detecting pathogens and, at the same time, appearing to play a role in mating as well.

Olfaction and Sexual Behaviour

People involved in romantic relationships often report smelling their partners' clothing during physical separations to facilitate feelings of closeness (McBurney et al. 2006). Indeed, sensing a partner's body odour has a soothing effect (Hofer et al. 2018), at least in individuals with a secure attachment (Granqvist et al. 2019). In fact, olfactory loss, even when undetected, undermines human wellbeing (Oleszkiewicz et al. 2020), including sexuality in men (Croy et al. 2013a; Ottaviano et al. 2013, 2015) and romantic attachment in women (Croy et al. 2013a). Interestingly, body odour is rated as more important for attraction by women (Herz and Inzlicht 2002).

I contend that smell is the most relevant sense in the experience of disgust, because of its usefulness to detecting invisible pathogen cues before they come into contact with our bodies. However, very few studies have used olfactory-induced disgust to study how sexual behaviour may be affected (see also Consedine, Chap. 2, this volume). In fact, among the studies mentioned above, only Laith Al-Shawaf et al. (2019) used olfactory induced pathogen disgust in their paradigm. Charmaine Borg et al. (2019) explicitly hypothesised that olfaction-induced disgust could have an impact on sexual arousal. To test their hypothesis, they recruited 78 male participants who were exposed to a pornographic video while being exposed to either a disgusting odour or a control odour. Measures of subjective and physiological disgust and arousal were collected before and after watching the video. The researchers found that exposure to a disgusting odour led to lower subjective and physiological arousal, as compared to the control condition. Building upon this relationship between disgust (mainly when olfactory induced) and sexual behaviour, Joshua Tybur et al. (2011) devised an ingenious experiment in which 99 participants were either exposed to a faeces-like smell or a control condition and then had to rate their intention to buy and use condoms in the next 6 months. The authors found that exposure to a pathogen cue (i.e., the faeces-like smell) significantly increased the likelihood of purchasing and using condoms (for more on disgust and purchase intentions see Powell, Chap. 15, this volume). Overall, from the few studies reviewed, it seems that inducing disgust through olfaction may have a meaningful impact on sexual behaviour.

So far, I have reviewed the role of olfaction as a sensory channel through which disgust can be induced. However, little is known about the role that olfactory function may have on sexual behaviour. On the one hand, we might hypothesise that an impairment in olfaction could reduce disgust, thus promoting sexual promiscuity. On the other hand, it might be that sensory impairment may promote a greater vigilance towards pathogen cues using other senses and/or greater motivation to avoid situations of pathogen threat in the absence of important sources of information about the risks. Additionally, some studies suggest that olfaction can also be used to promote negative assortative (or disassortative) mating (Roberts and Little 2008), namely a mating strategy that promotes greater MHC diversity in order to increase the fitness of the species (but see Havlicek and Roberts 2009, for a more nuanced account on how assortative and disassortative mating may interact). Besides, body odours can be used to evaluate attractiveness (Roberts et al. 2011; Groyecka et al. 2017) and personality (Sorokowska 2013; Sorokowska et al. 2011).

Olfaction, Disgust, Sexual Behaviour, and Aging

Seminal research has shown that olfactory function declines with age (Doty et al. 1984a; Larsson et al. 2000), and sexual attitudes and behaviours face a similar decline as a function of aging (Chao et al. 2011; Laumann et al. 2005). Data from adult males (Ottaviano et al. 2013) showed an association between olfactory sensitivity to *butanol* and sexual desire among young men, whereas the association was absent in older adults, perhaps due to floor effects in both variables. This result was replicated on young men by the same lab when measuring sensitivity to *burgeonal* (Ottaviano et al. 2015). More recently, a study found that olfactory impairment caused by chronic rhinosinusitis correlates with erectile dysfunction (Tai et al. 2016). However, the authors acknowledge that several pathophysiological factors may cause erectile dysfunction in rhinosinusitis patients. For instance, hypoxia can lead to a physiological alteration that hampers erection.

In a recent study on a sample of patients with olfactory disorders, Laura Schäfer et al. (2019) assessed qualitatively and quantitatively the effect of olfactory loss on sexual desire. They found that more than one-fourth of their participants reported changes in sexual desire, and these changes were predicted, among other factors, by olfactory impairment, as assessed by the SST total score. Taken together, these results point towards a negative effect of olfactory deficits on sexuality. However, it is not clear whether disease avoidance mechanisms play a role in the association between olfactory impairment and a decrease in sexual desire. In fact, the negative association between olfactory function and sexuality might be due to biological changes related to aging that affect both phenomena, along with other sensory processes. Another possible explanation is that olfaction plays an important role in sexuality (Bendas et al. 2018) and its reduction would negatively affect sexual desire.

Aging processes cause a decline in sensory and cognitive functions, but it is not clear to what extent this process impacts on affective processes. In particular, from our perspective, it is relevant to assess whether aging affects disgust, both in terms of recognition and experience. Indeed, one may argue that sensory and cognitive impairment should dampen the experience of sensory-induced disgust. But, from a compensatory prophylaxis standpoint, the elderly should compensate for their lessening reliance on olfactory function by being more vigilant (e.g., actively monitoring the environment using the other senses) and reactive to pathogen cues, given that the elderly are more vulnerable to diseases and/or diseases are more life-threatening to them. Results from self-report measures suggest that disgust-sensitivity decreases with age (Curtis et al. 2004; Fessler and Navarrete 2005; Quigley et al. 1997; Rozin et al. 2000), although in a scale more attuned to measure disgust propensity in clinical populations the opposite effect was found (Olatunji et al. 2007a). But, in terms of behavioural measures of emotion recognition, meta-analytic evidence (Gonçalves et al. 2018; Ruffman et al. 2008) suggests that elders do not decline in their ability of recognising disgust, whereas a decline is observed for all other discrete emotions, including negative emotions such as anger, sadness and, to a lesser extent, fear.

Furthermore, evidence from olfactory research shows that older participants—but not younger ones—display a more appropriate response to disgusting stimuli (a higher electromyographic activity of the *corrugator* muscle) as a function of their nutritional status, namely whether they were in a state of malnutrition (Joussain et al. 2017). Even more compelling evidence in favour of the compensatory prophylaxis hypothesis comes from a study that showed anosmic patients perform better in recognition of facial expressions of disgust (Lemogne et al. 2015). Furthermore, among patients with acquired anosmia, there was a positive correlation between the duration of anosmia and recognition performance (although it must be noted that these results lack specificity, as they extend to fear recognition as well; Lemogne et al. 2015). Overall, these results can be interpreted as potential compensation in the emotional and socio-cognitive component of our disease avoidance mechanism for the decline of the sensory part.

Future studies should aim to build a bridge between the investigation on changes in olfaction and changes in disgust sensitivity in aging populations. Of particular interest would be bridging these two fields of investigation in the domain of sexual behaviour, because aging comes with a greater vulnerability to disease, thus changing the balance between reproductive benefits and pathogen risks that come with sexual intercourse.

Disgust and Olfaction: Resolving Measurement Issues

One of the outstanding issues that remains to be explored more thoroughly is the evaluation of both olfactory function and disgust sensitivity. Self-reported disgust sensitivity seems to be prone to gender differences, as compared to behavioural measures. It is therefore unclear whether these differences reflect real differences in the underlying construct or rather different response biases that might be a function of cultural differences. In fact, when it comes to pathogen disgust, gender differences tend to be tiny (Liuzza et al. 2016; Tybur et al. 2009), or not statistically significant, when psychophysiological (Smith et al. 2011) and neural (Schienle et al. 2005) responses to disgust elicitors are analysed. This highlights a potential measurement issue that is worthy of further investigation for both the constructs under scrutiny here. Some work has been done in this regard (Liuzza et al. 2016; Olatunji et al. 2007b; Tybur et al. 2009), and Tybur and Karinen (2018) recently provided an insightful review on measurement issues in disgust sensitivity (see also Tybur, Chap. 6, this volume). In their review, Tybur and Karinen provided evidence to suggest that disgust sensitivity is not just an epiphenomenon of other higher-order traits. However, there are other remaining measurement issues that I contend could be better addressed using an Item Response Theory (IRT) approach to measurement along with a more often-used Classical Test Theory (CTT) approach.

IRT is "a model-based measurement in which trait level estimates depend on both persons' responses and on the properties of the items that were administered (Embretson and Reise 2013, p. 13). One of the core features that distinguish this approach to measurement from CTT (Gulliksen 1950; Spearman 1907, 1913), is that the reliability (or information in IRT *parlance*) of a test changes for different levels of the latent trait. This is a feature that may allow testing, for instance, the reliability of the different disgust scales for different levels of respondents' latent ability, and choosing the most appropriate test, or the most appropriate items from an item bank as a function of the trait(s) that are going to be assessed (e.g., in a clinical vs. non-clinical population). Another feature of the IRT approach is to test whether each single item functions in a similar way (DIF) in different populations where it is a priori expected to function homogenously, namely whether the relationship between the latent trait and the probability to respond to an item holds the same in different populations (e.g., women *vs.* men or elders *vs.* youngsters). This analysis may help to better understand to what extent current gender or age differences found in the disgust sensitivity literature are due to genuine differences or, perhaps, to different response biases.

A few studies—if any—have validated measures of disgust sensitivity within an IRT framework, a more extensive cross-cultural validation study using IRT analyses would provide better insights than standard psychometrics. Also, a study that investigates gender differences in disgust sensitivity and olfaction functions, and the differential moderating role of gender in different countries as a function of existing gender gap(s), would help elucidate whether part of the lack of validity could be attributed to different gender norms, in a similar way to other personality traits (Mac Giolla and Kajonius 2018). Although such an endeavour seems daunting at first glance, it is nonetheless an inevitable step that has to be taken in response to the reproducibility credibility crisis (Open Science Collaboration 2015), and to a related, but often overlooked, measurement crisis (Flake et al. 2017) in Psychology. For instance, the Psychological Science Accelerator (PSA, Moshontz et al. 2018) is an initiative that allows cross-cultural replication and novel studies in a distributed lab network that encompasses five continents. Such a call for a change in research practices in the domain of chemical senses is echoed by a similar stance by Tristram Wyatt (Wyatt 2019), who recently warned against the high risk of false positive findings in chemo-sensory research, for instance in the case of the investigation on the existence of putative "human pheromones." This call for change extends to research in disgust and, in particular, its measurement.

Self-ratings of olfactory functions are hardly accurate (Landis et al. 2003; Philpott et al. 2006; Shu et al. 2009; Soter et al. 2008). Behavioural-based assessments (e.g., SST and UPSIT) seem to be more valid, although I contend that more validation studies should be conducted if we want to reliably estimate the relationship that olfactory function holds with disgust sensitivity.

The full examination of the SST encompasses three different abilities: detection, discrimination, and identification. The scores on each of these tests are then summed to provide a unique measure of olfactory function. In order to meaningfully interpret a sum-score under the assumption of a reflective model of measurement, it should be shown that: (1) the data are compatible with a unidimensional model, or at least with a hierarchical model where olfactory function acts as a higher order latent variable; and (2) the factor loadings and the variances are equal for each of the subtests (i.e., threshold, discrimination, identification; McNeish and Wolf 2020). In other words, the data should be compatible with what is known using a parallel form of measurement (Graham 2006). However, there are good reasons for suspecting that the assumption that qualifies a parallel form of measurement is not met, starting from the observation that the subtests tap into different processes, since the threshold assesses a sensory process. At the same time, identification and discrimination are more intertwined with other cognitive processes (Hedner et al. 2010). Therefore, more research has to be conducted in order to provide compelling evidence that a sum score could be used. Such evidence is particularly warranted, given that the sum-score of the SST is currently used to establish cut-off values for the diagnosis of hyposmia and anosmia (see also Fried and Nesse 2015, for a related concern on the use of sum scores for the diagnosis of Depression).

A second measurement issue that was highlighted when reviewing the topic of gender differences in olfactory function is related to the possibility that single items could work differently in different populations, for instance, in women vs. men, in elders vs. youngsters, or western vs. eastern cultures. Again, assessing DIF within either within an IRT framework, or within a CTT by conducting a multi group confirmatory factor analysis, could provide important insights that may guide a revision of the tests to lead to scores that reflect latent traits in similar ways across genders, ages, and cultures. Otherwise, any differences observed could be due to just a measurement artefact.

In one of the few studies using an IRT approach, Kelly Minor et al. (2004) tested whether schizophrenic and control participants could be compared in terms of latent traits. To this purpose, they used the one-parameter IRT model, also known as the Rasch model, named after its founder, the Danish Mathematician Georg Rasch (1980). Under the Rasch model, all the items have the same discrimination parameter, but different difficulty parameters. Minor and colleagues estimated the Rasch model for schizophrenic patients—who typically have poorer performance in olfactory identification tests—and controls separately. Although findings confirm that the UPSIT adequately defines a single construct of olfactory ability for healthy and schizophrenic respondents, when looking at the test information (reliability) for different levels, the authors noticed that the UPSIT falls short for higher levels of the latent variable. In other words, it reliably measures people with reduced olfaction, but not people with high levels of olfactory function. Furthermore, Minor et al. (2004) found that some items seem to work differently in the two groups, which raises the question of whether schizophrenia patients' UPSIT scores represent a pure odour identification measure.

Similar results to Minor et al. (2004) were achieved in another study where the authors utilised a Rasch measurement model for the UPSIT (Lange et al. 2002). However, in this study, the authors found a problematic amount of items that did not seem to fit with the Rasch model, thus questioning the sum-score as a statistic that sufficiently represents the latent ability. To better understand the source of the misfit, they conducted DIF analyses. Even though they did not find evidence for DIF as a function of gender and age, they found that items worked differently for controls and patient groups (Alzheimer and Parkinson patients in this case). Although the authors conclude that, overall, the UPSIT provides a reliable measure, nonetheless, the richness of the IRT analyses provided many more insights on olfactory functioning measures than CTT-based approaches typically provide. I, therefore, conclude that more IRT-based analyses, and in general, more rigorous and better-powered validation studies on diverse samples will be of use to better ascertain the intimate relationship that occurs between disgust and olfaction. As it has been shown that disgust assessment can be undermined by measurement issues (Tybur and Karinen 2018), addressing these issues is of paramount importance in order to draw reliable conclusions on the relationship between emotional and sensory processes.

Conclusion

To summarise, this chapter has explored how disgust and olfaction, an affective and a sensory process respectively, can work hand in hand in promoting disease avoidance in humans. In particular, I focused on two domains where evidence on disgust reactivity and olfactory function can shed light on the functioning of our disease avoidance mechanism: gender differences and sexual behaviour, with consideration of differences related to aging. Although there is some indirect evidence on the synergy of affective and sensory processes in promoting disease avoidance, a surprisingly small amount of studies, if any, have directly addressed this research question and this represents a key future avenue for sensory disgust research. Finally, I reviewed how additional psychometric considerations should be taken into account in order to build more reliable evidence in the domain of olfaction and disgust via improved self-report measurement, and linked my considerations with similar concerns that arise in other fields of psychology.

Acknowledgment I wish to thank Mr. Roberto Giorgini for his precious comments.

References

Adolph, D., & Pause, B. M. (2012). Different time course of emotion regulation towards odors and pictures: are odors more potent than pictures? *Biol Psychol, 91*(1), 65-73.

Alaoui-Ismaïli, O., Robin, O., Rada, H., Dittmar, A., & Vernet-Maury, E. (1997a). Basic emotions evoked by odorants: comparison between autonomic responses and self-evaluation. *Physiology & Behavior, 62*(4), 713-720.

Alaoui-Ismaïli, O., Vernet-Maury, E., Dittmar, A., Delhomme, G., & Chanel, J. (1997b). Odor hedonics: connection with emotional response estimated by autonomic parameters. *Chemical senses, 22*(3), 237-248.

Al-Shawaf, L., & Lewis, D. M. (2013). Exposed intestines and contaminated cooks: Sex, stress, & satiation predict disgust sensitivity. *Personality and Individual Differences, 54*(6), 698–702.

Al-Shawaf, L., Lewis, D. M., & Buss, D. M. (2014). Disgust and mating strategy. *Evolution and Human Behavior, 36*(3), 199–205.

Al-Shawaf, L., Lewis, D. M., Alley, T. R., & Buss, D. M. (2015). Mating strategy, disgust, and food neophobia. *Appetite, 85*, 30-35.

Al-Shawaf, L., Lewis, D. M., & Buss, D. M. (2018). Sex differences in disgust: Why are women more easily disgusted than men?. *Emotion review, 10*(2), 149-160.

Al-Shawaf, L., Lewis, D. M., Ghossainy, M. E., & Buss, D. M. (2019). Experimentally inducing disgust reduces desire for short-term mating. *Evolutionary Psychological Science, 5*(3), 267-275.

Axelsson, J., Sundelin, T., Olsson, M. J., Sorjonen, K., Axelsson, C., Lasselin, J., & Lekander, M. (2018). Identification of acutely sick people and facial cues of sickness. *Proceedings of the Royal Society B: Biological Sciences, 285*(1870), 3–9.

Bendas, J., Hummel, T., & Croy, I. (2018). Olfactory Function Relates to Sexual Experience in Adults. *Archives of Sexual Behavior, 47*(5), 1333–1339.

Bensafi, M., Rouby, C., Farget, V., Bertrand, B., Vigouroux, M., & Holley, A. (2002). Psychophysiological correlates of affects in human olfaction. *Neurophysiologie Clinique/Clinical Neurophysiology, 32*(5), 326–332.

Blacker, K.-A., & LoBue, V. (2016). Behavioral avoidance of contagion in childhood. *Journal of Experimental Child Psychology, 143*, 162–170.

Borg, C., & De Jong, P. J. (2012). Feelings of disgust and disgust-induced avoidance weaken following induced sexual arousal in women. *PloS one, 7*(9), e44111.

Borg, C., Oosterwijk, T. A., Lisy, D., Boesveldt, S., & De Jong, P. J. (2019). The influence of olfactory disgust on (Genital) sexual arousal in men. *PloS one, 14*(2), e0213059.

Bryce, J., Boschi-Pinto, C., Shibuya, K., Black, R. E., & the WHO Child Health Epidemiology Reference Group. (2005). WHO estimates of the causes of death in children. *Lancet, 365*, 1147–1152.

Chan, K. Q., Holland, R. W., van Loon, R., Arts, R., & van Knippenberg, A. (2016). Disgust and fear lower olfactory threshold. *Emotion, 16*(5), 740.

Chan, K. Q., van Dooren, R., Holland, R. W., & van Knippenberg, A. (2019). Disgust lowers olfactory threshold: a test of the underlying mechanism. *Cognition and Emotion*, 1-7.

Chao, J. K., Lin, Y. C., Ma, M. C., Lai, C. J., Ku, Y. C., Kuo, W. H., & Chao, I. C. (2011). Relationship among sexual desire, sexual satisfaction, and quality of life in middle-aged and older adults. *Journal of Sex & Marital Therapy, 37*(5), 386-403.

Croy, I., Olgun, S., & Joraschky, P. (2011). Basic emotions elicited by odors and pictures. *Emotion, 11*(6), 1331.

Croy I., Bojanowski V., & Hummel T. (2013a). Men without a sense of smell exhibit a strongly reduced number of sexual relationships, women exhibit reduced partnership security—a reanalysis of previously published data. *Biol. Psychol. 92*, 292–294.

Croy, I., Maboshe, W., & Hummel, T. (2013b). Habituation effects of pleasant and unpleasant odors. *International Journal of Psychophysiology, 88*(1), 104-108.

Croy, I., Bendas, J., Wittrodt, N., Lenk, M., Joraschky, P., & Weidner, K. (2017). Gender-specific relation between olfactory sensitivity and disgust perception. *Chemical senses*, *42*(6), 487-492.

Curtis, V., Aunger, R., & Rabie, T. (2004). Evidence that disgust evolved to protect from risk of disease. *Proceedings of the Royal Society B: Biological Sciences, 271*, S131–S133. https://doi.org/10.1098/rsbl.2003.0144

Darwin, C. (1872). *The expression of emotions in man and animals*. London: Murray.

DeJesus, J. M., Shutts, K., & Kinzler, K. D. (2015). Eww she sneezed! Contamination context affects children's food preferences and consumption. *Appetite*, *87*, 303–309.

Doty, R. L. (Ed.). (1976). *Mammalian olfaction, reproductive processes and behavior*. Academic.

Doty, R. L., Shaman, P., Applebaum, S. L., Giberson, R., Siksorski, L., & Rosenberg, L. (1984a). Smell identification ability: changes with age. *Science, 226*(4681), 1441-1443.

Doty, R. L., Shaman, P., & Dann, M. (1984b). Development of the University of Pennsylvania Smell Identification Test: a standardized microencapsulated test of olfactory function. *Physiology & behavior, 32*(3), 489-502.

Economou, A. (2003). Olfactory identification in elderly Greek people in relation to memory and attention measures. *Arch. Gerontol. Geriatrics, 37*, 119–130.

Embretson, S. E., & Reise, S. P. (2013). *Item response theory*. Psychology Press.

Ferdenzi, C., Roberts, S. C., Schirmer, A., Delplanque, S., Cekic, S., Porcherot, Cayeux, I., Sander, D., & Grandjean, D. (2013). Variability of affective responses to odors: culture, gender, and olfactory knowledge. *Chem Senses*, *38*(2), 175-186.

Ferdenzi, C., Richard Ortegón, S., Delplanque, S., Baldovini, N., & Bensafi, M. (2020). Interdisciplinary challenges for elucidating human olfactory attractiveness. *Philosophical Transactions of the Royal Society B, 375*(1800), 20190268.

Fessler, D. M., & Navarrete, C. D. (2005). The effect of age on death disgust: Challenges to terror management perspectives. *Evolutionary Psychology*, *3*(1), 147470490500300120.

Fessler, D. M., Pillsworth, E. G., & Flamson, T. J. (2004). Angry men and disgusted women: An evolutionary approach to the influence of emotions on risk taking. *Organizational behavior and human decision processes, 95*(1), 107-123.

Fessler, D. M., Eng, S. J., & Navarrete, C. D. (2005). Elevated disgust sensitivity in the first trimester of pregnancy: Evidence supporting the compensatory prophylaxis hypothesis. *Evolution and Human Behavior, 26*(4), 344-351.

Flake, J. K., Pek, J., & Hehman, E. (2017). Construct validation in social and personality research: Current practice and recommendations. *Social Psychological and Personality Science*, 8(4), 370–378.

Fleischman, D. S., & Fessler, D. M. (2011). Progesterone's effects on the psychology of disease avoidance: Support for the compensatory behavioral prophylaxis hypothesis. *Hormones and Behavior, 59*(2), 271-275.

Fleischman, D. S., Hamilton, L. D., Fessler, D. M., & Meston, C. M. (2015). Disgust versus lust: Exploring the interactions of disgust and fear with sexual arousal in women. *PloS one, 10*(6), e0118151.

Fried, E. I., & Nesse, R. M. (2015). Depression sum-scores don't add up: why analyzing specific depression symptoms is essential. *BMC medicine, 13*(1), 72.

Gonçalves, A. R., Fernandes, C., Pasion, R., Ferreira-Santos, F., Barbosa, F., & Marques-Teixeira, J. (2018). Effects of age on the identification of emotions in facial expressions: a meta-analysis. *PeerJ, 6*, e5278.

Gottfried, J. A. (2006). Smell: central nervous processing. In T. Hummel & A. Welge-Lüssen (Eds.). *Taste and smell* (Vol. 63, pp. 44-69). Karger Publishers.

Gottfried, J. A., & Wilson, D. A. (2011). 5 Smell. *Neurobiology of sensation and reward*, 99.

Graham, J. M. (2006). Congeneric and (essentially) tau-equivalent estimates of score reliability: What they are and how to use them. *Educational and psychological measurement, 66*(6), 930-944.

Granqvist, P., Vestbrant, K., Döllinger, L., Liuzza, M. T., Olsson, M. J., Blomkvist, A., & Lundström, J. N. (2019). The scent of security: Odor of romantic partner alters subjective discomfort and autonomic stress responses in an adult attachment-dependent manner. *Physiology & behavior, 198*, 144-150.

Groyecka, A., Pisanski, K., Sorokowska, A., Havlíček, J., Karwowski, M., Puts, D., Roberts, S. C., & Sorokowski, P. (2017). Attractiveness Is Multimodal: Beauty Is Also in the Nose and Ear of the Beholder. *Frontiers in Psychology, 8*, 778.

Gulliksen, H. (1950). *Theory of mental tests*. Wiley.

Haidt, J., McCauley, C., & Rozin, P. (1994). Individual differences in sensitivity to disgust: A scale sampling seven domains of disgust elicitors. *Personality and Individual Differences, 16*(5), 701-713.

Havlicek, J., & Roberts, S. C. (2009). MHC-correlated mate choice in humans: a review. *Psychoneuroendocrinology, 34*(4), 497-512.

Havlicek, J., Roberts, S. C., & Flegr, J. (2005). Women's preference for dominant male odour: effects of menstrual cycle and relationship status. *Biology letters, 1*(3), 256-259.

Hedner, M., Larsson, M., Arnold, N., Zucco, G. M., & Hummel, T. (2010). Cognitive factors in odor detection, odor discrimination, and odor identification tasks. *Journal of Clinical and Experimental Neuropsychology, 32*(10), 1062-1067.

Herz, R. S., & Inzlicht, M. (2002). Sex differences in response to physical and social factors involved in human mate selection: The importance of smell for women. *Evolution and Human Behavior, 23*(5), 359-364.

Hofer, M.K., Collins, H.K, Whillans, A.V., & Chen, F.S. (2018). Olfactory cues from romantic partners and strangers influence women's responses to stress. *Journal of Personality and Social Psychology, 114*, 1-9.

Hoshika, Y., Imamura, T., Muto, G., Van Gemert, L. J., Don, J. A., & Walpot, J. I. (1994). International comparison of odor threshold values of several odorants in Japan and in the Netherlands. In Araki, S (Ed.) *Neurobehavioral Methods and Effects in Occupational and Environmental Health* (pp. 283-288). Academic.

Hummel, T., Sekinger, B., Wolf, S. R., Pauli, E., & Kobal, G. (1997). 'Sniffin'sticks': olfactory performance assessed by the combined testing of odor identification, odor discrimination and olfactory threshold. *Chem. Senses 22*, 39–52.

Hummel, T., Kobal, G., Gudziol, H., & Mackay-Sim, A. J. E. A. (2007). Normative data for the "Sniffin'Sticks" including tests of odor identification, odor discrimination, and olfactory thresholds: an upgrade based on a group of more than 3,000 subjects. *European Archives of Oto-Rhino-Laryngology, 264*(3), 237-243.

Ille, R., Wolf, A., Tomazic, P. V., & Schienle, A. (2016). Disgust-related personality traits in men with olfactory dysfunction. *Chem Senses. 41*(5), 427–431.

Ille, R., Wolf, A., Tomazic, P. V., & Schienle, A. (2017). Hyposmia and disgust: Gender-specific effects. *Chemical senses, 42*(6), 493-497.

Jones, B. C., Hahn, A. C., Fisher, C. I., Wang, H., Kandrik, M., Lee, A. J., Tybur, J., & DeBruine, L. M. (2018). Hormonal correlates of pathogen disgust: testing the compensatory prophylaxis hypothesis. *Evolution and Human Behavior, 39*(2), 166-169.

Joussain, P., Ferdenzi, C., Djordjevic, J., & Bensafi, M. (2017). Relationship between psychophysiological responses to aversive odors and nutritional status during normal aging. *Chemical senses, 42*(6), 465-472.

Landis, B. N., Hummel, T., Hugentobler, M., Giger, R., & Lacroix, J. S. (2003). Ratings of overall olfactory function. *Chem Senses, 28*(08), 691–694.

Lange, R., Donathan, C. L., & Hughes, L. F. (2002). Assessing olfactory abilities with the University of Pennsylvania smell identification test: a Rasch scaling approach. *Journal of Alzheimer's Disease*, *4*(2), 77-91.

Larsson, M., Finkel, D., & Pedersen, N.L. (2000). Odor identification: influences of age, gender, cognition and personality. *J. Gerontol. Psychol. Sci. B, 55*(5), 304–310.

Larsson, M., Nilsson, L. G., Olofsson, J. K., & Nordin, S. (2004). Demographic and cognitive predictors of cued odor identification: evidence from a population-based study. *Chem. Senses 29,* 547-554.

Larsson, M., Öberg, C., & Bäckman, L. (2005). Odor identification in old age: Demographic, sensory and cognitive correlates. *Aging, Neuropsychology, and Cognition, 12*(3), 231-244.

Laumann, E. O., Nicolosi, A., Glasser, D. B., Paik, A., Gingell, C., Moreira, E., & Wang, T. (2005). Sexual problems among women and men aged 40–80 y: prevalence and correlates identified in the Global Study of Sexual Attitudes and Behaviors. *International Journal of Impotence Research, 17*(1), 39.

Lemogne, C., Smadja, J., Zerdazi, E., Soudry, Y., Robin, M., Berthoz, S., Limosin, F., Consoli, S. M., & Bonfils, P. (2015). Congenital anosmia and emotion recognition: A case-control study. *Neuropsychologia, 72*, 52–58.

Liuzza, M. T., Lindholm, T., Hawley, C., Sendén, M. G., Ekström, I., Olsson, M. J., Larsson, M., & Olofsson, J. K. (2016). The body odor disgust scale (BODS): development and validation of a novel olfactory disgust assessment. *Chemical Senses, 42*(6), 499-508.

Liuzza, M. T., Olofsson, J. K., Cancino-Montecinos, S., & Lindholm, T. (2019). Body Odor Disgust sensitivity predicts stronger moral harshness towards moral violations of purity. *Frontiers in psychology, 10*, 458.

Lobmaier, J. S., Fischbacher, U., Probst, F., Wirthmüller, U., & Knoch, D. (2018). Accumulating evidence suggests that men do not find body odours of human leucocyte antigen-dissimilar women more attractive. *Proceedings. Biological sciences, 285*(1878), 20180566.

Mac Giolla, E., & Kajonius, P. J. (2018). Sex differences in personality are larger in gender equal countries: Replicating and extending a surprising finding. *International Journal of Psychology: Journal International de Psychologie, 54*(6), 705–711.

Mancini, F., Gragnani, A., & D'Olimpio, F. (2001). The connection between disgust and obsessions and compulsions in a non-clinical sample. *Personality and Individual Differences, 31,* 1173–1180.

McBurney, D. H., Shoup, M. L., & Streeter, S. A. (2006). Olfactory comfort: Smelling a partner's clothing during periods of separation. *Journal of Applied Social Psychology, 36*(9), 2325-2335.

McNeish, D., & Wolf, M. G. (2020). Thinking twice about sum scores. *Behavior Research Methods*, *52*(6), 2287–2305.

Minor, K. L., Wright, B. D., & Park, S. (2004). The Smell Identification Test as a measure of olfactory identification ability in schizophrenia and healthy populations: a Rasch psychometric study. *Journal of Abnormal Psychology*, *113*(2), 207–216.

Moshontz, H., Campbell, L., Ebersole, C. R., IJzerman, H., Urry, H. L., Forscher, P. S., Grahe, J. E., McCarthy, R. J., Musser, E. D., Antfolk, J., Castille, C. M., Evans, T. R., Fiedler, S., Falke J. K., Forero, D. A., Janssen, S. M. J., Keene, J. R., Protzko, J., Aczel, B., Solas, S. A., . . . Chartier, C. R. (2018). The Psychological Science Accelerator: Advancing psychology through a distributed collaborative network. *Advances in Methods and Practices in Psychological Science*, *1*(4), 501-515.

Murray, D. R., Jones, D. N., & Schaller, M. (2013). Perceived threat of infectious disease and its implications for sexual attitudes. *Personality and Individual Differences, 54*, 103–108.

Nesse, R. M. (2005). Natural selection and the regulation of defenses: A signal detection analysis of the smoke detector principle. *Evolution and human behavior*, *26*(1), 88-105.

O'Shea, K. J., DeBruine, L. M., & Jones, B. C. (2019). Further evidence for associations between short-term mating strategy and sexual disgust. *Personality and Individual Differences, 138*, 333-335.

Oaten, M., Stevenson, R. J., & Case, T. I. (2009). Disgust as a disease-avoidance mechanism. *Psychological Bulletin*, *135*(2), 303–321.

Oaten, M., Stevenson, R., Tapp, C., Case, T. I., & Cousins, A. (2018). The role of disgust in male sexual decision-making. *Frontiers in Psychology, 9*, 2602.

Olatunji, B. O., Cisler, J. M., Deacon, B. J., Connolly, K., & Lohr, J. M. (2007a). The Disgust Propensity and Sensitivity Scale-Revised: Psychometric properties and specificity in relation to anxiety disorder symptoms. *Journal of Anxiety Disorders*, *21*(7), 918-930.

Olatunji, B. O., Williams, N. L., Tolin, D. F., Abramowitz, J. S., Sawchuk, C. N., Lohr, J. M., & Elwood, L. S. (2007b). The Disgust Scale: item analysis, factor structure, and suggestions for refinement. *Psychological Assessment, 19*(3): 281–297.

Oleszkiewicz, A., Kunkel, F., Larsson, M., & Hummel, T. (2020). Consequences of undetected olfactory loss for human chemosensory communication and well-being. *Philosophical Transactions of the Royal Society B*, 375(1800), 20190265.

Olsson, M. J., Lundström, J. N., Kimball, B. A., Gordon, A. R., Karshikoff, B., Hosseini, N., Sorjonen, K., Höglund, C. O., Solares, C., Soop, A., Axelsson, J., & Lekander, M. (2014). The scent of disease: human body odor contains an early chemosensory cue of sickness. *Psychological Science*, *25*(3), 817-823.

Open Science Collaboration. (2015). Estimating the reproducibility of psychological science. *Science*, *349*(6251), aac4716.

Ottaviano, G., Zuccarello, D., Frasson, G., Scarpa, B., Nardello, E., Foresta, C., Marioni, G., & Staffieri, A. (2013). Olfactory sensitivity and sexual desire in young adult and elderly men: an introductory investigation. *American Journal of Rhinology & Allergy, 27*(3), 157-161.

Ottaviano, G., Marioni, G., Frasson, G., Zuccarello, D., Marchese-Ragona, R., Staffieri, C., Nardello, E., Frigo, A.C., Foresta, C., & Staffieri, A. (2015). Olfactory threshold for bourgeonal and sexual desire in young adult males. *Medical Hypotheses, 84*(5), 437-441.

Pfeiffer, C. A., & Johnston, R. E. (1994). Hormonal and behavioral responses of male hamsters to females and female odors: Roles of olfaction, the vemeronasal system, and sexual experience. *Physiology and Behavior, 55*, 129–138.

Philpott, C. M., Wolstenholme, C. R., Goodenough, P. C., Clark, A., & Murty, G. E. (2006). Comparison of subjective perception with objective measurement of olfaction. *Otolaryngol Head Neck Surgery, 134*(03), 488–490.

Quigley, J. F., Sherman, M. F., & Sherman, N. C. (1997). Personality disorder symptoms, gender, and age as predictors of adolescent disgust sensitivity. *Personality and Individual Differences, 22*(5), 661-667.

Rasch, G. (1980). *Probabilistic models for some intelligence and attainment tests*. University of Chicago Press.

Regenbogen, C., Axelsson, J., Lasselin, J., Porada, D. K., Sundelin, T., Peter, M. G., Lekander, M., Lundström, J. N., & Olsson, M. J. (2017). Behavioral and neural correlates to multisensory detection of sick humans. *Proceedings of the National Academy of Sciences, 114*(24), 6400-6405.

Roberts, S. C., & Little, A. C. (2008). Good genes, complementary genes and human mate preferences. *Genetica, 132*(3), 309–321.

Roberts, S. C., Kralevich, A., Ferdenzi, C., Saxton, T. K., Jones, B. C., DeBruine, L. M., Little, A. C., & Havlicek, J. (2011). Body Odor Quality Predicts Behavioral Attractiveness in Humans. *Archives of Sexual Behavior, 40*(6), 1111–1117.

Rottman, J., DeJesus, J., & Greenebaum, H. (2019). Developing Disgust: Theory, Measurement, and Application. In V. LoBue, K. Pérez-Edgar, & K. A. Buss (Eds.), *Handbook of Emotional Development* (pp. 283-309). Springer.

Rozin, P., & Fallon, A. E. (1987). A perspective on disgust. *Psychological Review*, *94*(1), 23–41.

Rozin, P., Hammer, L., Oster, H., Horowitz, T., & Marmora, V. (1986). The child's conception of food: Differentiation of categories of rejected substances in the 16 months to 5 year age range. *Appetite*, *7*(2), 141–151.

Rozin, P., Haidt, J., & McCauley, C. R. (2000). Disgust. In Lewis, M., and Haviland, J. (Eds.), *Handbook of Emotions* (2nd ed., pp. 637-653). Guilford Press.

Ruffman, T., Henry, J. D., Livingstone, V., & Phillips, L. H. (2008). A meta-analytic review of emotion recognition and aging: Implications for neuropsychological models of aging. *Neuroscience & Biobehavioral Reviews*, *32*(4), 863-881.

Sabiniewicz, A., Wrage, F., & Hummel, T. (2021). The impact of cognitive and noncognitive factors on odor discrimination performance. *Journal of Sensory Studies, 36*(5), e12683. https://doi.org/10.1111/joss.12683

Santos, D. V., Reiter, E. R., DiNardo, L. J., & Costanzo, R. M. (2004). Hazardous events associated with impaired olfactory function. *Archives of Otolaryngology – Head & Neck Surgery*, *130*(3), 317-319.

Sarolidou, G., Axelsson, J., Sundelin, T., Lasselin, J., Regenbogen, C., Sorjonen, K., Lundström, J. N., Lekander, M., & Olsson, M. J. (2019). Emotional expressions of the sick face. *Brain, Behavior, and Immunity, 80*, 286–291.

Schäfer, L., Mehler, L., Hähner, A., Walliczek, U., Hummel, T., & Croy, I. (2019). Sexual desire after olfactory loss: Quantitative and qualitative reports of patients with smell disorders. *Physiology & Behavior*, *201*, 64-69.

Schaller, M., & Murray, D. R. (2008). Pathogens, personality, and culture: Disease prevalence predicts worldwide variability in sociosexuality, extraversion, and openness to experience. *Journal of Personality and Social Psychology*, *95*, 212-221.

Schienle, A., Schäfer, A., Stark, R., Walter, B., & Vaitl, D. (2005). Gender differences in the processing of disgust-and fear-inducing pictures: an fMRI study. *Neuroreport*, *16*, 277–280.

Shu, C. H., Hummel, T., Lee, P. L., Chiu, C. H., Lin, S. H., & Yuan, B. C. (2009). The proportion of self-rated olfactory dysfunction does not change across the life span. *Am J Rhinol Allergy, 23*(04), 413–416.

Simpson, J. A., & Gangestad, S. W. (1991). Individual differences in sociosexuality: evidence for convergent and discriminant validity. *Journal of Personality and Social Psychology, 60*(6), 870-883.

Smith, K. B., Oxley, D., Hibbing, M. V., Alford, J. R., & Hibbing, J. R. (2011). Disgust sensitivity and the neurophysiology of left-right political orientations. *PloS one*, *6*(10), e25552.

Sorokowska, A. (2013). Assessing personality using body odor: Differences between children and adults. *Journal of Nonverbal Behavior, 37*(3), 153–163. https://doi.org/10.1007/s10919-013-0152-2

Sorokowska, A., Sorokowski, P., & Szmajke, A. (2011). Does personality smell? Accuracy of personality assessments based on body odour. *European Journal of Personality, 26*(5), 496–503.

Sorokowski, P., Karwowski, M., Misiak, M., Marczak, M. K., Dziekan, M., Hummel, T., & Sorokowska, A. (2019). Sex differences in human olfaction: a meta-analysis. *Frontiers in Psychology, 10*, 242.

Soter, A., Kim, J., Jackman, A., Tourbier, I., Kaul, A., & Doty, R. L. (2008). Accuracy of self-report in detecting taste dysfunction. *The Laryngoscope, 118*(4), 611-617.

Soudry, Y., Lemogne, C., Malinvaud, D., Consoli, S. M., & Bonfils, P. (2011). Olfactory system and emotion: common substrates. *European Annals of Otorhinolaryngology, Head and Neck Diseases, 128*(1), 18-23.

Soussignan, R., Schaal, B., Marlier, L., & Jiang, T. (1997). Facial and autonomic responses to biological and artificial olfactory stimuli in human neonates: Re-examining early hedonic discrimination of odors. *Physiology & Behavior*, 62(4), 745–758.

Spearman, C. (1907). Demonstration of formulae for true measurement of correlation. *American Journal of Psychology, 18*, 161-169.

Spearman, C. (1913). Correlations of sums and differences. *British Journal of Psychology*, 5, 417-426.

Stafford, L. D. (2017). The role of the chemical senses in disgust's disease avoidance. *Chemical Senses*, *42*(6), 455-456.

Stafford, L. D., Fleischman, D. S., Le Her, N., & Hummel, T. (2018). Exploring the emotion of disgust: differences in smelling and feeling. *Chemosensors, 6*(1), 9.

Steiner, J. E. (1979). Human facial expressions in response to taste and smell stimulation. *Advances in Child Development and Behavior*, *13*, 257–295.

Stevenson, R. J. (2010). An initial evaluation of the functions of human olfaction. *Chemical Senses*, *35*(1), 3-20.

Stevenson, R. J., Oaten, M. J., Case, T. I., Repacholi, B. M., & Wagland, P. (2010). Children's response to adult disgust elicitors: Development and acquisition. *Developmental Psychology*, *46*(1), 165–177.

Stevenson, R. J., Case, T. I., & Oaten, M. J. C. (2011). Effect of self-reported sexual arousal on responses to sex-related and non-sex-related disgust cues. *Archives of Sexual Behavior, 40*(1), 79-85.

Susskind, J. M., Lee, D. H., Cusi, A., Feiman, R., Grabski, W., & Anderson, A. K. (2008). Expressing fear enhances sensory acquisition. *Nature Neuroscience, 11*(7), 843.

Tai, S. Y., Wang, L. F., Tai, C. F., Huang, Y. T., & Chien, C. Y. (2016). Chronic rhinosinusitis associated with erectile dysfunction: A population-based study. *Scientific Reports, 6,* 32195.

Temmel, A. F., Quint, C., Schickinger-Fischer, B., Klimek, L., Stoller, E., & Hummel, T. (2002). Characteristics of olfactory disorders in relation to major causes of olfactory loss. *Archives of Otolaryngology–Head & Neck Surgery*, *128*(6), 635-641.

Thornhill, R., & Gangestad, S. W. (1999). The scent of symmetry: a human sex pheromone that signals fitness?. *Evolution and Human Behavior*, *20*(3), 175-201.

Tybur, J. M., & Karinen, A. K. (2018). Measurement and theory in disgust sensitivity. In V. Zeigler-Hill & T. K. Shackelford, (Eds.), *Handbook of Personality and Individual Differences* (pp. 159-179). SAGE.

Tybur, J. M., Lieberman, D., & Griskevicius, V. (2009). Microbes, mating, and morality: individual differences in three functional domains of disgust. *Journal of Personality and Social Psychology, 97*(1), 103–122.

Tybur, J. M., Bryan, A. D., Magnan, R. E., & Hooper, A. E. C. (2011). Smells like safe sex: Olfactory pathogen primes increase intentions to use condoms. *Psychological Science, 22*(4), 478-480.

Tybur, J. M., Inbar, Y., Güler, E., & Molho, C. (2015). Is the relationship between pathogen avoidance and ideological conservatism explained by sexual strategies?. *Evolution and Human Behavior*, *36*(6), 489-497.

Tybur, J. M., Çınar, Ç., Karinen, A. K., & Perone, P. (2018). Why do people vary in disgust?. *Philosophical Transactions of the Royal Society B: Biological Sciences, 373*(1751), 20170204.

Walliczek-Dworschak, U., & Hummel, T. (2017). The human sense of olfaction. *Facial Plastic Surgery*, *33*(04), 396-404.

Wyatt, T. D. (2019). Reproducible research into human semiochemical cues and pheromones: learning from psychology's renaissance. *PeerJ Preprints*, e27908v1.

Chapter 8
The Paradox of Moral Disgust and Three Possible Resolutions

Roger Giner-Sorolla

The phrase "moral disgust" comes to us as a paradox. "Moral" is one of the most positively valued adjectives in the English language, while "disgust" is one of the most negative states, reminding us that high standards must at times be guarded by bad feelings. Morality refers to a realm of human experience that is tied to the highest expressions of reason, while disgust reacts to things that are "rank and gross in nature" (to use the Shakespearean phrase cited in Chapman and Anderson 2013). To be moral, one should have self-control over instinctive reactions, but is not disgust one of the most instinctive of reactions? Indeed, the phrase "moral disgust" flirts with oxymoron status, a paradoxical marriage of angel and worm.

In the spirit of the "man-bites-dog" headline, paradox can make a topic counter-intuitive, fascinating, and inspirational for research. In a recent Google Scholar search, "moral disgust," at about 5000 hits, beat "moral anger" (1960 hits), "moral shame" (1900 hits), and "moral contempt" (668 hits) in popularity, among negative emotions of condemnation. "Moral guilt" had more hits, but the phrase might refer to a state of culpability rather than to an emotion; indeed, when the word "emotion" is required, "moral guilt" goes down to 2880, while "moral disgust" plus "emotion" still comes out on top at 3440 hits. At the same time, this field of research has attracted many scholars who, to follow the metaphor along, doubt that the man really did bite the dog, or have questions about definitions—maybe the "dog" was only a hot dog? Claims that emotion is a necessary and sufficient element of moral judgment in general have been questioned, as have claims that disgust in moral contexts can be distinguished from other emotions. Apparent effects of disgust on morality have been ascribed to other factors than the emotion itself. Meanwhile, further research has brought up the possibility that moral disgust is a performative

R. Giner-Sorolla (✉)
School of Psychology, University of Kent, Canterbury, UK
e-mail: R.S.Giner-Sorolla@kent.ac.uk

P. A. Powell, N. S. Consedine (eds.), *The Handbook of Disgust Research*,
https://doi.org/10.1007/978-3-030-84486-8_8

expression rather than a real feeling; or that it responds to more general cues than the violation of bodily taboos with which it seems to have an affinity.

We currently stand in the middle of a re-evaluation of the very concept of moral disgust. In this chapter, I review the original moral intuitionist claims; the debate over the extent of disgust's involvement in moral condemnation; and the more critical perspectives that, conceptually and empirically, seek to resolve the paradox by uncoupling disgust in part or whole from morality in part or whole. That is, it seems we can have the wholehearted emotion of disgust, but only applying to some forms of moral disapproval. Alternatively, we can apply some but not all characteristic features of disgust to most kinds of moral faults, ending up with a nominal "disgust" that may just be drawn upon for its ability to evoke shunning, social coordination, or other responses appropriate to a contagious bad idea. In summing up, I focus on the implications of different views of the flexibility of disgust for moral judgment, as one more way of resolving the paradox. That is, what disgust and moral standards do have in common is a sense that they are and should be inflexible to some degree.

Intuitionism and Maximal Views of Moral Disgust

Moral psychology theory in the twentieth century tended to be cognitive in nature, led by Kohlberg's lifelong focus on the thought process rather than the outcome of moral judgment. But the twenty-first century began with Haidt's (2001) introduction into psychology of the philosophical concept of moral intuitionism, including such writers as Hume and Adam Smith. The claims made for intuition in Haidt's article were sweeping: most moral judgments are made intuitively; dilemmas such as Kohlberg's in which competing principles inform reasoned judgment are the exception, not the rule, in everyday life; and moral reasoning exists primarily to justify intuitive judgments and convince other people of them. In Haidt's view, intuition is a rapid, unreflective, and emotionally "hot" process providing judgments without much initial access to the reasons underlying them.

Theoretical work in behavioural economics and evolutionary psychology has gone beyond intuitionism to focus on the specific characteristics of emotions. As compared to mere evaluative intuitions (i.e., good or bad feelings), emotions have a number of additional traits. They are "hot", motivating processes that, to use the insight of Scherer (1984), "'decouple' the behavioral reaction from the stimulus event" (p. 295); that is, like motivations, they allow for flexible and adaptive responding to meet a goal. They are also more specific than evaluations, and this specificity has been tied to additional adaptive differentiation within the kind of threats that can arouse a negative emotion, or the kind of opportunities that can arouse a positive one. Fear, for example, is adapted to react to obvious threats to bodily safety, with appropriate behavioural responses ranging from freezing in place to fleeing. Disgust, while it is also a negatively valenced emotion, defends against contagious pathogens that threaten the body in a different way. Thus, disgust reacts to likely signs and contexts of those pathogens (e.g., rotten smells, dead animals), and prepares behaviours that defend against those (e.g., avoidance, wrinkling the nose). Finally, specific emotions are expressive and have been studied since Darwin in terms of facial signals (Ekman 2006), whose uncontrollable, communicative nature helps coordinate social action and show cooperation with other individuals and collectives (Frank 1988).

Disgust became a star witness early on in the case for moral intuitionism, due to the inaccessibility to reasoning of specifically disgust-based moral reactions. Studies on "moral dumbfounding" (Björklund et al. 2000) have an important place in the intuitionist arguments of Haidt (2001). These studies presented several disgusting harmless violations of moral taboos (such as consensual incest) that were written to exclude any harmful consequences, and found that respondents had trouble articulating reasons why these acts were wrong (see also McHugh et al. 2017). Along the way, scholars have entertained the further notion that the specific emotion of disgust underlies nearly all instances of moral condemnation, even when harm to another person is explicit. Evidence in favour of this maximal connection between disgust and morality has come from studies showing that disgust inductions tend to sharpen moral condemnation judgments, from correlations between dispositional disgust and moral judgment, and from other examples of non-biological immoral elicitors heightening disgust, including physiological reactions (Chapman et al. 2009; Chapman and Anderson 2013).

Why would disgust play such a maximal role in morality? The most common answer (Chapman and Anderson 2013; Hutcherson and Gross 2011) is that, while physical disgust and moral disgust inputs have little in common, the output actions of disgust as contamination defence—avoiding, expelling, and abhorring the offending object—are also useful actions to take towards moral offenders. Anger is relatively more common than disgust when one's own rights are violated (Hutcherson and Gross 2011; Molho et al. 2017), and it cues directly aggressive actions that bring to attention the rights and power of the wronged individual (Molho et al. 2017; Sell et al. 2009). However, disgust is the more appropriate response when collective moral norms are violated, because its associated actions fortify the position that the offender has no place anywhere in the community. It says "I am offended for US" more loudly than it says "I am offended for MYSELF."

A related account of disgust's maximal presence in morality relies on the communication function of disgust expressions. According to Tybur et al. (2013), disgust at moral offenses is best understood as a coordination signal that—in line with a rapid and intuitive understanding of disgust, bringing us back to intuitionism—compels a unified collective response to moral offenses, avoiding the collectively undesirable situation of a feud over disagreement. Whatever the kind or context of the wrongdoing, feeling disgust helps us only insofar as it allows us to credibly express disgust. In support of this notion, Kupfer and Giner-Sorolla (2017) found that people understand that disgust, more than anger, communicates moral as opposed to self-interested motivations. The final study showed that disgust expressions can be strategically chosen to signal moral motives, even when the personal feeling described is more strongly anger. Molho et al. (2017) also conducted studies that supported the specific role of disgust in moral situations. In their findings, disgust encouraged coordinated social action through lateral information-sharing (gossip, in not so many words) and avoidant sanctioning, as opposed to anger's more direct approach. One might also note here that lateral and avoidant actions need to be collective and total in order to be effective. Otherwise, they just partition the group into two hostile factions that support and oppose the behaviour, which thereby becomes merely controversial rather than actively taboo.

It is worth bringing up two more relatively recent explanations for why disgust might be applied to a wide variety of moral wrongs. These are less focused on the utility of the outputs of disgust (i.e., avoidant action, coordinated expression) and more focused on the utility of cognitive aspects of disgust. To be specific, they are: disgust's object focus, useful for tracking moral character; and the contagious and dose-insensitive nature of disgust, useful in thinking about the spreading of immoral ideas and values.

In one of many contemporary appraisal accounts of emotion, Ortony et al. (1990) characterise disgust, like love, as an emotion that primarily responds to the nature of objects, rather than to the tenor of a situation or to the evaluation of the self. Later observations (Haidt 2003) raise the possibility that in the moral sphere, disgust functions to appraise the "object" of a person's moral character, rather than the "situation" of a moral wrong. Vignette experiments directly testing this conjecture found support for it (Giner-Sorolla and Chapman 2017). When a person's desire to commit wrong and the harmful outcome of their action were manipulated independently, desire influenced levels of disgust and degree of harm influenced levels of anger. The effect of desire on disgust was mediated by judgments of the actor's bad character. Parallel findings emerge from studies of judgments of immoral actions embedded in media, fiction, or the imagination (Rolfe and Giner-Sorolla 2020; Sabo and Giner-Sorolla 2017). Without literal harm to arouse anger, disgust apparently prevails when any kind of immoral action is referred to in, for example, song lyrics or a video game. Such immoral depictions, even though harmless, reveals the desires and tolerances of creators and consumers, which signal laxity of moral character.

An interesting but under-examined possibility proposes that disgust is morally relevant because beliefs that challenge moral norms bear functional similarity to the viruses, bacteria, and parasites disgust evolved to control. That is, ideas can start small and multiply, like a viral contagion or the spread of genetic traits, as referenced in Dawkins' "meme" metaphor (e.g., Dawkins 1993). If immoral acts are to be controlled, they must be categorically rejected no matter how small or symbolic the "dose," and without much concern for the risk of false-positive reactions. Indirect evidence comes from recent studies (Rottman and Young 2019) that used moral violation types associated in other research with disgust and anger, and found that "purity" (disgust-related) violations, such as incest and animal-human genetic engineering, were judged less in accordance with their magnitude than "harm" (anger-related) violations, such as animal and human cruelty.

Relatedly, some recent findings from our own lab indicate that symbolic moral contamination from association with extremist symbols (cf. Rozin et al. 1986) is not likely to occur through literal contact, but rather through concerns about one's own reputation (Kupfer and Giner-Sorolla 2021). The current battles about what ideas should be permissible in the academy, media, and workplace hinge on the rejection not only of problematic ideas but also of people who express them and of others who associate with them. The so-called "cancel culture" is proof on a large scale (if not yet from the lab) that modelling reactions on contagion and association can have weighty functions in regulating the moral nature of a society.

Paradox Resolution 1: All of Morality, Not All of Disgust

Whether moral disgust is explained in terms of action, communication, character evaluation, or social contagion management, each of these explanations applies disgust to a maximal model of morality, covering several different moral domains and types of wrongs. In doing so, they resolve the paradox between high-minded morality and lowly disgust by claiming the exaptation of a limited set of disgust-related responses or appraisals to cover the moral sphere. We can speculate that all four ideas just covered are simultaneously true, and imagine a schematic narrative: "people of bad character/will spread bad ideas, no matter how many there are/and we exclude them from our society/by letting everyone know we think they are disgusting."

Up to a certain point, this narrative resembles a skeleton of the biological disgust emotion: "things that are basically gross/will spread bad stuff, no matter how many there are/and we exclude them from our life/by letting everyone know we think they are disgusting." Yet it is not at all clear whether or not moral disgust possesses all four of these features, or overlapping completely with the disgust felt towards pathogen inducers. Just one of these features seems sufficient to produce higher endorsement of disgust in most of the self-report paradigms used in research thus far. Also, experiments testing disgust reactions to immoral actions that do not involve bodily or pathogen cues tend to show that morally elicited disgust in its broadest sense lacks some features that "core" pathogen disgust has, such as involvement of the gastric and taste system (Herz 2011), parasympathetic rather than sympathetic activation (Ottaviani et al. 2013), or rapid time course (Simpson et al. 2006). If people say they are "disgusted" at a wide range of moral transgressions, then it is possible to keep that as a concept of moral disgust, but only by treating as unnecessary some parts of the core emotion of disgust. Excluding some hallmarks of disgust to allow all parts of morality to be covered is, then, one way to resolve the paradox of moral disgust.

Paradox Resolution 2: All of Disgust, Not All of Morality

A different resolution of the moral disgust paradox keeps the full emotion of disgust, but limits its influence to certain kinds of moral situations. This solution is made possible by theories of morality that propose multiple domains of wrongs with different defining features and characteristic emotions, initially the three-domain model (Shweder et al. 1997) later extended into the Moral Foundations model with five or more moral domains (Graham et al. 2009). Some earlier writers on morality proposed it to be universally concerned with regulating harm to others (Turiel 1977), an idea more recently defended in views that harm is a central and necessary concept to morality (e.g., Schein and Gray 2018). But, theories that have proposed multiple moral domains explicitly include moral concerns that do not involve harm or victims. Such concerns include universal taboos, such as those against incest, as well as cultural norms that regulate consensual sexuality in other ways, and miscellaneous observances that are given moral weight, such as avoiding certain foods or modifying the body in certain ways. Shows of allegiance or communal solidarity, such as respect for national symbols or observing social norms of dress, also are phenomena explained as separate domains of morality.

In particular, a certain kind of moral violation seems to be intuitively disgusting: the ones labelled "divinity" by Shweder et al. (1997) and "purity" by Graham et al. (2009) whose most salient examples are violations of sexual and food taboos. Although these violations were initially presumed to be seen as wrong because they transgressed "sacred" values which religion and culture often seek to regulate, later research has shown that this is not the whole story. For example, symbolic desecrations of sacred religious objects, provided they do not involve materials that are pathogenic and disgusting in themselves, evoke more anger than disgust (Kollareth and Russell 2019; Royzman et al. 2014). Other studies have found that, religious believers who consider "heretical" thoughts produce self-reports of disgust that predict moral judgment more strongly than anger, and feelings of contamination, but no characteristic disgust facial expressions (Ritter et al. 2016).

The role of actual "divine" or "purity" concepts in sacred acts not involving the body or infection is still unclear from this small literature. Perhaps the differences are due to desecration being seen as an external act of aggression against one's religion, whereas impure thoughts conform more to a disease model, being located within one's own mind and liable to spread if not checked. In a previous review of the literature, Russell and Giner-Sorolla (2013) were not satisfied with these sacredness-based terms to characterise violations such as incest, and instead presented evidence for a "bodily-moral" view of moral transgressions that primarily elicit disgust. These are transgressions against norms that regulate how the body is used, whether sexually; in eating; or in other ways, such as altering the body for decoration. When disgust is properly separated from anger, the review found, bodily-moral transgressions are more likely to arouse disgust, while socio-moral transgressions such as harming someone else or cheating are more likely to arouse anger.

Reviewing research on this topic can be frustrating at times because of the lack of a standard experimental protocol that takes care to separate truly moral anger from truly moral disgust. The slipperiness of the definition of "purity" or "divinity" concepts is a case in point, starting with early research on the divinity-disgust link in which not all stimuli used to represent divinity violations had a moral component (Rozin et al. 1999). Even more recent attempts to create normed material for moral research have the same potential problem. Clifford et al. (2015), for example, presented vignettes for which the "sanctity" (purity) category was normed on judgments of how "degrading or disgusting" it is, but the categories were not tested on how immoral the violation is, and possibly non-moral hygiene violations such as "using someone else's toothbrush" were included.

There are other frequent failings of research to clearly categorise moral acts and emotions, some of which we have identified in reviews of the literature (Giner-Sorolla et al. 2018; Russell and Giner-Sorolla 2013). Some studies compare disgust to low-arousal emotions like sadness but not to anger, so that residual anger could influence reactions attributed to disgust. Others treat offenses to the self as moral issues, when the involvement of self-interest makes this more ambiguous and third-party judgments would be a clearer test of morality (e.g., Skarlicki et al. 2013). Still other studies treat harmful actions involving pathogens, like putting insects in someone's food, as examples of "purity" violations, when these are wrong because they hurt other people, not wrong because they are categorically forbidden as, for example, consensual incest is (e.g., Kupfer et al. 2020). Eating bugs yourself is gross, but it is not a moral violation of the same kind as making someone else eat bugs (for more on eating bugs see Powell, Chap. 15, this volume).

Among studies that do clearly define moral violations and emotions, some findings of bodily-moral disgust in response to harm or unfairness have been found (e.g., Cannon et al. 2011, in which unfair actions also elicited facial signals of disgust), and these might be explicable by the previously mentioned role of moral character in disgust. However, there are also a number of findings that visceral or physiological signs of disgust are especially prone to follow bodily-moral violations. Royzman et al. (2008) found that oral inhibition measures were linked to negative moral judgments of sibling incest, as compared to more normal sexual activity. Another measure of oral sensitivity, the PROP test to detect "supertaster" status, showed correlations with pathogen and sexual disgust scale items, but not with disgust towards non-sexual sociomoral transgressions (Herz 2011). Also, a recent set of studies using the antiemetic properties of ginger to quell disgust-related nausea found that the treatment reduced the severity of moral judgments, but only for bodily-moral violations (Tracy et al. 2019).

In addition, we should keep in mind that any kind of moral transgressions bring with them a fair amount of anger, further distinguishing them from pathogen disgust violations that have no moral component (Russell and Giner-Sorolla 2013). Studies comparing moral transgressions to pathogen transgressions have found differences in keeping with this observation. Across both quantitative and qualitative measures, there is more anger accompanying disgust at moral transgressions, with a slower and longer-duration response pattern (Abitan and Krauth-Gruber 2015; Simpson et al. 2006; Rubenking and Lang 2014). Even bodily-moral transgressions (incest) show a physiological signature in which sympathetic arousal predominates, as opposed to mere pathogen or "visceral" disgust, in which parasympathetic systems are more active (Ottaviani et al. 2013). When considering the same kind of moral transgressions, feelings of disgust related to a lower heart rate (typical of parasympathetic activation), and feelings of anger to a higher heart rate (typical of sympathetic activation; Konishi et al. 2020). Research on facial expressions, too, has found differences between a slack-lipped "sickened" face that is judged more appropriate for physically disgusting situations, and the classic wrinkled-nose "disgust" face that has structural similarities to anger expressions and is judged more appropriate for morally objectionable situations (Yoder et al. 2016).

One point of disagreement about linking disgust to a special form of morality has involved the role of harm. The kind of moral violations that most clearly elicit more disgust than anger can be, and initially were, described as "harm-free" because they often involve individuals following their own bliss into incest, necro-bestiality (with the infamous chicken carcass of Haidt et al. 1993), or ethically sourced cannibalism via cloning. Even in the newer, character-based takes on moral disgust, the kind of vignettes that bring out more disgust than anger have involved characters who are thinking of doing some kind of nefarious deed but do not try to carry it out, or try without success (Giner-Sorolla and Chapman 2017). However, the extension of morality to acts that do not directly harm individuals has been challenged by theory and findings that emphasise the centrality of harm to negative moral judgment (e.g., Schein et al. 2016). The most recent version of this theory allows for the moralisation of apparently "harmless" acts, but maintains that such acts are subjectively identified with harm to communal or spiritual goods (Schein and Gray 2018).

In the process of moralisation, an issue that was previously seen as a personal choice or pragmatic matter becomes a moral mandate. There is reason to believe that adopting harm into the narrative is key to the moralisation process (Rozin 1999). For example, moralisation of smoking over the twentieth century intensified with evidence that it was harmful to the self, and later with further evidence about the harm to others of second-hand smoke. In moralisation, concepts of harm can become generalised and disconnected from the specific issue of harm reduction to individuals, as research shows that moral concerns have become associated more with disgust towards smoking than with concerns about health (e.g., Alderman et al. 2010; Helweg-Larsen 2014; Rozin and Singh 1999).

Reframing the debate over different kinds of morality into one about qualitatively different kinds of harm can help in reinterpreting the existing literature. Disgust reacts preferentially to harms that are distal and diffuse, rather than immediate and individual. The "injury" that someone causes by masturbating in an unusual way, or by sharing mere thoughts of deviant and sadistic activities, is inflicted metaphorically upon social norms about the use of the body and respect for others. As a whole, many of these norms and standards do protect against individual harm, although some are moralised conventions. Thus, someone who is willing to go against societal disgust and gleefully take part in some forbidden activity, is likely to be seen as also capable of breaking norms against interpersonal harm with a grin; an inference confirmed by research on perceptions of immoral actors (Chakroff et al. 2017). Whether the perception of harm in apparently harmless acts occurs through an effortful post-hoc process, or immediately and automatically, is a question where the literature gives conflicting and as yet unresolved answers (Gray et al. 2014; Gutierrez and Giner-Sorolla 2007). Perhaps more strongly contextualising our concept of harm—harm to whom, and when?—might help give answers.

Finally, the experimental literature showing effects of induced incidental states of disgust on moral judgment has been cited in favour of "true" disgust driving all kinds of moral judgment, not just bodily-moral. As discussed in a review of this literature (Giner-Sorolla et al. 2018), many of these studies lack tight experimental controls, and cannot rule out the influence of co-activated states, such as anger and irritation, at being exposed to disgusting smells or sights. What's more, the numerous failures to replicate incidental disgust studies (Landy and Goodwin 2015) suggest a theoretical retrenchment (Schnall et al. 2015), in which feelings of disgust need to be noticeable but not overwhelming, and need to be misattributed to moral evaluations, in order to have an effect. We argued in our review that perhaps the object-bound nature of disgust makes transfer of disgust from non-moral to moral objects a less convincing mechanism than the transfer of other moods and emotions that are more tied to situations and goals (Giner-Sorolla et al. 2018).

No Morality in Moral Disgust Examples? The Pathogen Explanation

Beyond these partial resolutions of the moral disgust paradox, there have been altogether more Gordian attempts to solve the paradox by questioning the existence of any such state as "moral disgust." One such line of argument points out that violations involving eating or sexual contact or other bodily-moral violations, by their nature, involve the body and thus expose it to infection. A general form of this argument is that the profile of moral disgust might be adapted to police the kind of violations of norms about the body that overall work to regulate infection (e.g., Inbar and Pizarro 2014). A more specific form of the argument, however, proposes that any special disgust felt at violations such as incest, faecal desecration of a church, or eating cloned human steak is a free-rider from concerns that the activity itself could spread disease (e.g., Kayyal et al. 2015; Royzman et al. 2014). This stronger form proposes that disgust relates to the infection potential of any given activity and does not intrinsically relate to the moral evaluation of the activity.

A first response to this argument is that, in some senses, controlling infection *is* a moral issue. As we are all now acutely aware in the shadow of COVID-19, what other people do with their bodies—sneezing, coughing, getting close—does bear on the welfare of the whole community. Disease-spreading acts fit a classic form of moral problem in which the careless behaviour of one person harms many more. Pathogen disgust as a form of community morality has been proposed more than studied (Curtis 2013). However, this interpretation risks reducing moral judgments of bodily acts to just another literal form of interpersonal harm: infection instead of injury. It does not capture the sense that some acts involving the body are seen as wrong in and of themselves, unnatural and disgusting.

Evidence that disgust applies primarily to bodily activities in which infection is possible has sometimes been presented as a reduction of disgust to pathogen rather than moral concerns (e.g., Kayyal et al. 2015; Kollareth and Russell 2017). However, development of materials for such studies has not tightly controlled infection potential versus perceived immorality, relying instead on general properties such as direct physical, sexual, or ingestive contact, which also covary with the bodily-moral category. A more precise comparison would acknowledge that there are activities that do involve bodily contact, thus have similar potential for infection, but are viewed morally in different ways: for example, in English-speaking countries, eating horse meat versus beef. As well, there are bodily activities viewed in many societies with moral disgust that do not involve literal transmission of germs, a prominent example being masturbation (surely a more hygienic activity infection-wise than the alternative).

In controlled studies that hold the infection potential constant, but vary the normative versus non-normative nature of the bodily act, we and other labs have found strong differences in disgust (Giner-Sorolla et al. 2012; Gutierrez and Giner-Sorolla 2007; Royzman et al. 2008; Russell and Giner-Sorolla 2011a). Furthermore, in regression analyses, this disgust even when pitted against anger does its part to predict moral judgment, working against the idea that disgust is just a free-rider. These findings do leave open a more general form of adaptive argument: that norms about the use of the body in general include the kind of norms that protect against disease, and thus the general disgust reaction to their violation is useful, even if it sometimes catches examples that are not literally related to the transmission of disease. Such an imprecise mechanism, in my view, would be more appropriate to a cultural or biological selection process that had to operate in the absence of any explicit theory of germ transmission. Indeed, similar ideas have been advanced to explain why food taboos concern themselves with meat more than vegetable matter, even if these involve rejecting meat products that are objectively no more infectious than the culturally allowed alternative (Navarrete and Fessler 2003).

However, it would be difficult to prove, without reaching back into the mists of time, that pathogen defence is the only reason for bodily-moral acts evoking disgust. Sex and eating, as the most biologically primary acts for preservation of the group and individual, respectively, are particular targets of cultural regulation and variation. Breaking their norms threatens collective survival in a way, perhaps, that breaking other cultural conventions about dress, manners, or religion does not. A related idea under the label "existential disgust" (e.g., Goldenberg 2005) is that we carry culturally elaborated, tacit motives to distinguish ourselves from animals and to deny our own mortality. Thus, breaking rules about the body threatens the wall that culture has put up to separate us from the animals and to give us symbolic immortality through belonging.

Though the specific evidence for existential disgust is mixed (see again our review, Giner-Sorolla et al. 2018), enforcing cultural bottom-lines about what to eat and how to have sex would more generally seem to be a job for the consensus-enforcing functions of disgust explained in the previous sections, rather than for the workings of anger, which seem more attuned to legitimate disputes between individuals. Although some challenges to the definition of the purity moral domain have tried to explain it away as being based on bizarre and unusual violations (e.g., Gray and Keeney 2015), violations of fundamental consensual norms would have to be seen as somewhat unusual by definition in order to engage the society-wide ostracism implied by disgust. Thus, covarying out the perceived weirdness of disgust-inducing acts that are seen as immoral, but not directly harmful, might throw out the conceptual baby with the bathwater.

No Disgust in Moral Disgust Examples? The Anger-Metaphor Explanation

Another challenge to research seeking to distinguish moral disgust from moral anger is the close connection between disgust and anger in moral and other social judgment domains (for a review, see again Giner-Sorolla et al. 2018). Although the two emotions have been posited as separate and "basic" negative states, a number of their features coincide. They are both high-arousal and unpleasant states; their facial expressions include similar components and are the two most commonly confused basic emotions, so that in recent empirically supported schemes of facial expression they are fused into one (Jack et al. 2016). If care in measurement and context-setting is not taken, anger and disgust responses towards social objects can become empirically indistinguishable.

In this context an early doubt about moral "disgust" was framed: given existing evidence about the use of the term "disgusted" to express anger, might verbal self-reports of disgust at moral violations simply be the deployment of a synonym for some other emotional state, such as anger (Nabi 2002)? If we accept emotion as a multi-faceted construct including language terms, the apparent relationship between disgust and moral judgment can be picked apart as only language-deep: the terminology corresponding to disgust, but everything else belonging to anger. Nabi's (2002) preferred distinctive term for disgust, "grossed-out," is akin to other measures that propose to capture more visceral forms of the emotion, tapping turned guts rather than hot heads; for example, the previously mentioned sick face, oral inhibition measures (Cusimano et al. 2018), or verbal measures incorporating physical sensations (e.g., Nummenmaa et al. 2014). In a way, to reduce moral disgust to the use of a term is simply a more extreme version of the first resolution we considered, in which a partial form of disgust can be accepted as applying to many moral situations.

As mentioned before, many studies using more visceral disgust measures, or taking other steps such as covarying out levels of expressed anger from expressed disgust, have produced a relatively narrow picture of moral disgust centred on bodily-moral violations and, at times, on expressions reflecting bad character. However, a stronger form of the mere-synonym conjecture has not been well supported. That is, verbal reports of disgust behave independently from verbal and facial reports of anger in predicting endorsement of disgusted facial expressions, even for non-bodily moral violations (but more so for bodily-moral; Gutierrez et al. 2012).

The other evidence we have reviewed already shows that while verbal disgust towards moral violations often coexists with anger and incorporates elements of anger, it maintains other characteristics of disgust distinct from anger, such as action tendencies supporting indirect social sanctions and withdrawal, rather than anger's impulse towards direct attack. Whether to call the anger-disgust blend "anger" on one side or "disgust" the other depends, ultimately, on one's standards for categorising emotions.

A Third and Final Resolution: Shared Irrationality

In summary, the empirical evidence psychologists have gathered on disgust and morality leads to a qualification of the bald statement that the emotion of disgust, in its purest form, is essentially entangled with morality, in its total form. Rather, it seems that in order to extend the emotion of disgust to all areas of morality, we have to give up something of the visceral nature of core or pathogen disgust, and treat disgust as a feeling of social disapproval of bad characters, one that facilitates expressions of revulsion, thoughts of contamination, and actions of coordinated shunning. On the other hand, disgust as a complete and visceral emotion seems most reliably related to moral violations, such as incest, that involve violating fundamental rules of bodily conduct—not merely acts that harm others or violate their rights. However, the few findings that stand in exception to this summary, as well as the plausible hypotheses that have not yet been tested (e.g., whether visceral disgust is absent from reactions to marks of bad character), lead me to express this two-handed resolution somewhat cautiously.

A final and even more speculative possibility for the coincidence of disgust and morality rests on the inaccessibility of the core concepts of why something is disgusting, just as the core concepts of morality, past a certain level of argumentation, stand on their own and do not admit of further reasoning. To jump directly into this analogy, it is just as useful to not know why we fight for the ideals of fairness, equality, benevolence, or tradition, as to not know why we are disgusted at incest or sadism. When arguing about moral injustices, the articulated reasons, if pursued long enough, ultimately reach some inarguable, near-tautological principle such as "harming people is wrong" or "people have a right to consent to decisions that affect them." In fact, terminal moral values have been shown to share many traits in common with social truisms (Maio and Olson 1998). People find it hard to articulate reasons why they are right, and are vulnerable to counterarguments attempting to shake these seldom-defended and self-evident truths.

Likewise, disgust can be a form of truism, as shown in the kind of moral dumbfounding results that support intuitionist views of morality, and in studies building on them to examine the rationalisation of disgust. These studies show specifically that when it comes to the same moral target violating bodily norms, disgust but not anger is difficult to spontaneously explain (Russell and Giner-Sorolla 2011b). In other words, even if participants can endorse plausible reasons for disgust when presented with them, such reasons are more difficult to produce in relation to the immoral example (see also McHugh et al. 2017). Disgust, unlike anger, also showed a small but meaningful number of tautological responses such as "it's disgusting." Other studies show that disgust is hard to reason out of, as well as into (Piazza et al. 2013). For example, mitigating circumstances that could reduce disgust felt towards moral violations, as opposed to anger, were found to be harder to imagine and less effective.

These results support a broader feature of disgust that seems intuitively right. It is hard to explain why something is really disgusting, just as it's hard to explain why something is fun or tasty. The reason for revulsion at incest is locked away in the millennia-old fog of ignorance of population genetics, just as the reason for revulsion at unfamiliar meat animals is locked away in the ignorance of germ theory. Beyond encoding strategies for long-term biological fitness, adhering unquestioningly to a symbolic practice might give benefits more generally in terms of social cohesion. While it is hard to study these processes in the lab, I can here draw on personal experience with attending highly liberal, "God-optional" religious services in which the sociological reason for every affirmation and practice was explained in full. The contrast between these and more traditional rites has shown me that understanding one's religion, as satisfying as it might be to one's self-image as a rational and tolerant person, gets in the way of really feeling and believing in a common, shared mystery.

In other words, following Frank's (1988) argument, locking a response behind a door of rational inaccessibility might be a way to ensure that it is not overridden by temporary self-interested concerns. Just as people find foods tasty or disgusting without always being able to explain why, some moral judgments might be beyond reasoning, and might thereby be strengthened. If I can explain why going to Mass is useful to me, I can abandon the practice when it is no longer useful to me as an individual. To give another example, condemning incest might conflict with benevolent desires to approve of people's happiness. These wishes are usually adaptive in a society. But, an unreasoning opposition to the act of incest works over many generations to preserve the fitness of the species. In the absence of scientific knowledge, only a seemingly irrational stance can defend the incest taboo against persuasive arguments to loosen it, "just this once."

Even when arguing about injustices that arouse more anger than disgust, the articulated reasons must ultimately reach some inarguable, near-tautological value principle such as "harming people is wrong" or "people have a right to consent to decisions that affect them" (Maio and Olson 1998). Being unable to disarm this principle, itself, also has value. Ultimately, we can see the relationship between disgust and morality not as a paradox of opposites, but as a matter of a shared faith. Revulsion at the worm, and longing for the state of the angels, are tied together by their common need to serve as the immovable limits of the moral range.

References

Abitan, A., & Krauth-Gruber, S. (2015). The two sides of disgust: A lexical and thematic content analysis of narratives of personally experienced physical and moral disgust. *Social Science Information*, *54*(4), 470-496.

Alderman, J., Dollar, K. M., & Kozlowski, L. T. (2010). Commentary: understanding the origins of anger, contempt, and disgust in public health policy disputes: applying moral psychology to harm reduction debates. *Journal of Public Health Policy*, *31*(1), 1-16.

Björklund, F., Haidt, J., & Murphy, S. (2000). Moral dumbfounding: When intuition finds no reason. *Lund Psychological Reports*, 1, 1–15.

Cannon, P. R., Schnall, S., & White, M. (2011). Transgressions and expressions: Affective facial muscle activity predicts moral judgments. *Social Psychological and Personality Science*, 2(3), 325–331.

Chakroff, A., Russell, P. S., Piazza, J., & Young, L. (2017). From impure to harmful: Asymmetric expectations about immoral agents. *Journal of Experimental Social Psychology*, *69*, 201-209.

Chapman, H. A., & Anderson, A. K. (2013). Things rank and gross in nature: A review and synthesis of moral disgust. *Psychological Bulletin*, *139*(2), 300-327.

Chapman, H. A., Kim, D. A., Susskind, J. M., & Anderson, A. K. (2009). In bad taste: Evidence for the oral origins of moral disgust. *Science*, *323*(5918), 1222-1226.

Clifford, S., Iyengar, V., Cabeza, R., & Sinnott-Armstrong, W. (2015). Moral foundations vignettes: A standardized stimulus database of scenarios based on moral foundations theory. *Behavior Research Methods*, *47*(4), 1178-1198.

Curtis, V. (2013). *Don't look, don't touch: The science behind revulsion*. Oxford University Press.

Cusimano, C., Royzman, E. B., Leeman, R. F., & Metas, S. (2018). Methodology is the core disgust problem: Response to Inbar and Scott (2018). *Judgment and Decision Making*, *13*(6), 639.

Dawkins, R. (1993). Viruses of the mind. In Dahlbom, B. (Ed.), *Dennett and his Critics* (pp. 13-27). Blackwell.

Ekman, P. (Ed.). (2006). *Darwin and facial expression: A century of research in review*. Malor Books.

Frank, R. H. (1988). *Passions within reason: The strategic role of the emotions*. WW Norton & Co.

Giner-Sorolla, R., & Chapman, H. A. (2017). Beyond purity: Moral disgust toward bad character. *Psychological Science*, *28*(1), 80–91.

Giner-Sorolla, R., Caswell, T. A., Bosson, J., & Hettinger, V. (2012). Emotions in sexual morality: Testing the separate elicitors of anger and disgust. *Cognition & Emotion*, *26*, 1208-1222.

Giner-Sorolla, R., Kupfer, T., & Sabo, J. (2018). What makes moral disgust special? An integrative functional review. *Advances in Experimental Social Psychology*, *57*, 223-289.

Goldenberg, J. L. (2005). The body stripped down: An existential account of the threat posed by the physical body. *Current Directions in Psychological Science*, *14*(4), 224-228.

Graham, J., Haidt, J., & Nosek, B. A. (2009). Liberals and conservatives rely on different sets of moral foundations. *Journal of Personality and Social Psychology*, *96*(5), 1029–1046.

Gray, K., & Keeney, J. E. (2015). Impure or just weird? Scenario sampling bias raises questions about the foundation of morality. *Social Psychological and Personality Science*, *6*(8), 859–868.

Gray, K., Schein, C., & Ward, A. F. (2014). The myth of harmless wrongs in moral cognition: Automatic dyadic completion from sin to suffering. *Journal of Experimental Psychology: General*, *143*(4), 1600–1615.

Gutierrez, R., & Giner-Sorolla, R. (2007). Anger, disgust, and presumption of harm as reactions to taboo-breaking behaviors. *Emotion*, *7*, 853-868.

Gutierrez, R., Giner-Sorolla, R., & Vasiljevic, M. (2012). Just an anger synonym? Moral context influences predictors of disgust word use. *Cognition & Emotion*, *26*, 53-64.

Haidt, J. (2001). The emotional dog and its rational tail: A social intuitionist approach to moral judgment. *Psychological Review*, *108*(4), 814–834.

Haidt, J. (2003). The moral emotions. In R. J. Davidson, K. R. Scherer, & H. H. Goldsmith (Eds.), *Handbook of affective sciences* (pp. 852–870). Oxford University Press.

Haidt, J., Koller, S. H., & Dias, M. G. (1993). Affect, culture, and morality, or is it wrong to eat your dog?. *Journal of personality and social psychology*, *65*(4), 613.

Helweg-Larsen, M. (2014). Does moralization motivate smokers to quit? A longitudinal study of representative samples of smokers in the United States and Denmark. *Nicotine & Tobacco Research*, *16*(10), 1379-1386.

Herz, R. S. (2011). PROP taste sensitivity is related to visceral but not moral disgust. *Chemosensory Perception*, *4*(3), 72.

Hutcherson, C. A., & Gross, J. J. (2011). The moral emotions: A social–functionalist account of anger, disgust, and contempt. *Journal of Personality and Social psychology*, *100*(4), 719-737.

Inbar, Y., & Pizarro, D. A. (2014). Pollution and purity in moral and political judgment. In H. Sarkissian, & J. Wright (Eds.), *Advances in experimental moral psychology* (pp. 111-129). Bloomsbury Academic.

Jack, R. E., Sun, W., Delis, I., Garrod, O. G., & Schyns, P. G. (2016). Four not six: Revealing culturally common facial expressions of emotion. *Journal of Experimental Psychology: General*, *145*(6), 708.

Kayyal, M. H., Pochedly, J., McCarthy, A., & Russell, J. A. (2015). On the limits of the relation of disgust to judgments of immorality. *Frontiers in Psychology*, *6*, 951.

Kollareth, D., & Russell, J. A. (2017). On the emotions associated with violations of three moral codes (community, autonomy, divinity). *Motivation and Emotion*, *41*(3), 322-342.

Kollareth, D., & Russell, J. A. (2019). Disgust and the sacred: Do people react to violations of the sacred with the same emotion they react to something putrid?. *Emotion*, *19*(1), 37.

Konishi, N., Himichi, T., & Ohtsubo, Y. (2020). Heart rate reveals the difference between disgust and anger in the domain of morality. *Evolutionary Behavioral Sciences, 14*(3), 284–298.

Kupfer, T. R., & Giner-Sorolla, R. (2017). Communicating moral motives: The social signaling function of disgust. Social *Psychological and Personality Science*, *8*(6), 632-640.

Kupfer, T. R., & Giner-Sorolla, R. (2021). Reputation management as an alternative explanation for the "contagiousness" of immorality. *Evolution and Human Behavior, 42*, 130-139.

Kupfer, T. R., Inbar, Y., & Tybur, J. M. (2020). Reexamining the role of intent in moral judgements of purity violations. *Journal of Experimental Social Psychology*, *91*, 104043.

Landy, J. F., & Goodwin, G. P. (2015). Does incidental disgust amplify moral judgment? A meta-analytic review of experimental evidence. *Perspectives on Psychological Science*, *10*(4), 518–536.

Maio, G. R., & Olson, J. M. (1998). Values as truisms: Evidence and implications. *Journal of Personality and Social Psychology*, *74*(2), 294.

McHugh, C., McGann, M., Igou, E. R., & Kinsella, E. L. (2017). Searching for moral dumbfounding: Identifying measurable indicators of moral dumbfounding. *Collabra: Psychology*, *3*(1), 23.

Molho, C., Tybur, J. M., Güler, E., Balliet, D., & Hofmann, W. (2017). Disgust and anger relate to different aggressive responses to moral violations. *Psychological Science*, *28*(5), 609-619. https://doi.org/10.1177/0956797617692000

Nabi, R. L. (2002). The theoretical versus the lay meaning of disgust: Implications for emotion research. *Cognition & Emotion*, *16*(5), 695-703.

Navarrete, C. D., & Fessler, D. (2003). Meat is good to taboo: Dietary proscriptions as a product of the interaction of psychological mechanisms and social processes. *Journal of Cognition and Culture*, *3*(1), 1-40.

Nummenmaa, L., Glerean, E., Hari, R., & Hietanen, J. K. (2014). Bodily maps of emotions. *Proceedings of the National Academy of Sciences*, *111*(2), 646-651.

Ortony, A., Clore, G. L., & Collins, A. (1990). *The cognitive structure of emotions*. Cambridge University Press.

Ottaviani, C., Mancini, F., Petrocchi, N., Medea, B., & Couyoumdjian, A. (2013). Autonomic correlates of physical and moral disgust. *International Journal of Psychophysiology*, *89*(1), 57–62.

Piazza, J., Russell, P. S., & Sousa, P. (2013). Moral emotions and the envisaging of mitigating circumstances for wrongdoing. *Cognition & Emotion, 27*(4), 707-722.

Ritter, R. S., Preston, J. L., Salomon, E., & Relihan-Johnson, D. (2016). Imagine no religion: Heretical disgust, anger and the symbolic purity of mind. *Cognition & Emotion, 30*(4), 778-796.

Rolfe, H., & Giner-Sorolla, R. (2020). Moral and non-moral anger and disgust in judgments of music. Unpublished manuscript. University of Kent.

Rottman, J., & Young, L. (2019). Specks of dirt and tons of pain: Dosage distinguishes impurity from harm. *Psychological Science, 30*(8), 1151-1160.

Royzman, E. B., Leeman, R. F., & Sabini, J. (2008). "You make me sick": Moral dyspepsia as a reaction to third-party sibling incest. *Motivation & Emotion, 32*(2), 100-108.

Royzman, E., Atanasov, P., Landy, J. F., Parks, A., & Gepty, A. (2014). CAD or MAD? Anger (not disgust) as the predominant response to pathogen-free violations of the divinity code. *Emotion, 14*(5), 892–907.

Rozin, P. (1999). The process of moralization. *Psychological Science, 10*(3), 218-221.

Rozin, P., & Singh, L. (1999). The moralization of cigarette smoking in the United States. *Journal of Consumer Psychology, 8*(3), 321-337.

Rozin, P., Millman, L., & Nemeroff, C. (1986). Operation of the laws of sympathetic magic in disgust and other domains. *Journal of Personality and Social Psychology, 50*(4), 703.

Rozin, P., Lowery, L., Imada, S., & Haidt, J. (1999). The CAD triad hypothesis: A mapping between three moral emotions (contempt, anger, disgust) and three moral codes (community, autonomy, divinity). *Journal of Personality and Social Psychology, 76*(4), 574–586.

Rubenking, B., & Lang, A. (2014). Captivated and grossed out: An examination of processing core and sociomoral disgusts in entertainment media. *Journal of Communication, 64*(3), 543-565.

Russell, P. S., & Giner-Sorolla, R. (2011a). Moral anger, but not moral disgust, responds to intentionality. *Emotion, 11*, 233-240.

Russell, P. S., & Giner-Sorolla, R. (2011b). Social justifications for moral emotions: When reasons for disgust are less elaborated than for anger. *Emotion, 11*, 637-646.

Russell, P. S., & Giner-Sorolla, R. (2013). Bodily-moral disgust: What it is, how it is different from anger and why it is an unreasoned emotion. *Psychological Bulletin, 139*, 328-351.

Sabo, J. S., & Giner-Sorolla, R. (2017). Imagining wrong: Fictitious contexts mitigate condemnation of harm more than impurity. *Journal of Experimental Psychology: General, 146*(1), 134.

Schein, C., & Gray, K. (2018). The theory of dyadic morality: Reinventing moral judgment by redefining harm. *Personality and Social Psychology Review, 22*(1), 32-70.

Schein, C., Ritter, R. S., & Gray, K. (2016). Harm mediates the disgust-immorality link. *Emotion, 16*(6), 862.

Scherer, K. R. (1984). On the nature and function of emotion: A component process approach. In K. R. Scherer & P. Ekman (Eds.), *Approaches to emotion* (pp. 293-318). Psychology Press.

Schnall, S., Haidt, J., Clore, G. L., & Jordan, A. H. (2015). Landy and Goodwin (2015) confirmed most of our findings then drew the wrong conclusions. *Perspectives on Psychological Science, 10*(4), 537-538.

Sell, A., Tooby, J., & Cosmides, L. (2009). Formidability and the logic of human anger. *Proceedings of the National Academy of Sciences, 106*(35), 15073-15078.

Shweder, R. A., Much, N. C., Mahapatra, M., & Park, L. (1997). The 'big three' of morality (autonomy, community, and divinity) and the 'big three' explanations of suffering. In A. M. Brandt & P. Rozin (Eds.), *Morality and health* (pp. 119–169). Routledge.

Simpson, J., Carter, S., Anthony, S. H., & Overton, P. G. (2006). Is disgust a homogeneous emotion?. *Motivation and Emotion, 30*(1), 31-41.

Skarlicki, D. P., Hoegg, J., Aquino, K., & Nadisic, T. (2013). Does injustice affect your sense of taste and smell? The mediating role of moral disgust. *Journal of Experimental Social Psychology, 49*(5), 852-859.

Tracy, J. L., Steckler, C. M., & Heltzel, G. (2019). The physiological basis of psychological disgust and moral judgments. *Journal of Personality and Social Psychology*, *116*(1), 15.

Turiel, E. (1977). Distinct conceptual and developmental domains: social convention and morality. *Nebraska Symposium on Motivation*, *25*, 77-116.

Tybur, J. M., Lieberman, D., Kurzban, R., & DeScioli, P. (2013). Disgust: Evolved function and structure. *Psychological Review*, *120*(1), 65-84.

Yoder, A. M., Widen, S. C., & Russell, J. A. (2016). The word disgust may refer to more than one emotion. *Emotion*, *16*(3), 301.

Part III
Disgust and Its Applications

Chapter 9
The Realm of Disgust in Sexual Behaviour

Charmaine Borg and Peter J. de Jong

The Goal of Pleasure and Reproduction Versus Remaining Disease and Contagion Free

Sex and sexual behaviour are a core part of life; besides reproduction, sex offers opportunities for pleasure and can be beneficial for bonding, intimacy and for many facets of one's life. However, along with these desirable features, sex and sexual behaviour can also have the clear down side of providing ample opportunities for contracting infectious diseases. To elaborate on this high risk of contagion, let us consider one of the simplest sexual activities, French kissing. This activity entails the sharing of saliva, with the capacity of more than 80 million bacteria in a single 10-s kiss transferred to the other person (Kort et al. 2014). Other sexual activities, such as intercourse, and/or coming into physical contact with the ejaculate and vaginal fluids, may similarly pose a high risk of contamination (Curtis 2013; Kort et al. 2014). Thus, the inherent contagious nature of sexual behaviours and sexual (by)-products may help explain why sexual stimuli, may also be generally considered as potent disgust elicitors (Rozin et al. 1995).

Disgust is thought to serve the evolutionary function of self-protection (Rozin et al. 2008) and disease avoidance (Oaten et al. 2009; see also Bradshaw and Gassen, Chap. 3, this volume). By the expressed disgust-driven inhibitory tendencies, operating via defensive reflexes, or by actively motivating avoidance (or withdrawal) of disgust-evoking stimuli, disgust is thought to protect us from contamination by non-visible pathogens (Curtis et al. 2011; Oaten et al. 2009). In line with the protective function of disgust, it has been shown that disgust propensity varies as a function of vulnerability to disease. For instance, Fessler et al. (2005) found a temporary

C. Borg (✉) · P. J. de Jong
Department of Clinical Psychology and Experimental Psychopathology, University of Groningen, Groningen, Netherlands
e-mail: c.borg@rug.nl; p.j.de.jong@rug.nl

P. A. Powell, N. S. Consedine (eds.), *The Handbook of Disgust Research*,
https://doi.org/10.1007/978-3-030-84486-8_9

increase in disgust propensity during the first trimester of pregnancy, during which the risk of afflicting diseases is highest. Clearly, disgust-induced avoidance of sexual stimuli may be helpful to reduce contamination risks, yet it seems incompatible with sexual pleasure and the functional goal of procreation. This raises the question of how the conflict between the opposing goals of disease avoidance and procreation can be resolved. How do people generally succeed in having pleasurable sex in the face of the disgust eliciting properties of the stimuli inherently involved in sexual behaviours?

The Effect of Sexual Arousal on Disgust and Its Characteristic Avoidance

Considering that both disease avoidance and procreation are of paramount evolutionary importance, there should be a mechanism that facilitates pleasurable and functional sexual experiences. One hypothesis that has been put forward is that sexual arousal may temporarily reduce feelings of disgust; thus, to the extent that sexual stimuli elicit arousal, this may counteract or neutralise the disgust eliciting properties of sex (Koukounas and McCabe 2001), thereby facilitating sexual approach (see also Fig. 9.1). Consistent with the assumption that sexual arousal might temporally reduce disgust, sexually aroused male students were found to report less subjective disgust in response to sex-related disgust elicitors than unaroused participants (Stevenson et al. 2011). A follow up study replicated and extended this pioneering work in female students (Borg and de Jong 2012). Again, experimentally heightened sexual arousal decreased disgust in response to sex-related disgust elicitors and reduced disgust-induced avoidance (Borg and de Jong 2012). Thus, sexual arousal not only reduced subjective appraisals but also behavioural avoidance of (sexual) disgust elicitors. In other words, it appears that sexual arousal may not only counteract the subjective perception of disgust but might also transform the disgust-induced avoidant and inhibitory tendencies into increased tendencies- and willingness to approach.

These findings suggest that sexual arousal has the power to override disgust-driven inhibitions/avoidance. It is thus possible that sexual arousal induces approach tendencies towards certain sexual stimuli and behaviours that would be inhibited in a non-sexually aroused state. In line with this argument, in a within-subjects study design, it has been shown that male participants ($N = 24$) in a sexually aroused state were significantly more open towards several sexual activities and behaviours that evoked repulsion in the absence of experimentally induced sexual arousal (e.g., "have sex with someone who is extremely fat", or "getting sexually excited by contact with an animal"; Ariely and Loewenstein 2006). This indicates that sexual arousal might also reduce the prerequisites that potential sex mates need to fulfil for sexual appeal.

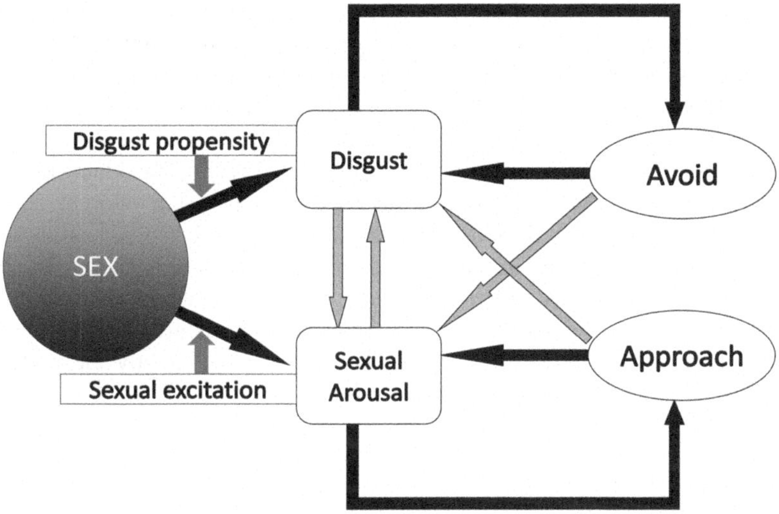

Fig. 9.1 Giving in to arousal or staying stuck in disgust? Disgust-based mechanisms in sex and sexual dysfunction. Model adapted from de Jong et al. (2013). Black arrows indicate excitation, whereas grey arrows refer to inhibition. The model illustrates how sexual arousal and sexual disgust are interrelated and jointly affect sexual behaviour resulting in either sexual avoidance, when disgust dominates, or sexual approach, when arousal outweighs disgust. Some sex-related stimuli are assumed to be inherently disgusting due to their associated contamination risk. Other sex stimuli, that become apparent somewhere around the process of puberty, are associated with sexual readiness (and a good fitness/healthy status). The latter sex stimuli have the potential to trigger sexual arousal that are expected to override disgust elicited by concurrently available sex stimuli that are somehow associated with contamination threat. Individual differences in trait disgust propensity will moderate the strength of the sex-disgust relationship, whereas individual differences in trait sexual excitation/arousability will moderate the relationship between sex and sexual arousal

The repulsion reported by the male participants in the unaroused condition is referred to as "sexual disgust", which, from an evolutionary perspective, has the function of assisting in the selection of an appropriate sexual partner, and to avoid sex partners of low intrinsic qualities and poor genetic make-up (Tybur et al. 2009; see also Tybur, Chap. 6, this volume). Sexual disgust is also thought to protect us from engaging in intercourse with someone that is genetically too close to us (e.g., father, mother, sister, or brother). Put simply, imagining sexual acts with each of these categories typically elicits disgust, and disgust promotes rejection and avoidance (Tybur et al. 2009). Thus, in combination, the available findings indicate that sexual arousal might temporally reduce both pathogen (or core) and sexual disgust.

The Weakening Influence of Disgust on Sexual Arousal

Consistent with the model depicted in Fig. 9.1, recent work indicates that the obstructive relationship between sex and disgust is bidirectional, meaning that not only sexual arousal reduces disgust, but induced disgust may also weaken sexual arousal (e.g., Borg et al. 2019b). It has, for example, been shown that sexually explicit images elicited less self-reported sexual arousal when primed by disgusting pictures (Andrews et al. 2015). Following this idea, a recent experimental study showed that sexual arousal elicited by an erotic movie could be weakened by prior exposure to an aversive, disgusting odour (Borg et al. 2019b). The odour (associated with rotten food, rotten eggs, and tetrahydrothiophene - the odorant added to cooking gas) resulted in a decrease of both subjective and genital sexual arousal compared to the participants in the odourless control condition. These two studies support the view that disgust has an inhibitory effect on sexual arousal. Together, the available evidence suggests that sexual arousal and disgust have a mutually inhibiting relationship, however, as yet it is unclear what exactly the underlying mechanism of this bidirectional relationship is.

The Paradox

The available evidence seems to indicate that prior to the surge in testosterones in prepubertal boys and girls; these are typically repelled by intimate sexual behaviours such as French kissing. However, in spite of the disgust eliciting properties of sex and sexual behaviours, yet consistent with the function of procreation and sexual pleasure, at some stage most people generally become interested and get involved in intimate and sexual physical contact. The key to solving this paradox can be found in the capacity to become sexually aroused around the age that children become physically ready for reproduction. Probably due to the surge in testosterones in early adolescence, children may at that time acquire sexual approach motivation, together with sensitivity for sexual cues indicative of high mating value (e.g., breasts, blushed skin, particular body shapes, and body postures). The emergence of sexual cues may accentuate the sexual appeal and elicit sexual arousal at the time proximate to the process of puberty due to the surge in the gonadal hormones (Borg et al. 2019a; Peper and Dahl 2013). The gonadal hormones, which are known to contribute to the typical physical and behavioural changes in puberty, only start to be released by the age of 9–10 years in girls and 10–12 years in boys (Peper and Dahl 2013). Surrounding this age, these sex-related stimuli signalling a healthy status and reproduction fitness, may then become motivationally salient.

This process of transition from childhood into adolescence leads to the onset of biologically triggered sexual motivation. When sexual interest and sexual motivation commence, relevant sexual cues can elicit sexual excitation. In turn, this sexual excitation may (temporarily) reduce the inhibitory forces and avoidance behaviour that may at least partly be driven by the default disgust response towards sexual stimuli and behaviours that may signal contamination risk (Borg et al. 2019a; Koukounas and McCabe 2001).

It has been argued that testosterones impact the ability to become sexually aroused, and on the willingness to approach stimuli previously considered disgusting or avoided. In the meantime, testosterones may also impact on people's general disgust propensity as a protective and compensatory mechanism for the increased risks to infections that comes as part and parcel with becoming autonomous and sexually active (Curtis 2013). In line with this notion, two cross-sectional studies (samples merged, $N = 248$, 6–17 years, 137 female students) using scenario-based measurements showed that general disgust propensity was overall higher in early adolescence (12–14 years) than in pre-adolescence (9–11 years; Borg et al. 2019a). Although this finding is consistent with the view that the heightened level of testosterones in early adolescence might be the motor behind the heightened general disgust propensity in this age group, it should be acknowledged that testosterone levels were not assessed in this study.

Consistent with the view that the ability to get sexually aroused may counteract the default sex-disgust response, it was found that disgust propensity towards sex-relevant disgust elicitors declined from pre-adolescence to early and middle adolescence (Borg et al. 2019a). In line with the view that this would serve the functional goal of procreation and sexual pleasure, such decline was restricted to contexts where peers and non-family members were the source of the sex-relevant disgust elicitors. Coherent with the view that sexual disgust might help prevent sexual approach of individuals with high genetic similarity (Tybur et al. 2009; see also Tybur, Chap. 6, this volume), disgust propensity increased from pre- to middle adolescence when parents were the primary source of the sex-relevant disgust elicitors (Borg et al. 2019a). The reduced disgust-eliciting properties of sexual behaviours may lower the threshold for adolescents to become involved in sexual behaviours. In turn, this weakened threshold may further contribute to a reduction of the disgust-eliciting properties of sexual stimuli that are involved in sex by habituation, thereby facilitating a healthy and typical sexual development. A recent online cross-sectional study ($N = 240$, 116 males, 9–17 years) provided further support for this idea. In this study we recorded the responses of youths towards 20 pictures (including inherently disgusting and sex-related stimuli) and six sexual behaviour scenarios (previously used in the study of Borg et al. 2019a), presented on visual analogue scales for the dimensions of liking and disgust. Preliminary data suggest that the pattern of findings of this online study is very close to the first two studies conducted in a classroom context (Oosterwijk et al. 2020). Self-reported disgust towards sex-related stimuli and behaviours (e.g., kissing) again followed a linear age trend, with sex-related disgust decreasing from the pre-adolescence group to the middle adolescence group.

An important next step would be to follow up these cross-sectional studies with a longitudinal approach. Longitudinal studies with repeated self-reported assessments of the disgust response towards sexual behaviours, together with repeated assessments of concrete sexual approach behaviours, are necessary to examine whether the current findings provide an adequate representation of people's actual developmental trajectories. Further, such studies may help to arrive at concrete conclusions in explaining the overall decline of (sex-relevant and sex-irrelevant) disgust from early to middle adolescence.

Besides assessing the disgust eliciting properties of particular sexual behaviours and sex stimuli that may signal contamination risk, it would also be important to concurrently assess the sexual arousing properties of stimuli that are proposed to be relevant for sexual appeal (e.g., body features such as breasts, and specific body postures). Such information would help determine what factors are critically involved in sexual approach and whether the decline in disgust eliciting properties comes as a result of the heightened arousing properties of the stimuli that are relevant for sexual appeal. Additionally, it would be relevant to know the extent to which involvement in sexual behaviour contributes to a further decline in sexual disgust, from one sexual encounter to the next, and/or from one partner to a new partner. Finally, it would be helpful to test whether changes in arousal/disgust/sexual approach can be mapped on to the surge in testosterone during early adolescence.

If indeed testosterone levels play a crucial role in the shift from disgust-driven avoidance to sexual approach, it would be useful to test whether blocked testosterones have an intensifying effect on the disgust properties of sex and non sex-related disgusting stimuli and whether any shifts in disgust are translated into actual avoidance behaviours. The treatment of patients that are diagnosed with prostate cancer, offers the opportunity to study this relationship in a naturalistic framework. These patients are typically subjected to testosterone blockers as part of Androgen Deprivation Therapy (ADT), which leads to chemical castration. This type of treatment follows a typical trajectory. Patients begin with normal testosterone levels, however, following administration of ADT treatment, testosterone peaks at 3 days and then drops to castration levels 4 weeks later. Thus, this group of patients offers the opportunity to study the impact of heightened and lowered testosterones on sexual disgust within participants. To test whether avoidance is indeed intensified in the context of low testosterone, automatic approach and avoidance tendencies may be measured in an Approach Avoidance Task (AAT) in response to stimuli that are typically considered as sexually appealing and sex stimuli that may signal contamination risk (Hinzmann et al. 2019). Such an approach has the potential to help in understanding the impact of testosterones on the disgust properties of some sexual stimuli and to inform patients and their partners about what to expect following ADT.

The Evolution of the Disgust-Evoking Properties of Some Sexual Stimuli

As alluded to earlier in this chapter, sexual behaviour entails two categories of stimuli; those that promote reproductive health and signal sexual appeal (e.g., breasts) and stimuli that signal contamination thereby supporting the function of disease avoidance (e.g., saliva). Besides, some stimuli might fall in between these two categories (e.g., lips). Based on this categorisation, some specific sex stimuli are in general likely to maintain their inherently disgusting status, whereas other sex stimuli are likely to attract sexual appeal and trigger sexual excitation. Furthermore, sex stimuli may change their conditioned status either to a positive one following a positive sexual experience/ pleasure/exposure, or to a negative/aversive one, after a negative sexual experience, or because sex/specific sexual behaviour occurred in the absence of sexual excitation.

Thus far, research on the relationship between developmental stage and the transition from sexual disgust to sexual desire (e.g., Borg et al. 2019a; Oosterwijk et al. 2020) has focused on potential changes in adolescents' disgust responses towards sex related stimuli and behaviours that may signal contamination risks (Curtis 2013). However, these studies have not focused on stimuli that may acquire sexual appeal. It thus remains to be tested how the arousal-eliciting properties of particular stimuli develop, and which stimuli are exactly involved/most critical in this acquisition. In addition, it would be important to examine how the change in sexual appeal of sex-relevant stimuli relates to the change in sexual disgust, as well as how these processes relate to and contribute to the rise of actual sexual behaviours.

To the extent that adolescents get involved in sexual approach behaviour, habituation of disgust is expected given that prolonged physical contact with disgusting stimuli is a potent way to reduce the disgust eliciting properties of otherwise initially disgusting stimuli (de Jong 2013; Rozin and Fallon 1987). Repeated exposure to sex-relevant disgust elicitors may even result in a more permanent reduction of disgust, thereby lowering the threshold for sexual approach. Repeated assessment of both sexual behaviour and the disgust eliciting status of particular sex-relevant stimuli would allow for the assessment of whether prolonged contact with such stimuli would indeed be followed by reduced disgust. Repeated exposure with the same partner would eventually also result in a pattern of equalisation of the bacteria thereby reducing contamination threat. Therefore, the decline in disgust eliciting properties of the sex-related stimuli will probably be most pronounced when repeated exposure occurs with the same sex-partner. Such a pattern would also be in line with the more general phenomenon that the disgust eliciting properties of stimuli highly depend on the stimulus' source; using a tooth brush of your sibling is typically less disgusting than using a tooth brush of an unknown person (Borg et al. 2019a; Peng et al. 2013).

It would be interesting and helpful to combine the responses to specific sex stimuli that are selected based on their contamination potential, with the testing of the appealing properties of non-contamination signal type (of gender specific) sex stimuli (such as breasts). This would allow identification of where an increase in attraction may promote sexual arousal and excitation and examination into how the change in sexual appeal relates to the disgust evoking properties of contamination-signalling sex stimuli. Perhaps it would be possible to identify trajectories that are involved in straightforward sexual development compared with trajectories that may relate to sexual problems and dysfunctions. Insight into such trajectories together with individual difference variables that may set people at risk for unfavourable trajectories might improve insights into factors involved in a healthy vs. more problematic sexual development. In addition, it may provide clues to improve current available treatment options for sexual dysfunctions.

Applications: Disgust Based Interventions and Gaps for Future Research

Involving disgust in the functional analyses of sexual problems and in promoting pleasurable sex might be especially important because disgust does not only influence the way we feel, but, as already discussed in this chapter, characteristically promotes avoidance and escape behaviour (Oaten et al. 2009). In light of what we have discussed earlier, the latter are very likely to weaken sexual pleasure, because every opportunity for contact with sexual stimuli is prevented. Consequently, this prevention of contact with sexual stimuli eliminates any possibility to focus and elaborate on, such stimuli for appraising it as sexually stimulating.

In the last decade, the role of disgust in several psychopathologies, has received increasing support (see Davey, Chap. 11, this volume), igniting several exposure-focussed interventions targeting disgust. Unfortunately, a prevalent finding has been that disgust is hard to unlearn (Bosman et al. 2016; see also Reynolds and Askew, Chap. 5, this volume). A possible explanation is that it is relatively difficult to disprove the contaminating properties of the stimuli in question, because the threat concerns pathogenic stimuli that cannot be directly detected by our senses. This might have geared the disease avoidance system to being overly conservative (de Jong and Borg 2019). This may help explain why exposure to stimuli that have been in contact with disgust elicitors are relatively ineffective in reducing people's avoidance to these stimuli (see e.g., Tolin et al. 2004). Furthermore, it may also help to explain why exposure is relatively ineffective in reducing disgust-induced avoidance of stimuli that previously acquired predictive value for disgusting outcomes (Borg et al. 2015).

Taking all of the above limitations into consideration, for over a decade it has nevertheless been acknowledged that prolonged exposure to inherently disgusting stimuli seems to be the most efficient manner to reduce their disgust-eliciting status (e.g., Meunier and Tolin 2009). Exposure tasks should be designed in such a way that they provide safety information and reduce the signal value of the target stimuli as a sign of contamination threat; prolonged physical contact seems therefore the most critical component (Borg et al. 2015; Bosman et al. 2016). In this context, it is especially important to reduce avoidance and safety behaviours, which in the long run maintain the associated disgust reactions to the stimuli. Research in the context of fear conditioning has shown that safety behaviours undermine the efficacy of extinction procedures to reduce learned fear responses (Lovibond et al. 2009), and it seems reasonable to assume that safety behaviours in the context of exposure to disgust eliciting stimuli will similarly undermine the efficacy of exposure to reduce disgust.

Safety behaviour in this context can be expressed as either completely avoiding all sexual stimuli or only participating in sexual situations with certain items, stimuli, or with behaviours that signify safety and temporarily reduce disgust (e.g., using fabrics to avoid contact with bodily fluids during sexual activities). Although such behaviour might momentarily decrease participants' levels of disgust, these safety behaviours preclude prolonged direct physical contact and thus hinder the effectiveness of exposure in the long-term (Borg et al. 2011, 2020).

To investigate further how to optimally reduce (sexual) disgust, we designed a lab model of sexual disgust (Pawlowska et al. 2020). Similar to the traditional lab model of phobic fear, we used a classical conditioning procedure to facilitate acquisition/learning of disgust associations towards initially sexually arousing stimuli. During the acquisition stage, erotic films served as the Conditioned Stimuli (CS+ and CS−), where the CS+ film was repeatedly paired with a disgust film (Unconditioned Stimulus, US), and the CS− was not. Supporting the validity of this lab model of sexual disgust learning, the CS+ was rated as significantly more disgusting, less sexually arousing, and less pleasant than the CS−. Following this differential disgust conditioning procedure, the CS+ also elicited lower genital arousal than the CS− as indexed by vaginal pulse amplitude (VPA). Using this lab model, we recently tested the impact of mere exposure versus counterconditioning (in which the CS+ was paired with stimuli of high positive valence including pictures of romantic couples, palatable food, puppies, babies, and young animals) on acquired sexual disgust (Pawlowska et al. 2020). Both CS− only exposure (extinction) and the counterconditioning procedure were effective in eliminating the conditioning-induced differential physiological responsivity elicited by the CS+ and CS− as indexed by VPA. This appeared however not to be due to an increase in physiological arousal elicited by the CS+ from post-acquisition to extinction, but seems to be due to a general decline of the VPA to the erotic clips that already became evident during acquisition. This might have reduced the sensitivity of the VPA as a measure to detect more subtle differences between the CS+ and CS- over the course of the experiment.

In line with this, sexual arousal as indexed by a self-report measure showed a slightly different pattern in response to the extinction/counterconditioning procedures. Although both procedures were effective in reducing the acquired difference in subjective sexual arousal elicited by the CS+ versus the CS−, post extinction/counterconditioning the CS+ remained less sexual arousing than the CS−. A similar pattern was evident for the self-reported feelings of disgust. Although the acquired disgusting properties of the CS+ again reduced following the extinction and counterconditioning procedure, the CS+ remained more disgusting than the CS− (Pawlowska et al. 2020). The failure to fully eliminate the acquired disgust to the CS+ was paralleled with a failure to fully restore the sexually arousing properties of the CS+, which is consistent with the view that feelings of disgust may counterforce the generation of sexual arousal (as also proposed in the model depicted in Fig. 9.1).

In addition, the failure of both the CS-only extinction and the counterconditioning procedure to fully eliminate the acquired disgusting properties of the erotic clip that was used as the CS+ is consistent with previous findings indicating that disgust is relatively resistant to extinction procedures (e.g., Bosman et al. 2016). The findings of Pawlowska et al. (2020) also point to the relevance of coming up with fresh candidate strategies to target sexual disgust next to exposure and counterconditioning, which can then be tested within the context of this lab model (Pawlowska et al. 2020).

One possible candidate approach to target sexual disgust might be conceptual reorientation (Rozin and Fallon 1987). This refers to a "cognitive switch" in the conceptualisation and understanding of objects in a way that they no longer elicit disgust, by changing their core appraisal into something that is pleasurable or healthy and functional. For example, conceptual reorientation was used for a woman who indicated that she felt disgusted by "this slimy stuff" in her vagina by explaining that, in the absence of lubrication, a penis would feel like sandpaper. The reorientation of vaginal lubrication from being a highly aversive (bad-smelling and bad-tasting) fluid to a very helpful and positively valanced supporting substance appeared quite successful in modifying the originally dominant disgust appraisals and in reducing sexual avoidance (for detailed examples see de Jong et al. 2010). Reframing and challenging techniques from cognitive behavioural therapy may generally be helpful in changing relevant sex stimuli or behaviours from a disgust frame to promote pleasurable/rewarding or functional interpretations of these stimuli.

Such interventions can also be included within an emotional regulation framework. Using emotional regulation techniques, we tested whether a sexual arousal up-regulation strategy (the instruction to let one's emotion roll whilst watching an erotic movie) might effectively enhance feelings of sexual arousal and decrease feelings of disgust in response to sexual stimuli (van Overveld and Borg 2015). Indeed, brief sexual arousal up-regulation training appeared helpful in enhancing sexual arousal. However, this increase in sexual arousal was not paralleled with a decrease in disgust (van Overveld and Borg 2015). It needs to be said, however, that this study was conducted with sexually asymptomatic individuals who showed very weak disgust when exposed to the erotic movie to begin with.

In a follow up study, we successfully replicated the finding that up-regulation of sexual arousal could be successfully applied to increase arousal when exposed to an erotic video (Pawlowska et al. 2021; study 1). However, again, this strategy appeared ineffective in reducing feelings of disgust. Thus at least in the context of women without sexual problems and limited feelings of disgust when exposed to erotic materials, we failed to find evidence for successful sexual arousal upregulation to concurrently reduce feelings of disgust. This follow up study did however find some evidence for the inhibitory influence of disgust on the generation of sexual arousal. That is, it was found that the arousal up-regulation appeared ineffective in increasing sexual arousal when prior to the presentation of the erotic clip, the contaminating properties of sex were primed. Thus, making the disgust-relevant properties of sex more salient interfered with the subsequent efficacy of arousal up-regulation instructions. Together this implies not only that arousal up-regulation is not effective in reducing feelings of disgust, but it also implies that such up-regulation interventions might not be very helpful in increasing sexual arousal when sexual arousal problems are due to relatively strong pre-existing feelings of disgust. These cases suggest instead that the disgust eliciting properties of sexual stimuli should be reduced first, before aiming to increase sexual arousal.

We therefore also tested if disgust down-regulation strategies might perhaps be more effective than sexual arousal up-regulating strategies to reduce feelings of disgust. Findings indicated that indeed an emotion regulation strategy designed to reduce disgust was effective in reducing feelings of disgust when exposed to an erotic video (Pawlowska et al. 2021; study 2). However, this study only included asymptomatic participants and thus it remains to be tested whether such an intervention would also be effective in reducing disgust in individuals with sexual problems related to sexual disgust. As an important next step, it is relevant to test these (or similar) interventions in symptomatic samples to explore their clinical relevance. Further, it needs to be mentioned, that reducing or even completely eliminating disgust may not be sufficient to lead to pleasurable sex. Such a strategy may just take away one barrier to sexual excitation; pleasurable experiences and sufficient stimulation may still be required for sexual enjoyment.

Weakening Sexual Approach

The interventions mentioned up until now are mainly focused on understanding and/or weakening disgust responses in favour of promoting sexual approach. However, as we already alluded to in a previous section, disgust can also be applied to do the opposite; to weaken sexual approach (e.g., Borg et al. 2019b). Such an approach may have practical relevance in cases where the sexual approach in question is unwarranted or undesired and necessitates to be stopped (e.g., rape). In a recent experimental study among male students ($N = 78$, $M = 21.1$ years, $SD = 2.1$), we showed that sexual arousal elicited by an erotic movie could be weakened by exposure to a highly disgust-eliciting odour (Borg et al. 2019b). The odour resulted in a decrease of both subjective and genital sexual arousal compared to the participants in the odourless control condition. This study provides initial support for the relevance of using disgusting odours to undermine undesired behavioural actions motivated by sexual arousal (driven by e.g., poor judgment, coercive sexual behaviour; Borg et al. 2019b).

An important extension of the current findings would be to examine whether odour induced disgust might also attenuate automatic sexual approach behaviour (e.g., Hinzmann et al. 2019), and whether the findings in the lab can be replicated in an actual sexual context. As a first step, and partially to replicate these findings, one may use a scenario-based study in which participants are asked about their willingness to approach their preferred sexual stimuli under high and low contagion. Next, by using a virtual reality lab approach in which participants are given choices about their inclination / motivation to (sexual) approach behaviours in various types of situations under conditions of high vs. low disgust, the actual approach behaviour could be tested. Finally, in the previously described study the sample was composed only of men, thus replicating this work with women would test the robustness and generalisability of the findings.

Synthesis and Conclusions

In this chapter, we described how sexual behaviours entail confrontation with stimuli that both promote reproductive health and signal sexual appeal, and stimuli that also signal contamination and disease. Sexual behaviours and some sexual stimuli may elicit disgust responses and consequently hinder sexual arousal, whereas other sex stimuli, generally those associated with sexual appeal, may help in generating sexual arousal and are expected to weaken the disgust response related to the contaminating properties of sex. We explored pathways in understanding how people become involved in sex and sexual behaviours and discussed when the central sexual stimuli signal contagion. We also explored interventions that can be used in accentuating sexual arousal and weakening disgust, as well as novel interventions to weaken sexual arousal when this is undesired. These theoretically driven interventions are novel, however, the evidence supporting them is preliminary and further work in the area is required. In this chapter, we outlined these interventions to provide a contemporary perspective about the possible applications of disgust-based interventions in the context of human sexual expression. We also outlined a lab model of sexual disgust (Pawlowska et al. 2020), which offers a platform to test the relative efficacy of interventions aimed at reducing disgust for sexual stimuli. We hope that this chapter is helpful in inspiring future research to arrive at more conclusive answers about the mechanisms involved in the development and persistence of sexual disgust and about the procedures that are most effective in adjusting the invalidating effects of sexual disgust.

References

Andrews, A. R., Travis, C., Cholka, B. C., Cooper, V. T., & Bridges, J. A. (2015). Correlational and experimental analyses of the relation between disgust and sexual arousal. *Motivation and Emotion, 39*, 766-779.

Ariely, D., & Loewenstein, G. (2006). The heat of the moment: The effect of sexual arousal on sexual decision making. *Journal of Behavioral Decision Making, 19*, 87-98.

Borg, C., & de Jong, P. J. (2012). Feelings of disgust and disgust-induced avoidance weaken following induced sexual arousal in women. *PLoS ONE, 9*, e44111.

Borg, C., de Jong, P. J., & Weijmar Schultz, W. C. M. (2011). Vaginismus and dyspareunia: Relationship with general and sex-related moral standards. *Journal of Sexual Medicine, 8*, 223–231.

Borg, C., Bosman, C. R., Engelhard, I., Olatunji, B. O., & de Jong, P. J. (2015). Is disgust sensitive to classical conditioning as indexed by facial EMG and behavioral responses? *Cognition and Emotion, 30*, 669-86.

Borg, C., Hinzmann, J., Heitmann, J., & de Jong, P. (2019a). Disgust toward sex-relevant and sex-irrelevant stimuli in pre-, early, and middle adolescence. *Journal of Sex Research, 56*, 102-113.

Borg, C., Oosterwijk, T. A. , Lisy, D., Boesveldt, S., & de Jong, P. J. (2019b) The influence of olfactory disgust on (Genital) sexual arousal in men. *PLoS ONE, 14*(2), e0213059.

Borg, C., Both, S. S., ter Kuile, M. M., & de Jong, P. (2020). Sexual aversion. In K. S. K. Hall & Y. M. Binik (Eds.), *Principles and Practice of Sex Therapy* (6th ed., pp. 224-238). Guildford Press.

Bosman, R., Borg, C., & de Jong, P. J. (2016). Optimizing the extinction of conditioned disgust. *PLoS ONE, 11*, e0148626.

Curtis, V. (2013). *Don't Look, Don't Touch - the Science Behind Revulsion* (1st ed.). Oxford University Press.

Curtis, V., de Barra, M., & Aunger, R. (2011). Disgust as an adaptive system for disease avoidance behaviour. *Philosophical Transactions of the Royal Society Biological Sciences, 366*, 389-401.

de Jong, P. J. (2013). Learning mechanisms in the acquisition of disgust. In D. Hermans, B. Rime, & B. Mesquita (Eds.), *Changing Emotions* (pp. 74–80). Psychology Press.

de Jong, P. J., & Borg, C. (2019). Disgust sensitivity. In J. S. Abramowitz & S. M. Blakey (Eds.), *The Clinical Handbook of Fear and Anxiety: Psychological Processes and Treatment Mechanisms.* American Psychological Association.

de Jong, P. J., van Lankveld, J., Elgersma, H. J., & Borg, C. (2010). Disgust and sexual problems: Theoretical conceptualization and case illustrations. *International Journal of Cognitive Therapy, 3,* 24-41.

de Jong, P. J., van Overveld, M., & Borg, C. (2013). Giving in to arousal or staying stuck in disgust: Disgust-based mechanisms in sex and sexual dysfunction. *Annual Review of Sex Research, 50,* 247-262.

Fessler, D. M., Eng, S. J., & Navarrete, C. D. (2005). Elevated disgust sensitivity in the first trimester of pregnancy: Evidence supporting the compensatory prophylaxis hypothesis. *Evolution and Human Behavior, 26*, 344-351.

Hinzmann, J., Borg, C., Verwoerd, J. R. L., & de Jong, P. (2019). The reciprocal relationship between sexual arousal and disgust as evidenced in automatic approach-avoidance behavior. *Journal of Sex Research. 57*, 384-396.

Kort, R., Caspers, M., van de Graaf, A., van Egmond, W., Keijser, B., & Roeselers, G. (2014). Shaping the oral microbiota through intimate kissing. *Microbiome, 2*, 1-8.

Koukounas, E., & McCabe, M. P. (2001). Sexual and emotional variables influencing sexual response to erotica: A psychophysiological investigation. *Archives of Sexual Behavior, 30*, 393-408.

Lovibond, P. F., Mitchell, C. J., Minard, E., Brady, A., & Menzies, R. G. (2009). Safety behaviours preserve threat beliefs: Protection from extinction of human fear conditioning by an avoidance response. *Behaviour Research and Therapy, 47*, 716–720.

Meunier, S. A., & Tolin, D. F. (2009). The treatment of disgust. In: B. O. Olatunji & D. McKay (Eds.). *Disgust and Its Disorders* (pp. 9-29). American Psychological Association.

Oaten, M., Stevenson, R. J., & Case, T. I. (2009). Disgust as a disease-avoidance mechanism. *Psychological Bulletin, 135*, 303-321.

Oosterwijk, T., Borg, C., & van Dijk, M. (2020). *Explorative study investigating the subjective disgust responses towards core disgust, sex, and food stimuli in children between the ages of 9-17.* Manuscript submitted for publication.

Pawlowska, A., Borg, C., de Jong, P. J., & Both, S. (2020). The effect of differential disgust conditioning and subsequent extinction versus counterconditioning procedures on women's sexual responses to erotic stimuli. *Behaviour Research and Therapy, 134,* 103714.

Pawlowska, A., Borg, C., & de Jong, P. J. (2021). Up-regulating sexual arousal and down-regulating disgust while watching pornography: Effects on sexual arousal and disgust. *Journal of Sex Research, 58*, 353-363.

Peng, M., Chang, L., & Zhou, R. (2013). Physiological and behavioral responses to strangers compared to friends as a source of disgust. *Evolution and Human Behavior. 34,* 94–98.

Peper, J. S., & Dahl, R. E. (2013). The teenage brain: Surging hormones-brain-behavior interactions during puberty. *Current Directions in Psychological Science, 22*, 134-139.

Rozin, P., & Fallon, A. E. (1987). A perspective on disgust. *Psychological Review, 94,* 23-41.

Rozin, P., Nemeroff, C., Horowitz, M., Gordon, B., & Voet, W. (1995). The borders of the self: contamination sensitivity and potency of the body apertures and other body parts. *Journal of Research in Personality*, *29*, 318–340.

Rozin, P., Haidt, J., & McCauley, C. R. (2008). Disgust: The body and soul emotion in the 21st century. In: B. O. Olatunji & D. McKay (Eds.). *Disgust and Its Disorders* (pp. 9-29). American Psychological Association.

Stevenson, R. J., Case, T. I., & Oaten, M. J. (2011). Effect of self-reported sexual arousal on responses to sex-related and non-sex-related disgust cues. *Archives of Sexual Behavior*, *40*, 79-85.

Tolin, D. F., Worhunsky, P., & Maltby, N. (2004). Sympathetic magic in contamination related OCD. *Journal of Behavior Therapy and Experimental Psychiatry*, *35*, 193–205.

Tybur, J. M., Lieberman, D., & Griskevicius, V. (2009). Microbes, mating, and morality: Individual differences in three functional domains of disgust. *Journal of Personality and Social Psychology*, *97*, 103–122.

van Overveld, M., & Borg, C. (2015). Brief emotion regulation training facilitates arousal control during sexual stimuli. *The Journal of Sex Research*, *52*, 996-1005.

Chapter 10
Disgust, Prejudice, and Stigma

Lenny R. Vartanian, Tayla B. McCutcheon, and Sarah A. Rubenstein

Prejudice is a significant social issue, perpetrated and experienced by many people around the world. In its most benign form, prejudice is merely a heuristic used to conserve mental energy; at its worst, prejudice can lead to biased treatment of individuals or groups, unfair social and political structures, and various forms of abuse, oppression, and violence. Researchers seeking to understand the root causes of prejudice have long considered the role that emotions play in prejudicial attitudes and discriminatory behaviours. In this chapter, we specifically focus on the role that disgust plays in prejudice. We first describe some prominent theoretical frameworks for understanding the role of emotions in intergroup relations and prejudice. We then provide evidence for the connection between disgust and prejudice, followed by a discussion of some mechanisms underlying this connection. Finally, we highlight some unanswered questions and areas of future exploration.

Intergroup Emotions Elicited by Social Groups

The Stereotype Content Model developed by Fiske and colleagues (e.g., Fiske et al. 2002) proposes that social groups can be categorised on two orthogonal dimensions—competence (e.g., competent, intelligent, capable) and warmth (e.g., friendly, warm, sincere)—and that the emotional reactions elicited by a given social group will vary as a function of people's perceptions of their competence and warmth. For example, groups perceived as being high in both competence and warmth (e.g., health care professionals) might elicit admiration and pride; groups perceived as being low in competence but high in warmth (e.g., elderly people) might elicit pity

L. R. Vartanian (✉) · T. B. McCutcheon · S. A. Rubenstein
School of Psychology, UNSW Sydney, Sydney, Australia
e-mail: l.vartanian@unsw.edu.au; tayla.mccutcheon@student.unsw.edu.au; sarah.rubenstein@student.unsw.edu.au

P. A. Powell, N. S. Consedine (eds.), *The Handbook of Disgust Research*,
https://doi.org/10.1007/978-3-030-84486-8_10

and sympathy; groups perceived as being high in competence but low in warmth (e.g., rich people) might elicit envy and jealousy; and groups perceived as being low in both competence and warmth (e.g., welfare recipients) might elicit disgust, contempt, and anger.

Other researchers have argued that emotions in an intergroup context are dependent on the nature of the threat posed by the group. For example, anger is the typical response when valuable resources (such as jobs) are taken, disgust is the typical response when there is the potential for contamination, and fear is the typical response when physical safety is threatened (Cottrell and Neuberg 2005). Related work has further shown that emotional states can influence judgments of social groups, but only when the specific emotion is relevant to the group in question. For example, Dasgupta et al. (2009) showed that eliciting anger in participants led to heightened negative implicit evaluations of Arabs but not homosexuals, presumably because Arabs are associated with threat. In contrast, eliciting disgust led to heightened negative implicit evaluations of homosexuals but not Arabs, presumably because homosexuality is associated with violating particular moral values, values which are associated with disgust (see also Giner-Sorolla et al. 2012; Giner-Sorolla, Chap. 8, this volume). Similarly, because obese people are not generally seen as threatening to others, disgust responses are more relevant than are anger responses in prejudice towards obese people (Vartanian et al. 2013, 2016).

Finally, Rozin et al. (1999) described contempt, anger, and disgust as moral emotions that function to maintain the integrity of social order, and suggested that each of those emotions is uniquely associated with violation of a particular moral code. Contempt is elicited when individuals violate their duties or responsibilities within the community or social hierarchy; anger is elicited when individuals harm others or infringe on the freedom of others; and disgust is elicited when individuals cause impurity or degradation to the self or to others. Hutcherson and Gross (2011) further showed that anger is specifically evoked when the threat or transgression is viewed as being relevant to the self, disgust is associated with intentional immoral behaviours, and contempt seems to be related to judgments of someone being incompetent.

Overall, then, we see that social groups can elicit a variety of emotions, but which specific emotions are elicited seems to depend on the characteristics of the group in question. Disgust seems to be most relevant to groups perceived as low in competence and warmth, and is most likely to be elicited by groups seen as violating certain moral standards.

Evidence that Disgust Is Related to Prejudice Towards Outgroups

Before we begin examining the research evidence connecting disgust to prejudice, a note about terminology is useful for the non-social-psychologist reader. The social groups to which one belongs are typically referred to as "ingroups," and all other social groups (to which one does not belong) can be considered "outgroups." What the literature generally shows is that people have more favourable attitudes towards their ingroup (as well as towards individual members of their ingroup) than they do towards outgroups (as well as towards individual members of those outgroups). So, for example, a White American might consider other White Americans as members of their ingroup and consider African Americans (or any other racial minority) as members of outgroups. Of course, the picture is vastly more complicated because of the fact that we belong to so many different social categories (related to sex, ethnicity, nationality, occupation, religion, and many other aspects that define who we are) and thus the lines between an ingroup and outgroup are not always that clear cut. However, understanding the conceptual distinction is important because much of the research described in the rest of this chapter will refer to judgments of outgroup members.

Disgust Reactions to Social Groups

There is evidence that disgust is more strongly associated with some groups than with others. For example, we had participants rate their level of disgust towards 16 different social groups on a 9-point scale (1 = *Not at all disgusted*, 9 = *Extremely disgusted*; Vartanian 2010). The results of this survey are shown in Fig. 10.1. The highest disgust ratings were (in descending order) for drug addicts, smokers, obese people, politicians, and homeless people. These findings are consistent with the theoretical frameworks outlined above in that these social groups can all be viewed,

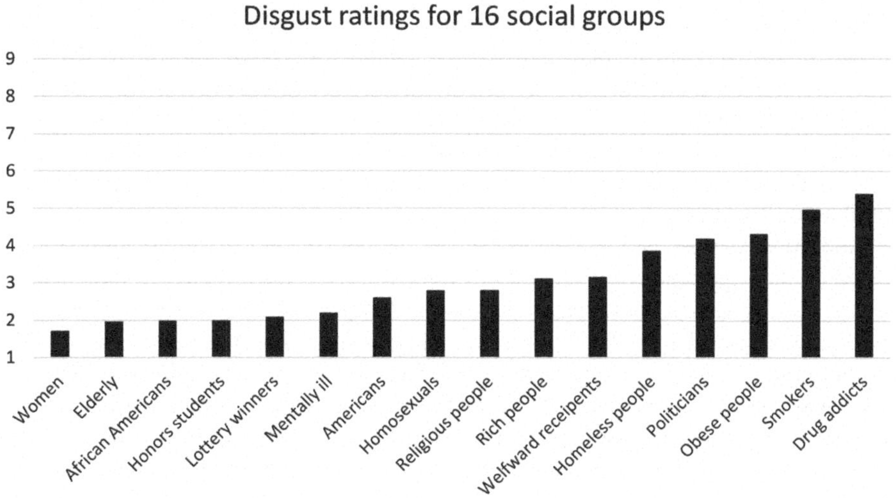

Fig. 10.1 Disgust ratings towards 16 different social groups. (Adapted from data reported in Vartanian (2010))

in some ways, as violating moral standards. In contrast, women, African Americans, the elderly, and individuals with mental illness received some of the lowest ratings of disgust. Other research has similarly shown that gay men received higher disgust ratings than did Mexican Americans and African Americans (Cottrell and Neuberg 2005). It is worth noting that disgust ratings in these studies, even among the groups rated as most disgusting, tend to be only moderately high. This means that people need not view a group as absolutely *repulsive* in order for prejudice to emerge.

Other research from our lab has shown that disgust is not just relevant to social groups at an abstract group level, but is also relevant to judgments of specific target individuals who are members of a particular social group. For example, participants reported more disgust towards a specific individual whose body weight was in the obese range than they did towards an individual whose body weight was in the normal-weight range (Vartanian et al. 2016, 2018). Because prejudice is often experienced at an individual level (i.e., specific individuals are the targets of prejudice), these findings highlight the potential relevance of disgust to interpersonal interactions.

Physiological Indicators of Disgust Towards Social Groups

The findings described in the section above are all based on self-report measures of disgust, and are thus dependent on individuals' willingness to acknowledge having these negative emotional reactions to a particular social group. Other research has attempted to capture emotional reactions to social groups using more direct assessments, specifically examining disgust at the physiological or neurobiological level. Findings from functional neuroimaging studies and brain injury research suggest the unique involvement of the insular cortex in disgust reactions. In particular, the insular cortex appears to play a role in regulating the parasympathetic nervous system and integrating sensory and visceral information in the processing of disgust (Calder et al. 2000, 2001). For example, activation of the anterior insula has been shown to occur in response to facial expressions of disgust (Phillips et al. 1997), as well as in response to photographs of disgusting scenes depicting contamination or mutilation (Wright et al. 2004).

Additionally, there is evidence to suggest that amygdala activation can be modulated by the experience of disgust. For example, fMRI was used to examine neural activity among female participants who viewed either disgusting or neutral images and found increased amygdala activation to disgusting images (Schienle et al. 2005). Furthermore, when disgusting images were presented, thereby inducing disgust, there was more pronounced activity in the right amygdala for participants who self-reported greater disgust sensitivity. Considering the amygdala is usually associated with fear responding and threat detection, this evidence might suggest that the amygdala responds to disgust, as well as fear, threats.

The medial prefrontal cortex (mPFC) has also been implicated in disgust reactions towards outgroups. For example, a number of studies have demonstrated that compared to familiar ingroups, stigmatised outgroups differentially activate the mPFC (Harris and Fiske 2006, 2007; Krendl et al. 2006). In these studies, participants were presented with photographs of different social groups that each represented one of the quadrants outlined in the Stereotype Content Model (i.e., either low or high competence and either low or high warmth). Participants had significantly reduced activation of the mPFC in response to photographs of outgroups characterised by low competence and low warmth when compared with photographs of all other social groups (Harris and Fiske 2006, 2007). The mPFC is often considered the social cognition region and has been implicated in person perception, for example, inferring individual attitudes, personality, and intent, and has also been implicated in theory of mind, which refers to the capacity to understand others mental and emotional states (Harris et al. 2005). Thus, reduced mPFC activity for groups falling into low competence/low warmth (e.g., homeless, poor, drug addicts) might suggest they are being perceived as so different that they do not experience the same complex emotions as people within the ingroups. That is, the feeling of disgust is facilitating a dehumanisation effect whereby people in the low competence/low warmth outgroups are not being considered to have complex emotions or thought processes thereby creating a "less than" human perception which can encourage prejudicial attitudes and behaviours (Fiske 2009).

Physiological measures such as salivary response (which is associated with experiences of disgust; Proctor and Carpenter 2007) have also been shown to increase when viewing stigmatised outgroups. For example, in one study, heterosexual male participants were exposed to images of either same-sex couples or mixed-sex couples (either kissing or engaging in public displays of affection [PDA]), neutral objects, or disgusting objects (O'Handley et al. 2017). The researchers found increased salivary alpha-amylase responses when participants viewed images of male same-sex couples kissing as well as when participants viewed disgusting images, compared to when they viewed images of same-sex couples engaging in PDA, mixed-sex couples kissing or engaging in PDA, or neutral images. These findings suggest that images that depict same-sex kissing are able to elicit a physiological response that is consistent with responses to disgusting stimuli.

Finally, some studies have used facial electromyography (EMG) to capture micro facial expressions of disgust towards specific social groups. Facial EMG can assess covert muscle activity, that is low levels of muscle contraction that do not necessarily lead to overt facial movements that can be observed visually (Tassinary and Cacioppo 1992). Research has shown that facial EMG responses can predict racial prejudice (e.g., Vanman et al. 1997), and there is evidence that moral transgressions elicit the same facial motor activity that disgusting tastes or exposure to disgusting images do (Chapman et al. 2009). Specifically, the levator labii muscles seem to be implicated in disgust responses. In a series of studies we conducted examining facial EMG responses to obese targets, images of obese individuals elicited slightly more levator activity than did neutral images, but we consistently failed to find any differences in levator activity towards obese versus non-obese individuals (Vartanian et al. 2018). One possible explanation for these findings is that facial EMG is not as sensitive as neurological measures in capturing disgust towards stigmatised outgroups. Recall that the ratings of disgust towards stigmatised outgroups are only moderately high in even the most extreme cases. It may be that the level of disgust needed to activate the levator muscles is considerably higher. Another possible explanation is that disgust responses to stigmatised groups (or at least obese individuals) are more a cognitive-conceptual response than they are a visceral reaction, such as the one might have in response to faeces or rotting food.

Disgust and Prejudice

In addition to documenting the emotional reactions to different social groups, research has also demonstrated an association between disgust responses and various forms of prejudice (including attitudes, stereotypes, and discriminatory or avoidant behaviour). In the study described earlier on disgust towards various social groups (Vartanian 2010), we found that ratings of disgust towards the 16 social groups (in aggregate) was negatively correlated with how favourable those groups were perceived (in aggregate). We have also found that ratings of disgust towards obese people in particular predicted negative attitudes towards obese people (Vartanian 2010; Vartanian et al. 2016). Research has similarly shown that disgust is associated with the stereotypical characteristics that are attributed to outgroups. For example, disgust towards obese people as a group is correlated with attributing characteristics such as "lazy," "sloppy," and "incompetent" to obese individuals (e.g., Vartanian et al. 2013, 2016).

Finally, studies have shown that disgust predicts discriminatory behaviour towards outgroups. For example, Cottrell et al. (2010) found that participants who expressed more disgust towards gays and lesbians also reported less support for gay rights. In our own work, we showed that disgust predicted a greater desire for social distance from individuals with obesity (Vartanian et al. 2016). In our study, we used two different measures of social distancing. One measure was a self-report questionnaire capturing participants' willingness or reluctance to engage with a target individual (e.g., willingness to work on the same project as the individual or rent a room in your home to the individual). The second measure was a computerised adaptation of the classic Seating Distance Task (Macrae et al. 1994). In the traditional version of this task, participants are led to a room where they believe the other participant has already selected a seat, and the measure of social distance is how far participants choose to sit from the other participant. In the computerised adaptation, participants were shown an image of a table and seven seats with the target individual's seat clearly marked, and they were asked to indicate which seat they would choose for themselves. The distance between the target's seat and the seat that the participants chose for themselves was taken as a measure of social distance. Results showed that disgust was a significant predictor of desire for social distance on both measures.

Disgust Sensitivity and Prejudice

Most research on the connection between disgust and prejudice has focused on individual differences in disgust responses given that there is likely to be substantial variation in the reaction to disgust-inducing stimuli like faeces, blood, or vomit. For example, some people might have a mild aversive reaction to blood, whereas other people might feel nauseous or faint. This variation in disgust response is referred to as *disgust sensitivity* (see Tybur, Chap. 6, this volume). A large body of correlational research has demonstrated that disgust sensitivity is associated with negative attitudes towards various groups, including immigrants, Muslims, homosexuals, and obese people (Choma et al. 2012; Hodson and Costello 2007; Lieberman et al. 2011; Olatunji et al. 2008; Tapias et al. 2007). Disgust sensitivity is also positively associated with attitudes towards one's ingroup (Crawford et al. 2014; Navarrete and Fessler 2006), suggesting that disgust not only generates avoidance of outgroups but may also heighten ingroup attraction, possibly in an attempt to rally support against a common threat (Navarrete and Fessler 2006). Disgust sensitivity has also been shown to predict negative behavioural intentions towards outgroups, such as increased support for discriminative policies and greater avoidance of intergroup contact (e.g., Aarøe et al. 2017). The heightened desire to avoid intergroup contact among those high in disgust sensitivity is important because it may prevent necessary experiences with outgroup members that could help foster acceptance and tolerance, and may thereby perpetuate the cycle of avoidance and prejudice.

Induced Disgust

In order to provide more causal evidence for a link between disgust and prejudice, a number of studies have induced feelings of disgust and then examined how this induced state influenced participants' responses to various outgroups (for more on inducing disgust see Consedine, Chap. 2, this volume). Asking participants to write about, read about, or imagine disgusting scenarios has been shown to influence prejudicial responding to outgroups. For example, Terrizzi Jr. et al. (2010) showed that asking participants to write about a scenario involving eating maggots (vs. eating lettuce in the control condition) can increase subsequent prejudicial attitudes towards homosexuals (this study is described in more detail below in the section on individual differences).

Unpleasant odours have also proven to be powerful elicitors of disgust. Inbar et al. (2012) randomly assigned participants to sit in one of two rooms, one with a noxious odour or one with no odour. Participants who were exposed to the noxious odour subsequently reported less warmth towards gay men than did those not exposed to the odour. Similarly, Cunningham et al. (2013) presented participants with a vial containing either a body odour scent, a parmesan cheese scent, or no odour. During a subsequent picture-viewing task, participants who were exposed to the body odour scent spent less time looking at images of gay couples compared to images of heterosexual couples. Further, these participants also reported less warmth towards gay (vs. heterosexual) men. These differences were not observed for participants in the other two conditions.

Another approach that researchers have used to examine the connection between disgust and prejudice is to increase the salience of disease or the perceived risk of infection. For example, Faulkner et al. (2004) found that participants who were shown a series of images designed to increase their perceived vulnerability to disease expressed less positive attitudes towards immigrant groups that were unfamiliar to them (but not immigrant groups that were familiar to them). Similarly, Park et al. (2007) manipulated the salience of contagious disease by showing participants a slide show that depicted either infectious disease threats or non-disease-related health threats. In a subsequent Implicit Associations Test, participants in the disease-salient condition showed more pronounced implicit associations between disease-connoting concepts and obese individuals. Moreover, Kenrick et al. (2013) found that parents who were made to feel vulnerable to infection reported more negative attitudes towards their own overweight children but not their normal weight children.

Is Disgust a Unique Emotional Response in Prejudice?

Consistent with the theoretical accounts outlined earlier in this chapter, there appears to be a unique association between disgust and prejudice towards certain outgroups. For example, disgust is more strongly associated with obesity than are either anger or contempt (Vartanian et al. 2013), and disgust is more strongly associated with gay men than are either anger or fear (Cottrell and Neuberg 2005). Furthermore, our research has shown that disgust, but not anger, is related to negative attitudes towards, stereotypes of, and a desire for social distance from, obese people (Vartanian et al. 2013, 2016). Another study by Matthews and Levin (2012) showed that disgust, but not anger, was associated with a desire to avoid Muslim culture (e.g., would consider preventing a child from reading books written by people of the Muslim world). Contempt is a bit more difficult to differentiate from disgust, possibly because of some conceptual overlap between the two constructs. We have found that disgust, but not contempt, was related to prejudice towards obese people (Vartanian et al. 2013, 2016), but others have found that both contempt and disgust are relevant to prejudice towards individuals with obesity (Wirtz et al. 2016).

Explanations for Disgust Towards Outgroups

We now turn our attention to a discussion of some of the possible mechanisms underlying (or explanations for) the association between disgust and prejudice towards an outgroup. Some of these have been discussed in previous chapters, so here we focus on their relevance to prejudice.

Behavioural Immune System

The behavioural immune system (BIS) works outside of conscious awareness utilising disgust to evoke avoidance of pathogens (see Bradshaw and Gassen, Chap. 3, this volume). Specifically, the BIS uses perceptual cues to indicate threat and activate disgust-related aversive emotions and avoidance behaviours (Schaller 2011), which can include distancing and social ostracism/exclusion (Park et al. 2003). Because pathogens themselves are not detectable by human senses, the BIS uses cues that might indicate a person could be carrying threatening pathogens (Tybur and Lieberman 2016). Some of the cues that might signify the presence of a pathogen include skin discoloration, facial and behavioural anomalies, body odour, and blemishes (Murray and Schaller 2016; Park et al. 2003). Thus, any outgroup who possesses one or more of these characteristics can activate the BIS and elicit disgust/avoidance reactions. Because the potential health costs of a false-negative (not detecting a pathogen when it is present) are greater than the costs of a false-positive (assuming the presence of a pathogen when none is present), the BIS has evolved to be overly sensitive. In this way, stigma towards outgroups can be seen as a signal-detection problem in an otherwise adaptive system (Park et al. 2007).

Evidence in support of the BIS explanation comes from a variety of sources. First, there is evidence that certain outgroups (e.g., obese people, people with physical disabilities) are associated with disease-related concepts at an automatic (implicit) level (e.g., Park et al. 2003, 2007). Second, as noted earlier, inducing disgust increases negative attitudes towards outgroups, including obese individuals and immigrants (Faulkner et al. 2004; Park et al. 2007). Finally, research has shown that when the subjective perceived vulnerability to disease is heightened (e.g., during the first trimester of pregnancy, or when participants were exposed to news coverage of the swine flu epidemic), there is increased prejudice towards outgroups (Huang et al. 2011; Navarrete et al. 2007). Furthermore, in this research, among participants who reported higher subjective ratings of perceived vulnerability to disease, prejudicial attitudes towards outgroups were reduced if they were given an opportunity to wash their hands (Huang et al. 2011).

Dehumanisation

Dehumanisation (or infra-humanisation) is considered the worst form of prejudice because it involves viewing outgroups as inhuman or as a lesser form of human (Dalsklev and Kunst 2015). Dehumanisation can include likening outgroup members to animals or machines, and consequently denying them uniquely "human" qualities or characteristics such as intelligence, cognitive abilities, and complex emotions, including love, hope, and guilt (Dalsklev and Kunst 2015). When individuals (or groups) are dehumanised, the failure to consider their mental and emotional experiences may facilitate severe discrimination and interpersonal violence (Harris and Fiske 2011). Dehumanising an outgroup removes that group from general societal moral protections and can work to justify the perpetration of hate crimes and interpersonal violence (Costello and Hodson 2010). In the same way that animals are often outside of moral protection and exploited by humans, dehumanisation works by rendering the outgroup as more animal-like and thus incapable of experiencing emotion or pain (Costello and Hodson 2010). Consequently, dehumanised groups are viewed as less deserving of the compassion, kindness, and respect that is given to other humans.

The psychological processes underlying dehumanisation can help us understand incidents of hate crimes and genocide (Schaller and Neuberg 2008). Throughout history, hate speech has been used to evoke fear and disgust towards outgroups and, in turn, to justify mass violence and genocide (Harris 2014). The choice of words within hate speech is important because these words are often used to incite disgust by comparing the outgroup to diseases or pathogen-threats such as cancer, faeces, parasites, or vermin. As Harris (2014) notes, these metaphors can also further expand the "us" and "them" divide. For example, the 1994 Rwanda genocide saw the Hutus refer to Tutsis as "cockroaches"; in Hitler's genocide during the Third Reich, the Nazis classified the Jewish people as "parasites" and "rats," relating them to vermin; and in the USA, Trump consistently referred to immigrants as "predators," "animals," and "killers" that "infest" the country (Harris and Fiske 2011; Warnock 2019). The consequence of this type of hate speech is that the outgroup is characterised as a disease that needs to be eradicated, and this mindset can lead to extreme violence towards members of that outgroup (Costello and Hodson 2010).

Even outside of the extreme examples just described, there is evidence that the language used to describe certain groups can impact perceptions of those groups. For example, one study showed how media portrayals can influence perceptions of immigrants. In this study, participants were asked to read a bogus newspaper article that included a cartoon of an immigrant arriving in Canada carrying several suitcases. Half of the participants saw a cartoon in which the suitcases were labelled with various diseases (thus depicting immigrants as carriers of disease); for the other half of participants, the cartoon contained no labels. Participants who viewed the cartoon with the disease labels subsequently showed a significantly higher belief that immigrants were pathogen threats, were more likely to dehumanise immigrants, and had less favourable views towards immigration more generally (Esses et al. 2013). Media representations of outgroups portrayed as vermin (using phrases such as "swarm," "invasion," or "flood") also increase disgust sensitivity and prejudicial attitudes in people who identified more strongly with their ingroup (Marshall and Shapiro 2018). Using this type of terminology facilitates dehumanised perceptions of outgroups, which has a direct impact on increasing prejudicial attitudes and stigmatisation (Esses et al. 2013; Marshall and Shapiro 2018).

Given the reference to disease and pathogens, it is not surprising that disgust is relevant in the context of dehumanisation. Intergroup disgust sensitivity is related to dehumanisation of outgroups (Hodson and Costello 2007), and in particular is associated with likening outgroup members to animals (Dalsklev and Kunst 2015). Disgust also appears to be more relevant to dehumanisation than other emotions. For example, one study used an experimental manipulation to prime either disgust, sadness, or neutral emotions. Although all participants showed a dehumanising bias towards outgroups, participants primed with disgust demonstrated the greatest associations between outgroups and animals, therefore adopting stronger socially dehumanising cognitions (Buckels and Trapnell 2013). Furthermore, the pattern of neural activity towards outgroups rated highest in disgust (the low-competence/low-warmth quadrant in the Stereotype Content Model) suggests that these groups are perceived as less than human (Harris and Fiske 2006).

Individual Differences in Social Conservatism

We have already discussed one set of individual differences—disgust sensitivity—related to the connection between disgust and prejudice. Here we outline another individual difference that can influence disgust and prejudice responses; namely, social conservatism.

Social conservatism includes a strong preference for ingroups, avoidance of outgroups, and strict obedience to social/cultural norms and rules (Shook et al. 2015). One study found that individuals higher in disgust sensitivity were also higher in social conservatism, which in turn predicted prejudicial attitudes towards homosexuals (Terrizzi Jr. et al. 2010). That is, the strength of the BIS, as measured by disgust sensitivity, was found to be predictive of socially conservative value systems; the more sensitive individuals are to disgust, the more conservative attitudes they favour, suggesting that disgust may promote social conservatism as a mechanism of pathogen avoidance (Shook et al. 2015; Terrizzi Jr. et al. 2010). Of course, the reverse causal path may be true as well, such that social conservatism promotes greater disgust sensitivity. Interestingly, in an eye-tracking study comparing attention to emotional faces among conservative and liberal participants, researchers found that the conservative participants paid significantly more attention to emotional faces depicting disgust (Oosterhoff et al. 2018). In order to avoid disgusting stimuli, greater attention is paid to cues (i.e., facial expressions) that might signal the presence of disgust threats.

This conservative belief system involves active avoidance of subjective pathogen threats and consequently promotes prejudicial behaviours (Terrizzi Jr. et al. 2010). Higher disgust sensitivity also predicts political conservatism, including authoritarianism, but seems to only apply to political issues associated with intergroup relations and pathogen avoidance, issues such as immigration, abortion, homosexuality, medicinal illicit drug use, and euthanasia (Inbar et al. 2009; Terrizzi Jr. et al. 2010). An experiment found that inducing disgust increased negative attitudes towards homosexuals for participants who were politically conservative but decreased negative attitudes for participants who were politically liberal (Terrizzi Jr. et al. 2010). The decrease in negative attitudes observed among the liberal participants may be explained by them having a less rigid definition of their ingroup, such that inducing disgust promotes ingroup preference and liberals include homosexuals as part of their broader ingroup (Terrizzi Jr. et al. 2010). The studies explained so far have contrasted disgust manipulations with neutral conditions. However, it is possible that these manipulations may have induced other, overlapping affective states, such as fear or anger. Future research that manipulates multiple negative and/or avoidance-promoting emotions would be important to examine these alternative explanations and broaden current understanding regarding the specificity of the disgust effect (see also Consedine, Chap. 2, this volume).

Pro-effort Bias

Our recent research in the domain of obesity stigma has identified a new pathway for understanding disgust and prejudice towards individuals with obesity. Earlier work had conceptualised stigma towards obese individuals in terms of beliefs about the controllability of body weight (e.g., Weiner et al. 1988). That is, believing that obesity is caused by overeating and lack of exercise is associated with more negative attitudes towards obese individuals. However, subsequent research has shown that changing people's beliefs about the controllability of body weight (such as by convincing them that obesity is caused by genetic factors) has very little impact on attitudes and stereotypes (Daníelsdóttir et al. 2010).

In our research, rather than focusing on the causes of the "onset" of obesity (i.e., how the problem developed in the first place), we focused on the "offset" of obesity (i.e., what people are doing to change the problem). We have shown that individuals with obesity elicit less disgust if they are described as putting in effort to lead a healthy lifestyle (i.e., eat healthy and exercise), even if there is no resulting weight loss (Beames et al. 2016; Vartanian et al. 2018). We similarly found that showing participants photographs of obese individuals engaging in healthy behaviours (e.g., eating vegetables, exercising) resulted in lower disgust ratings than did photographs of obese individuals engaging in unhealthy behaviours (e.g., eating junk food, sitting on the couch; Vartanian et al. 2018). We termed this process a "pro-effort bias" because it seemed to operate independently of beliefs about the controllability of the onset of the problem (Beames et al. 2016). It is plausible that this pro-effect bias also exists for other social groups for whom beliefs about their effort to offset the "problem" could be separable from beliefs about the onset of the problem. For example, drug users, smokers, and homeless people could potentially be conceptualised as putting in varying degrees of effort to remedy their situation, regardless of the perceived cause of the onset of the problem. Just as with obesity, perceptions of effort among individuals from these groups should be associated with lower ratings of disgust. If this is the case, then it may be possible to correct misperceptions about effort as a means of reducing prejudice towards these groups.

Unanswered Questions and Future Directions

Although research has provided numerous insights into the role of disgust in prejudice towards outgroups, there is a lot that we still do not fully grasp. For example, the "special" or "unique" place for disgust needs stronger support. Much of the evidence in this regard has come from studies showing that disgust is more strongly correlated with prejudice towards particular groups than are other emotions, such as anger or fear. However, there is little in the way of experimental work showing a causal link between specific emotions and prejudice. Furthermore, most studies in this area (as with much of psychology) has focused on participants from Western cultures, often university students, and the findings might therefore not be generalisable to other populations. If part of the reason that people experience disgust towards a particular social group is that the group is seen as violating a cultural norm, then cultures that have different norms would presumably view the group differently. Thus, further work is needed in these respects. Here we outline two additional areas in need of further exploration.

Differentiating Different Types of Disgust

Throughout this chapter, we have described disgust as though it was more or less a unitary construct. However, Tybur et al. (2009; see also Tybur, Chap. 6, this volume) provide evidence for the functional heterogeneity of disgust. Specifically, they outline three different functions of disgust, each of which solves a qualitatively different adaptive problem. Pathogen disgust, which is most similar to traditional conceptions of disgust, is related to the desire to avoid disease-causing agents. Moral disgust is related to avoidance of individuals who can potentially inflict social harms on oneself or one's social network. Sexual disgust relates to the avoidance of behaviours or partners that could reduce one's long-term reproductive success. Tybur et al. (2009) also suggest that different behavioural responses might be associated with each functional domain of disgust. For example, pathogen disgust might motivate the desire to clean; moral disgust might motivate the desire to punish the offending agent; and sexual disgust might motivate rejection in the context of a romantic relationship, but not in the context of a business partnership. Earlier in this chapter, we noted that different outgroups elicit different emotional reactions. In a similar way, we might expect that different outgroups would elicit different types of disgust reactions. Identifying the domain(s) of disgust that is (are) most central to particular outgroups is an important direction for future research because it could improve our understanding of what drives prejudice towards these groups.

Reducing Disgust as a Vehicle for Reducing Prejudice

A potential benefit of understanding the emotional underpinnings of prejudice is that it could help explain why prejudicial reactions are so difficult to change. Cognitive approaches to eliminating prejudice (e.g., convincing people that obesity is not under the person's control) are typically not very effective (Daníelsdóttir et al. 2010). If prejudice is based on emotional reactions, then we should not expect that challenging cognitions about a particular group would do much to reduce prejudice. Instead, it would be beneficial to identify ways to modify people's emotional reactions to those groups. One example of an approach that could be relevant in this context is increasing contact with outgroup members. Intergroup contact (Pettigrew and Tropp 2006) and even *imagined* intergroup contact (Crisp and Turner 2009) have both been shown to reduce prejudice towards outgroups. One of the mechanisms through which contact reduces prejudice is by reducing intergroup anxiety (Turner et al. 2008). There is also some recent evidence that disgust mediates the effect of intergroup contact on prejudice towards gay men, but not towards Whites, Blacks, or Asians (Seger et al. 2017). These findings suggest that contact with outgroup members may represent a useful way to reduce prejudice, possible by reducing negative emotional reactions to those groups. Further exploring the benefits of intergroup contact for reducing disgust and prejudice is an important direction for future research.

Conclusions

Disgust plays a key role in prejudice, and in particular prejudice towards groups that are viewed as violating moral standards. These disgust reactions could be driven by a highly sensitive behavioural immune system that views outgroup members as potential vectors of disease, by a tendency to view the groups as "less than human," and perhaps in some cases by a tendency to view these groups as not exerting sufficient effort to change their situations. Modifying people's emotional reactions to particular social groups could be a useful means of reducing prejudice towards those groups.

References

Aarøe, L., Petersen, M., & Arceneaux, K. (2017). The behavioral immune system shapes political intuitions: Why and how individual differences in disgust sensitivity underlie opposition to immigration. *American Political Science Review*, *111*, 277-294.

Beames, J. R., Black, M. J., & Vartanian, L. R. (2016). Prejudice toward individuals with obesity: Evidence for a pro-effort bias. *Journal of Experimental Psychology: Applied*, 22, 184-195.

Buckels, E. E., & Trapnell, P. D. (2013). Disgust facilitates outgroup dehumanization. *Group Processes & Intergroup Relations*, *16*, 771-780.

Calder, A., Keane, J., Manes, F., Antoun, N., & Young, A. (2000). Impaired recognition and experience of disgust following brain injury. *Nature Neuroscience*, *3*, 1077-1078.

Calder, A., Lawrence, A., & Young, A. (2001). Neuropsychology of fear and loathing. *Nature Reviews Neuroscience*, *2*, 352-363.

Chapman, H. A., Kim, D. A., Susskind, J. M., & Anderson, A. K. (2009). In bad taste: Evidence for the oral origins of moral disgust. *Science, 323,* 1222-1226.

Choma, B., Hodson, G., & Costello, K. (2012). Intergroup disgust sensitivity as a predictor of islamophobia: The modulating effect of fear. *Journal of Experimental Social Psychology*, *48*, 499-506.

Costello, K., & Hodson, G. (2010). Exploring the roots of dehumanization: The role of animal—human similarity in promoting immigrant humanization. *Group Processes & Intergroup Relations*, *13*, 3-22.

Cottrell, C. A., & Neuberg, S. L. (2005). Different emotional reactions to different groups: A sociofunctional threat-based approach to 'prejudice'. *Journal of Personality and Social Psychology, 88,* 770–789.

Cottrell, C. A., Richards, D. A., & Nichols, A. L. (2010). Predicting policy attitudes from general prejudice versus specific intergroup emotions. *Journal of Experimental Social Psychology*, *46*, 247-254.

Crawford, J., Inbar, Y., & Maloney, V. (2014). Disgust sensitivity selectively predicts attitudes toward groups that threaten (or uphold) traditional sexual morality. *Personality and Individual Differences*, *70*, 218-223.

Crisp, R. J., & Turner, R. N. (2009). Can imagined interactions produce positive perceptions? Reducing prejudice through simulated social contact. *American Psychologist, 64,* 231–240.

Cunningham, E., Forestell, C., & Dickter, C. (2013). Induced disgust affects implicit and explicit responses toward gay men and lesbians. *European Journal of Social Psychology*, *43*, 362-369.

Dalsklev, M., & Kunst, J. R. (2015). The effect of disgust-eliciting media portrayals on outgroup dehumanization and support of deportation in a Norwegian sample. *International Journal of Intercultural Relations*, *47*, 28-40.

Daníelsdóttir, S., O'Brien, K. S., & Ciao, A. (2010). Anti-fat prejudice reduction: A review of published studies. *Obesity Facts, 3,* 47–58.

Dasgupta, N., DeSteno, D., Williams, L. A., & Hunsinger, M. (2009). Fanning the flames of prejudice: The influence of specific incidental emotions on implicit prejudice. *Emotion, 9,* 585-591.

Esses, V. M., Medianu, S., & Lawson, A. S. (2013). Uncertainty, threat, and the role of the media in promoting the dehumanization of immigrants and refugees. *Journal of Social Issues*, *69*, 518-536.

Faulkner, J., Schaller, M., Park, J., & Duncan, L. (2004). Evolved disease-avoidance mechanisms and contemporary xenophobic attitudes. *Group Processes & Intergroup Relations*, *7*, 333-353.

Fiske, S. T. (2009). From dehumanization and objectification, to rehumanization: Neuroimaging studies on the building blocks of empathy. *Annals of the New York Academy of Sciences, 1167,* 31-34.

Fiske, S. T., Cuddy, A. C., Glick, P., & Xu, J. (2002). A model of (often mixed) stereotype content: Competence and warmth follow from perceived status and competition. *Journal of Personality and Social Psychology, 82,* 878–902.

Giner-Sorolla, R., Bosson, J. K., Caswell, T. A., & Hettinger, V. E. (2012). Emotions in sexual morality: Testing the separate elicitors of anger and disgust. *Cognition & Emotion, 26,* 1208-1222.

Harris, A. (2014). The psychological mechanisms of oppression: Empathy, disgust, and the perception of group membership. *Oregon Undergraduate Research Journal*, *7*, 38-47.

Harris, L. T., & Fiske, S. T. (2006). Dehumanizing the lowest of the low: Neuroimaging responses to extreme out-groups. *Psychological Science*, *17*, 847-853.

Harris, L. T., & Fiske, S. T. (2007). Social groups that elicit disgust are differentially processed in mPFC. *Social Cognitive and Affective Neuroscience, 2,* 45-51.

Harris, L. T., & Fiske, S. T. (2011). Dehumanized perception: A psychological means to facilitate atrocities, torture, and genocide? *Zeitschrift für Psychologie/Journal of Psychology, 219,* 175–181.

Harris, L. T., Todorov, A., & Fiske, S. T. (2005). Attributions on the brain: neuro-imaging dispositional inferences, beyond theory of mind. *Neuroimage, 28,* 763-769.

Hodson, G., & Costello, K. (2007). Interpersonal disgust, ideological orientations, and dehumanization as predictors of intergroup attitudes. *Psychological Science, 18,* 691-698.

Huang, J. Y., Sedlovskaya, A., Ackerman, J. M., & Bargh, J. A. (2011). Immunizing against prejudice: Effects of disease protection on attitudes toward out-groups. *Psychological Science, 22,* 1550-1556.

Hutcherson, C. A., & Gross, J. J. (2011). The moral emotions: A social-functionalist account of anger, disgust, and contempt. *Journal of Personality and Social Psychology, 100,* 719–737.

Inbar, Y., Pizarro, D., Knobe, J., & Bloom, P. (2009). Disgust sensitivity predicts intuitive disapproval of gays. *Emotion, 9,* 435-439.

Inbar, Y., Pizarro, D., & Bloom, P. (2012). Disgusting smells cause decreased liking of gay men. *Emotion, 12,* 23-27.

Kenrick, A., Shapiro, J., & Neuberg, S. (2013). Do parental bonds break anti-fat stereotyping? *Social Psychological and Personality Science, 4,* 721-729.

Krendl, A. C., Macrae, N. C., Kelley, W. M., Fugelsang, J. A., & Heatherton, T. F. (2006). The good, the bad, and the ugly: An fMRI investigation of the functional anatomic correlates of stigma. *Social Neuroscience, 1,* 5–15.

Lieberman, D., Tybur, J., & Latner, J. (2011). Disgust sensitivity, obesity stigma, and gender: Contamination psychology predicts weight bias for women, not men. *Obesity, 20,* 1803-1814.

Macrae, C. N., Bodenhausen, G. V., Milne, A. B., & Jetten, J. (1994). Out of mind but back in sight: Stereotypes on the rebound. *Journal of Personality and Social Psychology, 67,* 808–817.

Marshall, S. R., & Shapiro, J. R. (2018). When "scurry" vs. "hurry" makes the difference: Vermin metaphors, disgust, and anti-immigrant attitudes. *Journal of Social Issues, 74,* 774-789.

Matthews, M., & Levin, S. (2012). Testing a dual process model of prejudice: Assessment of group threat perceptions and emotions. *Motivation and Emotion, 36,* 564-574.

Murray, D. R., & Schaller, M. (2016). The behavioral immune system: Implications for social cognition, social interaction, and social influence. *Advances in Experimental Social Psychology, 53,* 75-129.

Navarrete, C., & Fessler, D. (2006). Disease avoidance and ethnocentrism: The effects of disease vulnerability and disgust sensitivity on intergroup attitudes. *Evolution and Human Behavior, 27,* 270-282.

Navarrete, C. D., Fessler, D. M., & Eng, S. J. (2007). Elevated ethnocentrism in the first trimester of pregnancy. *Evolution and Human Behavior, 28,* 60-65.

O'Handley, B., Blair, K., & Hoskin, R. (2017). What do two men kissing and a bucket of maggots have in common? Heterosexual men's indistinguishable salivary α-amylase responses to photos of two men kissing and disgusting images. *Psychology & Sexuality, 8,* 173-188.

Olatunji, B., Haidt, J., McKay, D., & David, B. (2008). Core, animal reminder, and contamination disgust: Three kinds of disgust with distinct personality, behavioral, physiological, and clinical correlates. *Journal of Research in Personality, 42,* 1243-1259.

Oosterhoff, B., Shook, N. J., & Ford, C. (2018). Is that disgust I see? Political ideology and biased visual attention. *Behavioural Brain Research, 336,* 227-235.

Park, J. H., Faulkner, J., & Schaller, M. (2003). Evolved disease-avoidance processes and contemporary anti-social behavior: Prejudicial attitudes and avoidance of people with physical disabilities. *Journal of Nonverbal Behavior, 27,* 65-87.

Park, J., Schaller, M., & Crandall, C. (2007). Pathogen-avoidance mechanisms and the stigmatization of obese people. *Evolution and Human Behavior, 28,* 410-414.

Pettigrew, T. F., & Tropp, L. R. (2006). A meta-analytic test of intergroup contact theory. *Journal of Personality and Social Psychology, 90,* 751–783.

Phillips, M., Young, A., Senior, C., Brammer, M., Andrew, C., Calder, A. J., Bullmore, E. T., Perrett, D. I., Rowland, D., Williams, S. C., Gray, J. A., & David, A. S. (1997). A specific neural substrate for perceiving facial expressions of disgust. *Nature, 389,* 495-498.

Proctor, G. B., & Carpenter, G. H. (2007). Regulation of salivary gland function by autonomic nerves. *Autonomic Neuroscience: Basis and Clinical, 133,* 3–18.

Rozin, P., Lowery, L., Imada, S., & Haidt, J. (1999). The CAD triad hypothesis: A mapping between three moral emotions (contempt, anger, disgust) and three moral codes (community, autonomy, divinity). *Journal of Personality and Social Psychology, 76,* 574–586.

Schaller, M. (2011). The behavioural immune system and the psychology of human sociality. *Philosophical Transactions of the Royal Society B: Biological Sciences, 366,* 3418-3426.

Schaller, M., & Neuberg, S. L. (2008). Intergroup prejudices and intergroup conflicts. In C. Crawford & D. Krebs (Eds.), *Foundations of evolutionary psychology* (pp. 401–414). Taylor & Francis Group/Lawrence Erlbaum Associates.

Schienle, A., Schäfer, A., Stark, R., Walter, B., & Vaitl, D. (2005). Gender differences in the processing of disgust-and fear-inducing pictures: An fMRI study. *Neuroreport, 16,* 277-280.

Seger, C. R., Banerji, I., Park, S. H., Smith, E. R., & Mackie, D. M. (2017). Specific emotions as mediators of the effect of intergroup contact on prejudice: Findings across multiple participant and target groups. *Cognition and Emotion, 31,* 923-936.

Shook, N. J., Terrizzi Jr, J. A., Clay, R., & Oosterhoff, B. (2015). In defense of pathogen disgust and disease avoidance: A response to Tybur et al. (2015). *Evolution and Human Behavior, 36,* 498-502.

Tapias, M., Glaser, J., Keltner, D., Vasquez, K., & Wickens, T. (2007). Emotion and prejudice: Specific emotions toward outgroups. *Group Processes & Intergroup Relations, 10,* 27-39.

Tassinary, L. G., & Cacioppo, J. T. (1992). Unobservable facial actions and emotion. *Psychological Science, 3,* 28-33.

Terrizzi, J. A., Jr., Shook, N. J., & Ventis, W. L. (2010). Disgust: A predictor of social conservatism and prejudicial attitudes toward homosexuals. *Personality and Individual Differences, 49,* 587–592.

Turner, R. N., Hewstone, M., Voci, A., & Vonofakou, C. (2008). A test of the extended intergroup contact hypothesis: The mediating role of intergroup anxiety, perceived ingroup and outgroup norms, and inclusion of the outgroup in the self. *Journal of Personality and Social Psychology, 95,* 843-860.

Tybur, J. M., & Lieberman, D. (2016). Human pathogen avoidance adaptations. *Current Opinion in Psychology, 7,* 6-11.

Tybur, J. M., Lieberman, D., & Griskevicius, V. (2009). Microbes, mating, and morality: individual differences in three functional domains of disgust. *Journal of Personality and Social Psychology, 97,* 103-122.

Vanman, E. J., Paul, B. Y., Ito, T. A., & Miller, N. (1997). The modern face of prejudice and structural features that moderate the effect of cooperation on affect. *Journal of Personality and Social Psychology, 73,* 941-959.

Vartanian, L. R. (2010). Disgust and perceived control in attitudes toward obese people. *International Journal of Obesity, 34,* 1302-1307.

Vartanian, L. R., Thomas, M. A., & Vanman, E. J. (2013). Disgust, contempt, and anger and the stereotypes of obese people. *Eating and Weight Disorders, 18,* 377-382.

Vartanian, L. R., Trewartha, T., & Vanman, E. J. (2016). Disgust predicts prejudice and discrimination toward individuals with obesity. *Journal of Applied Social Psychology, 46,* 369-375.

Vartanian, L. R., Trewartha, T., Beames, J. R., Azevedo, S. M., & Vanman, E. J. (2018). Physiological and self-reported disgust reactions to obesity. *Cognition and Emotion, 32,* 579-592.

Warnock, A. (2019). The dehumanization of immigrants and refugees: A comparison of dehumanizing rhetoric by all candidates in three U.S. presidential elections. *Journal of Purdue Undergraduate Research*, *9*, 49-59.

Weiner, B., Perry, R. P., & Magnusson, J. (1988). An attributional analysis of reactions to stigmas. *Journal of Personality and Social Psychology, 55,* 738–748.

Wirtz, C., van der Pligt, J., & Doosje, B. (2016). Derogating obese individuals: The role of blame, contempt, and disgust. *Journal of Applied Social Psychology, 46,* 216-228.

Wright, P., He, G., Shapira, N., Goodman, W., & Liu, Y. (2004). Disgust and the insula: fMRI responses to pictures of mutilation and contamination. *Neuroreport, 15,* 2347-2351.

Chapter 11
Mechanisms of Disgust in Psychopathology

Graham C. L. Davey

Over the previous two to three decades there has been accumulating evidence that the disease-avoidance emotion of disgust is closely associated with and regularly experienced in a number of different psychopathologies and mental health problems (Davey 2011; Knowles et al. 2018; Olatunji et al. 2010). Some of the symptoms of these mental health problems have obvious links to disgust and the kinds of events and stimuli that elicit the emotion (e.g., contamination fears in obsessive compulsive disorder [OCD]; Knowles et al. 2018; Melli et al. 2019; Poli et al. 2019). Some other conditions appear to have an indirect link with disgust through perceived harbingers of disease, such as small animal phobias (Davey 1994a; Mulkens et al. 1996; Ware et al. 1994). While still other psychopathologies have characteristics similar to the avoidance responses typical of disgust, such as fear of oral incorporation (e.g., "picky eating" and eating disorders; Davey et al. 1998; Harris et al. 2019; Harvey et al. 2002).

But, perhaps most perplexing, is the fact that there is a growing list of psychopathologies whose severity is directly associated with levels of disgust even though those conditions appear to have little or no theoretical or functional link with the disgust emotion. These disorders include separation anxiety (Muris et al. 1999), agoraphobia (Muris et al. 2000), symptoms of psychosis (Schienle et al. 2003), and height anxiety and claustrophobia (Davey and Bond 2006). This list of disgust-related psychopathologies raises a number of theoretical questions about how the disgust emotion has become associated with these conditions, and what role, if any, disgust plays in the development and maintenance of these psychopathologies. Such questions are not just of theoretical interest, they may also provide vital information about the kinds of effective interventions we might develop to alleviate these psychopathologies.

G. C. L. Davey (✉)
School of Psychology, University of Sussex, Brighton, UK
e-mail: grahamda@sussex.ac.uk

P. A. Powell, N. S. Consedine (eds.), *The Handbook of Disgust Research*,
https://doi.org/10.1007/978-3-030-84486-8_11

While there have been significant advances in identifying the psychopathologies associated with disgust, and advances to some extent in understanding some of the developmental and cognitive characteristics associated with the disgust emotion (Knowles et al. 2019; Rottman et al. 2019), there has been less progress in clearly identifying the mechanisms by which the disgust emotion influences psychopathology symptoms. Indeed, there remains the question if disgust has a causal influence at all, and is not simply an epiphenomenon of the psychopathology symptoms it is associated with. In this chapter, I will describe a number of putative mechanisms through which the emotion of disgust might influence the symptoms of mental health problems and examine how these putative mechanisms might fit the existing evidence on disgust-relevant psychopathologies. But, first it is necessary to review the characteristics and functionality of the disgust emotion, and discuss how these features relate to more commonly experienced psychopathology emotions such as anxiety and fear.

Disgust, Anxiety, and Fear

Disgust is regularly described as a specially evolved natural defensive emotion whose purpose is to protect against harm (Davey 1994b; Rozin and Fallon 1987), with the proposed primary biological function of protecting the organism specifically against pathogens, illness, disease, and contamination (Curtis and de Barra 2018; Stevenson et al. 2019; Tybur et al. 2013). As documented elsewhere (see Bradshaw and Gassen, Chap. 3, this volume), the main physiological, experiential, and behavioural response features of the disgust emotion are consistent with this disease-avoidance explanation and—in more specific cases—to prevent the oral incorporation of items that might potentially be causes of illness and disease.

In addition, some theorists have extended this basic "disease-avoidance" view of the disgust response to suggest that the emotion has evolved more sophisticated adaptive functions that protect not just the physical body from contamination and harm, but also protect the "psychological" body from harm in response to reminders of our own mortality and inherent animalistic nature (Haidt et al. 1994; Rozin et al. 2016). Proponents of this latter view support this account by arguing that most "primary" disgust items can be characterised by their animal origin and their ability to elicit fear of oral incorporation (e.g., phlegm/mucus, diarrhoea, slugs, maggots; Marzillier and Davey 2004). As a result, disgust is constantly a reminder of our animal origins and—because disgusting objects are usually universally revulsive—disgust has become an emotion frequently used to imbue socially and culturally unacceptable activities and attitudes with negative affect. Viewed in this broader, psychological perspective, disgust is not just a food-rejection response serving to avoid disease; it has also come to regulate behaviour in social and interpersonal contexts, and will affect our social attitudes by conveying culturally and morally unacceptable views. In effect, disgust can be used to convey the strong unacceptability of ideas and behaviours, and can also be elicited by violations of moral or social norms and by feelings of being "wronged" or betrayed (e.g., Giner-Sorolla et al. 2018; Plakias 2018; see also Giner-Sorolla, Chap. 8, this volume). Beyond its origins in physical disease avoidance, the psychological protection mechanism of disgust could also be linked to people's mental health, and is manifest in conditions such as OCD, where an inflated disgust response may be observed to moral violations.

However, rather than being a biologically pre-wired adaptation, it is worth remembering that disgust is an acquired emotion that is shaped during the first 5–6 years of life (DeJesus et al. 2015). It begins with a born dislike of bitter flavours, which is observed in human neonates. The infant then appears to embark on a developmental process in which they learn which foods are edible through both associative learning (Brown and Harris 2012) and the learning of culturally acceptable practices (Feder 2015; Stevenson et al. 2010). By around the age of 5 or 6 years, children have then acquired a concept of disgust that embraces notions of contamination; a learnt catalogue of objects that evoke disgust (including primary disgust objects such as faeces and mucus); and knowledge of the relationship of disgust with the avoidance of contaminants, pathogens, and illness (Blacker and LoBue 2016; Rottman et al. 2019; for more on how children learn to be disgusted see Reynolds and Askew, Chap. 5, this volume).

If we now compare the evolutionary role of disgust with that of fear and anxiety, there appear to be some clear and distinct evolutionary purposes for these different emotions. Fear is largely a biologically pre-wired reaction to imminent threats, with a network of reflexive responses reacting to stimuli such as loud noises, looming shadows, sudden sharp pain, rapid movements towards you, and staring eyes (all generalised characteristics of potential predators!; Russell 1979). In contrast, anxiety is a more flexible emotion for managing potential future threats and challenges that may not be so imminent. It is an emotion that develops after birth by recruiting some responses from the fear network (e.g., startle reactions and physiological arousal processes) but adds on some higher-level cognitive processes that influence attentional focus and defensive interpretations of both on-going and future events (Davey 2018, Chapter 1; Ouimet et al. 2009). The fact that anxiety is a hybrid emotion built out of a combination of fear reflexes and higher-level cognitive processes suggests that anxiety, like disgust, is almost certainly a learnt emotion rather than an innate one, and can be characterised as a cognitive schema (or "emotion schema", Izard 2007). A cognitive schema is a high-level conceptual structure or network in the brain that enables rapid, effective, and efficient responding to important events such as everyday threats and challenges (Britton and Davey 2014; Davey 2018; Neuberg et al. 2011). In this sense, both anxiety and disgust are learnt emotions whose eventual form will be determined by the early experiences and development of the individual. In contrast, fear is biologically pre-wired and relatively consistent in its reflexive features across individuals.

Disgust, Anxiety, and Fear in Psychopathology

Measures of disgust sensitivity and propensity have been shown to be highly correlated with measures of anxiety (Davey and Bond 2006; Ware et al. 1994), and with measures of specific anxiety-based problems such as small animal fears (Davey 1994a; Matchett and Davey 1991), spider phobia (Mulkens et al. 1996; Thorpe and Salkovskis 1998), blood-injury injection phobia (BII; de Jong and Merckelbach 1998; Olatunji et al. 2007b), contamination fear in OCD (Mancini et al. 2001; Olatunji et al. 2007a), and eating disorders (Davey et al. 1998). These correlations are instructive, but do not tell us too much about the role of disgust in these conditions, because in many cases these correlations may be inflated by a confounding of disgust-relevant items in the measures of both disgust and the psychopathology (Davey 2011). However, comparisons of individuals diagnosed with disgust-relevant psychopathologies do show that they experience higher levels of disgust than individuals without a diagnosis (Cisler et al. 2009a; Olatunji et al. 2010), which does suggest at the very least that disgust is experienced more intensely in individuals suffering these psychopathologies than in healthy individuals. Psychopathologies where this is the case include OCD contamination fears, small animal phobias, and BII phobia (Olatunji et al. 2010). It is perhaps not so surprising that in these psychopathologies disgust is experienced more because the objects of fear and anxiety in these conditions are also objects that would naturally elicit disgust (e.g., dirty toilets, pathogens, blood and injury, disease-relevant animals such as rats, slimy animals such as slugs and snails, and animals associated with contamination such as maggots).

But, there is also considerable overlap between the emotions of disgust and the emotions of anxiety and fear, so it is quite possible that experiencing one may facilitate the other. Both disgust and fear/anxiety have a dominant behavioural tendency of avoidance (Izard 1993). However, fear- and anxiety-motivated avoidance protects the person from almost any perceived threat (Woody and Teachman 2000), whereas disgust-motivated avoidance has the more specific function of avoiding pathogens and contaminants (Olatunji et al. 2010; Rozin and Fallon 1987).

To summarise, much of the evidence for the involvement of the disgust emotion in psychopathology is correlational in nature, so there is a real need for controlled experimental studies that enable us to identify causal relationships between disgust, anxiety, fear, and other symptoms of psychopathology.

Disgust as a Causal Agent in Psychopathologies

There is experimental evidence that disgust may play a causal role in facilitating symptoms in a number of anxiety-related conditions. For example, Webb and Davey (1993) asked a non-clinical population to rate fear to four categories of animals (predatory, fear-relevant small mammals and reptiles, fear-relevant invertebrates, and fear-irrelevant animals) before and after watching a violent, disgusting, or neutral video. They found that participants who watched the violent video showed increased fear ratings to larger predatory animals, whereas participants who watched the disgusting video showed increased fear to fear-relevant small mammals and invertebrates. The effects of disgust inductions on symptoms of contamination fear also suggest a causal role for disgust. In a study by Olatunji and Armstrong (2009) a disgust induction caused a significantly greater increase in distress to low contagion stimuli in high contamination fear individuals than low contamination fear individuals. Interestingly, inducing disgust evoked distress levels to potentially contagious stimuli in low contamination fear participants that paralleled levels found in high contamination fear participants, suggesting that high levels of disgust may be involved in the development of clinical levels of contamination fear. A further study by Davey et al. (2008) investigated the effect of an experimental disgust induction on subsequent anxiety to scenarios of fear-relevant, disgust-relevant, and fear-irrelevant stimuli in a non-clinical population. The results indicated that disgust facilitated levels of self-reported anxiety to a range of scenarios regardless of whether they were disgust-relevant, fear-relevant, or fear-irrelevant, and suggested that disgust can have a general facilitating effect on self-reported anxiety to a range of stimuli regardless of their disgust-relevant status.

While there is good evidence that disgust can facilitate anxiety in a number of experimental procedures, there are still some significant failures to demonstrate this effect—failures suggesting that disgust may only enhance anxiety in certain contexts and procedures. One significant failure to find an effect of induced disgust on anxiety was reported by Marzillier and Davey (2005). In a series of three studies and using a range of mood induction procedures, Marzillier and Davey (2005) demonstrated that induced anxiety produced increases in self-reported disgust, but there was no evidence that induced disgust facilitated self-reported anxiety. These results were independent of the type of mood induction procedure used and the type of dependent measure used, and suggest that if disgust does have a causal influence on anxious psychopathology then this influence is not mediated simply by experienced disgust facilitating experienced anxiety.

At this point, we can begin to speculate about how disgust facilitates anxiety in those procedures where it does, and why it fails in some other situations and procedures. First, because both disgust and anxiety facilitate avoidance responses, disgust may increase self-reported anxiety by priming avoidance goals that facilitate feelings of fear and anxiety (Olatunji and Armstrong 2009; Olatunji et al. 2004)—but only if there is something that this avoidance tendency can be directed at (such as a target stimulus or event within the procedure). However, if disgust facilitates only disgust-relevant avoidance responses, then we would expect disgust to selectively enhance anxiety to disgust-relevant stimuli and events rather than to disgust-irrelevant stimuli and events.

Secondly, because disgust is a negatively-valenced emotion, some of the negativity experienced through disgust may become associated with any proximal stimuli or events and facilitate associated levels of fear and anxiety, perhaps through a process of evaluative conditioning (De Houwer et al. 2001; see also Reynolds and

Askew, Chap. 5, this volume). But again, this is only likely to be detected in disgust induction procedures if there is a stimulus or event to be evaluated. Disgust is unlikely to facilitate anxiety or fear if there is nothing to which the disgust-induced negativity can be attached. And this is not withstanding the view that, at best, evaluative conditioning is a relatively elusive phenomenon (Field and Davey 1997, 1998).

A third possibility is that disgust as a negative emotion may facilitate information processing biases that affect attentional focus, decision-making, and interpretations of proximal events and stimuli. Indeed, there is good evidence that disgust does induce these negative information processing biases (and I'll talk more about these later in the chapter; Davey et al. 2006; Knowles et al. 2019; Leathers-Smith and Davey 2011). Once again, we would expect disgust to enhance anxiety and fear via its biasing effect on information processing only if there are events and stimuli on which these information-processing biases can operate. But, because these information-processing biases can potentially influence any event or stimulus, we would expect disgust in this respect to be able to facilitate fear and anxiety to any event or stimulus regardless of its disgust-relevant status. Such a view would be consistent with the findings of Davey et al. (2008) described earlier, and also explains why induced disgust does *not* enhance anxiety when there are no stimuli or events in the procedure that can be influenced by biased information processing (e.g., Marzillier and Davey 2005).

However, these putative explanations of how disgust might influence fear and anxiety allude primarily to cognitive processes involved in information processing and stimulus evaluation, but disgust also has a social and cultural role through which it might influence levels of fear and anxiety. For example, disgust can be used to imbue any thoughts and behaviours with negative affect (Rozin and Fallon 1987), can be evoked by feelings of moral violation or feelings of being "wronged" (Giner-Sorolla et al. 2018; Plakias 2018; see also Giner-Sorolla, Chap. 8, this volume), or by facilitating feelings of self-disgust (Powell et al. 2015a). Are these evolved "social" contributions of disgust also implicated in facilitating psychopathology in any way? It is reasonable to think so. There is, for instance, good reason to suppose that disgust directed at the self may influence feelings of shame, guilt, low self-esteem, and self-harm—all feelings and actions relevant to a number of common psychopathologies (Ille et al. 2014; Smith et al. 2015).

Putative Mechanisms of Disgust in Psychopathology

There are now a significant number of studies published linking disgust to psychopathology, and this evidence provides a background for discussing the possible mechanisms by which disgust influences mental health problems. In this section, I will discuss four possible processes by which disgust may have a direct or indirect causal effect on psychopathology symptoms. These are: (1) response driven effects of disgust, (2) disgust directed towards the self, (3) disgust-related effects of

Table 11.1 A summary of the four putative mechanisms of disgust described in this chapter alongside some of the types of psychopathology those mechanisms may causally influence

Putative mechanism	Potential psychopathologies
Response-driven effects of disgust	Avoidant psychopathologies influenced by decision-making under risk
Disgust directed towards the self (self-disgust)	Depression Borderline personality disorder Post-traumatic stress disorder Self-harm
Disgust-related effects of violations of moral or social norms	Obsessive-compulsive disorder Contamination fears
Stimulus-driven and strategic information-processing biases	Any psychopathology acquired and maintained via threat-related attentional and interpretation biases

violations of moral or social norms, and (4) stimulus-driven and strategic information-processing biases (see Table 11.1).

Response Driven Effects of Disgust

There is a clear pathway for disgust to influence psychopathology when the focus of those psychopathologies is disgust-relevant stimuli or events. We have already mentioned relevant candidates such as contamination fear in OCD, fear of small disease-relevant animals (including spider phobia), BII fears, and to some extent a range of eating disorders (such as anorexia nervosa and "picky eating"). When associated with high levels of disgust these psychopathologies may to some extent develop directly out of the disgust emotion's ability to enable the sufferer to quickly classify the focus of the psychopathology as dangerous and facilitate the deployment of strong avoidance, withdrawal, and rejection responses, as well as inhibiting habituation (Mason and Richardson 2010; Olatunji et al. 2009). Once these avoidance responses are established, this may then lead to anxiety and fear becoming predominant responses as the disgust-relevant stimuli and events are anticipated and coping strategies developed to avoid contact and contamination.

But, the responses that have evolved to characterise disgust may have some broader implications for psychopathology development than just facilitating anxious reactions to disgust-relevant stimuli. Most emotions that have evolved to protect against harm will have benefits (in terms of protection against harm), but will probably do so at the cost of foregoing other opportunities and benefits. For example, the pathogen-avoidance function of disgust must be traded-off against other important functional goals such as acquiring food or engaging in mating activities (e.g., Fleischman et al. 2015; see also Borg and de Jong, Chap. 9, this volume). In this sense, disgust acts to *regulate* exposure to potentially harmful pathogens rather than avoid them unconditionally, and so may contribute to decision-making under risk (e.g., weighing up the costs and benefits of interacting with something that may be potentially harmful; Sparks et al. 2018; Tybur et al. 2013). Sparks et al. (2018) argue that considering the role of disgust at this ultimate functional level, rather than at a proximal mechanism level, can help to explain some of the characteristics of the emotion, such as the higher levels of disgust propensity and sensitivity found in women as opposed to men (Al-Shawaf and Buss 2017). This higher-level functionality of disgust may influence risk-taking and protective decision-making, with higher levels of disgust being associated with greater risk avoidance that may trigger anxiety and fear when risk avoidance strategies are activated. In fact, all harm-avoidance emotions, including disgust, are likely to have their own effects on risky decision-making and contribute these inputs to risk taking and risk avoidance across diverse domains of behaviour (Sparks et al. 2018).

One implication of a higher-level functional view of disgust is that individual differences in risk-taking across all harm-related domains may in part be driven by disgust levels and its input into avoidant decision-making, with higher disgust levels leading to greater avoidance. Avoidance is a key feature of very many psychopathologies, and one that contributes directly to the symptoms of psychopathology. For example, avoidance detrimentally affects emotional regulation, which means you become less emotionally reactive the more you avoid. This then has a number of negative knock-on consequences. You'll be less privy to a range of useful information conveyed by an emotional response (an emotion will tell you whether you should be frightened of something, angered by it, or merely surprised by it); you'll become less adaptive in your interactions with your environment (because your emotional reaction would normally signal which adaptive responses are most appropriate to deploy in those circumstances); you'll be more likely to make less effective actions in the future (and thereby likely to experience more emotional distress); and the interpersonal value of the emotion is lost (your friends, family, and colleagues will be less likely to understand your behaviour—or they may even entirely misinterpret it; Salter-Pedneault et al. 2004).

Disgust Directed Towards the Self

Many researchers are beginning to identify self-disgust as a form of the disgust emotion that is directed inwards at core and stable features of the self (e.g., Powell et al. 2015a). Self-disgust may be generated by negative evaluation of the individual's own features or actions by using disgust to imbue attributes of the self with negative affect—either as a form of self-punishment or self-evaluation. Alternatively, disgust elicited by an external stimulus or disgust directed at the individual by others may become associated with attributes of the self, and through processes of iterative self-appraisal may become incorporated as a core attribute of the self, which influences how the individual perceives the world and the self over the longer term (i.e., has a "schematic" quality).

But, if self-disgust is to be a useful construct that can influence mental health symptoms, then it needs to be clearly differentiated from other similar negative emotions such as shame, guilt, self-criticism, and self-hatred. In effect, a definition of self-disgust should include disgust or contamination-based appraisals of the self, a strong physical sense of revulsion at the self along with other visceral aspects of disgust, and even attempts to cleanse the disgusting self (Gilbert et al. 2004; Simpson et al. 2010). These characteristics should help to distinguish self-disgust from other competing emotions. After conducting a review of the similarities and differences between self-disgust and other self-oriented negative constructs, Clarke et al. (2018) define self-disgust as "a psychologically destructive emotion, sometimes latent but easily triggered, with visceral content and resulting in a desire/need to avoid the disgusting aspect of the self psychologically and behaviourally and to attempt to expunge the self" (Clarke et al. 2018, p. 124).

If we are to successfully investigate whether self-disgust contributes to psychopathologies, we need to have valid and reliable psychometric measures of self-disgust that can identify the core features of the emotion and differentiate it from other similar negative emotional constructs. Examples are the Self-Disgust Scale (Overton et al. 2008) and the Questionnaire for the Assessment of Self-Disgust (Schienle et al. 2014). However, there is still some doubt about whether these measures capture all of the relevant features of self-disgust across different populations (see Clarke et al. 2018). But, even in the absence of truly valid measures of self-disgust, we can still speculate about the mechanisms by which self-disgust might have a direct or indirect causal influence on mental health symptoms.

Clarke et al. (2018) describe a number of possible pathways by which self-disgust might directly or indirectly influence mental health symptoms. For example, self-disgust may develop during childhood as a result of adverse childhood experiences (such as childhood neglect or physical or sexual abuse), and self-disgust then becomes a predictor or mediator of mental health problems in adulthood, such as depression or borderline personality disorder (see also Powell et al. 2015a). However, in examples such as this, it is important to be able to identify self-disgust as a genuine contributor to the psychopathology rather than just an unrelated correlate of the condition, or a consequence rather than cause, or a correlate of other constructs (such as shame or guilt) that may be the genuine causes of the psychopathology. A combination of mediation analyses and prospective studies may be needed to identify causal effects of self-disgust, as will the development of ecologically-valid lab-based procedures in which self-disgust can be directly manipulated (or alleviated if more ethically palatable, see e.g., Powell et al. 2015b) and its causal effect on potential mental health symptoms measured.

In their systematic review of self-disgust, Clarke et al. (2018) found varying levels of evidence of a *potential* causal role for self-disgust in depression, trauma-related problems such as posttraumatic stress disorder (PTSD), body image and eating disorders, OCD, and also found evidence for a bidirectional relationship between self-disgust and self-harm. However, in many of these cases it was difficult to effectively rule out the possibility that self-disgust was either a consequence of the psychopathology or merely a correlate of another construct that may have been a genuine causal factor.

If self-disgust does play a direct or indirect causal role in psychopathology symptoms, what is the nature of the mechanisms that may mediate these effects? These processes are likely to be complex—not least because there may be a bidirectional relationship between self-disgust and symptoms, but also because self-disgust may affect symptoms directly and also indirectly by affecting other internal states that in turn affect symptoms (e.g., self-disgust may directly affect self-harm, but also increase levels of depression, which in turn affects self-harm; Abdul-Hamid et al. 2014; Smith et al. 2015). In other cases, the role of self-disgust may be to either create or mediate biases in the way that information is processed that can help to develop or maintain common mental health problems, such as depression (Powell et al. 2013), but more direct laboratory-based experimental studies may be needed to determine whether self-disgust itself has a direct effect on information processing biases, including attentional biases and interpretation biases.

Disgust-Related Effects of Violations of Moral or Social Norms

There is considerable research suggesting that the disease-avoidance emotion of disgust has also evolved to convey the strong unacceptability of ideas and behaviours, and can also be elicited by violations of moral or social norms and by feelings of being "wronged" or betrayed (e.g., Giner-Sorolla et al. 2018; Plakias 2018; see also Giner-Sorolla, Chap. 8, this volume). This view has been particularly prominent in what has come to be known as the "body-to-soul preadaptation theory" of disgust in which it is hypothesised that there has been an expansion in the function of disgust from guarding the "body" to guarding the "soul" (Haidt 2012; Rozin et al. 2009). As such, people will react to violations of the sacred (e.g., violations of moral norms) with the same disgust emotion that is involved in reactions to pathogens, contamination, and putrid objects, such as faeces (e.g., Lieberman et al. 2003; Plakias 2018; Ritter and Preston 2011). If violations of moral norms and the sacred do elicit disgust, then might this influence the acquisition and maintenance of psychopathology, and if so, by what processes would this happen?

It is worth making clear that once visceral disgust has been elicited (whether it be by violation of moral norms or any other means) it could facilitate psychopathology symptoms through any of the putative mechanisms discussed in this chapter. But, might there be some psychopathologies that are particularly vulnerable to the elicitation of moral disgust? One extension of the view that disgust is elicited by violations of moral norms and the sacred is that disgust is also highly associated with religiosity. For example, Terrizzi Jr. et al. (2012) found that the disease-avoidance component of disgust was highly correlated with religious conservatism, and this in turn mediated religious prejudices. Religious societies are frequently concerned with the cleanliness of their followers' minds and bodies (Graham and Haidt 2010), and religious rituals throughout the world often involve purification practices (Sica et al. 2002). Thus, any violation of the moral order required by particular religious beliefs is likely to threaten purity and elicit disgust (Ritter and Preston 2011), and a bidirectional relationship may exist between religiosity and the emotion of disgust (Olatunji et al. 2005).

What is significant about the relationship between disgust and religiosity is that disgust, morality, and religiosity are regularly implicated in the development of many forms of OCD (Inozu et al. 2014). In particular, disgust plays a causal role in the contamination fears that often underlie compulsive washing and purifying rituals in OCD (Olatunji 2010), and religiosity has been estimated to account for almost a quarter of all obsessive concerns in clinical samples (Summerfeldt et al. 1998). Indeed, many religions may espouse everyday customs and rituals that are risk factors for compulsive behaviour or obsessive thoughts (e.g., orthodox Judaism emphasises cleanliness related to dietary restrictions, family purity, praying correctly etc., Huppert et al. 2007). In addition, a more radical possibility is that adopting high moral or religious standards is itself caused by disgust, with the former then acting as a risk factor for OCD. This proposition is supported by a range of studies demonstrating that a disgust induction leads to more severe moral judgments (Horberg et al. 2009; Wheatley and Haidt 2005). Accordingly, as religiosity and high moral standards are a significant feature of OCD conditions, disgust may be one of the causal drivers for these standards rather than simply being an outcome of them (particularly when applied to cleanliness and purity).

However, while this evidence suggests that moral and sacred violation disgust may be a vulnerability factor for OCD in religious individuals, is moral disgust the same emotion as that elicited by pathogens, illness, and disease? There is recent evidence to suggest it may not always be (see also Giner-Sorolla, Chap. 8, this volume). First, there is emerging evidence that visceral disgust may only be elicited by bodily-related moral violations, such as incest or religious food taboos, while negative emotions such as anger are more likely to be elicited by sociomoral violations such as theft (Giner-Sorolla et al. 2018). Indeed, studies have demonstrated that the word "disgust" is often used to express anger or a negative reaction of intense disapproval, rather than featuring the full range of visceral and pathogen-avoidance components of core disease-avoidance disgust (Herz and Hinds 2013; Royzman and Kurzban 2011). Furthermore, a recent study by Kollareth and Russell (2019) found that the emotional reaction to a sacred religious violation was not the same as that to simple threats to health. Pathogen exposure led to a predominantly "grossed out" reaction similar to that caused by core disgust, but this "grossed out" reaction was not differentially found with exposure to sacred religious violations. These findings are very similar to those of Royzman et al. (2014), who reported that participants' reaction to pathogen-free sacred violations was predominantly anger and did not show core disgust-related features such as nausea, gagging, loss of appetite, or a desire to avoid.

So any mechanisms that mediate the association between disgust elicited by the violation of moral/social norms and psychopathology symptoms are likely to be complex and be contaminated by other negative emotions (such as anger) elicited by these violations. However, the relevance to the core disgust emotion of purity rituals common to many religions may be a significant risk factor for the development of OCD compulsions and obsessions in highly religious individuals—mainly as a result of the disgust emotion fuelling contamination fears if purity rituals are not fully and properly adhered to (see Davey et al. 2014, Chapter 2).

Stimulus-Driven and Strategic Information-Processing Biases

Disgust is a negatively experienced emotion and as such has been shown to have effects on information processing similar to many other negative emotions (e.g., Knowles et al. 2019). In particular, experiencing disgust creates biases in both attention processes and interpretation processes—effects that are likely to influence the detection and threatening interpretation of stimuli and events, regardless of their disgust-relevance. This potentially provides the basis for a mechanism that could explain why disgust is associated with psychopathologies that, *prima facie*, appear to have little or no disgust-relevance, such as separation anxiety (Muris et al. 1999), agoraphobia (Muris et al. 2000), symptoms of schizophrenia (Schienle et al. 2003), and height anxiety and claustrophobia (Davey and Bond 2006) (cf. Davey 2011; although it should be made clear that there is as yet no clear experimental evidence that disgust plays a *causal* role in these disgust-irrelevant psychopathologies).

Knowles et al. (2019) have provided a comprehensive review of information processing biases in disgust, with the most robust information-processing bias seeming to be at the attentional level, with disgust-prone individuals exhibiting an initial enhanced orienting response and attentional bias towards disgusting cues compared to neutral cues (Cisler et al. 2009b; Vogt et al. 2011). Further studies show a more general attentional and orienting bias to threat generally in disgust-prone individuals (Stevenson et al. 2014). However, while disgust may be associated with an initial stimulus-driven attentional bias towards disgust-relevant or threat-relevant stimuli, this appears to be subsequently followed by significant avoidance of these stimuli (Armstrong et al. 2014; Bradley et al. 2015). Knowles et al. (2019) argue that this subsequent avoidance of disgust and threat-related material, after initial orientation to the material, may represent a strategic activity in disgust-prone individuals where the individual chooses to attend away from disgusting or threatening stimuli as a form of emotional regulation to reduce distress.

What appears to be a downstream consequence of these attentional biases is that subsequent ambiguous information is then interpreted in a threatening rather than benign manner. This has been called the cognitive bias hypothesis (Hirsch et al. 2006), in which pre-conscious attentional biases to threat determine congruent biases in interpretation. Studies that have examined this interpretation bias following disgust mood inductions suggest that experienced disgust results in a threat-interpretation bias (in which ambiguous stimuli and events are interpreted as threatening rather than benign). This disgust-induced bias is not dissimilar to the threat-interpretation bias caused by anxious mood, and is not moderated by measures of trait anxiety or anxiety sensitivity (Davey et al. 2006; Leathers-Smith and Davey 2011; Mayer et al. 2009). The combined attention-interpretation bias caused by disgust can act to impart stimuli and events with threatening meaning and cause congruent biases in recall and recognition (Everaert et al. 2013). It is now well known that experimentally inducing interpretation biases in individuals leads to congruent effects on state anxiety (Mathews and Mackintosh 2000; Salemink et al. 2007), and this suggests that the effect that disgust has on interpretation biases would have a direct causal effect on experienced anxiety via biased processing. This is likely to impact on any anxiety-related psychopathology regardless of whether it shares a disease-avoidance functionality with the disgust emotion or not—an assumption borne out by the fact that recent studies have indicated that information-processing biases can be identified as mediators between disgust and subsequent measures of psychopathology (Ólafsson et al. 2019; Zanjani et al. 2018).

If such a mechanism is operating regularly and allows disgust to causally facilitate psychopathology symptoms, then we need further experimental studies that will clearly demonstrate that: (1) disgust's role is causal in its effect on psychopathology measures; (2) that information processing biases such as attentional and interpretation biases play a mediating role in this process; and (3) that these factors can affect any psychopathology symptom regardless of its disgust relevance.

Conclusions

Disgust is no longer "the forgotten emotion of psychiatry" (Phillips et al. 1998). Its broad involvement in psychopathology is now well known and well documented. It is experienced more intensely in those suffering a range of psychopathologies than in healthy individuals, and experimental lab-based studies have indicated that its effect on psychopathology symptoms can be causal. But, the involvement of disgust in psychopathology is likely to be complex and often difficult to unravel, and we are still some distance away from developing predictive models of the role of disgust in psychopathology. More prospective studies are needed that identify disgust as a genuine risk factor for psychopathologies, and we cannot develop models of disgust in psychopathology until we have clearly established the causal effects of disgust through controlled experimental manipulation of its features, and objective measurement of the outcomes of those manipulations, on potential psychopathology symptoms.

In this chapter I have discussed a number of putative mechanisms by which the disgust emotion may exert its effect on psychopathology symptoms, and this range of putative mechanisms testifies to the complexity of the disgust emotion and the many different levels at which it may influence cognition, behaviour, and mental health. Evidence that validates some or all of these mechanisms will play an important role in helping to develop either preventative or ameliorative interventions for disgust-relevant psychopathologies.

References

Abdul-Hamid, S., Denman, C., & Dudas, R. B. (2014). Self-relevant disgust and self-harm urges in patients with borderline personality disorder and depression: a pilot study with a newly designed psychological challenge. *PLoS One*, *9*(6), e99696.

Al-Shawaf, Lewis, & Buss (2017) Sex differences in disgust: Why are women more easily disgusted than men? *Emotion Review*, *10*, 149-160.

Armstrong, T., McClenahan, L., Kittle, J., & Olatunji, B. O. (2014). Don't look now! Oculomotor avoidance as a conditioned disgust response. *Emotion*, *14*(1), 95–104.

Blacker, K.-A., & LoBue, V. (2016). Behavioral avoidance of contagion in childhood. *Journal of Experimental Child Psychology*, *143*, 162–170.

Bradley, M. M., Costa, V. D., & Lang, P. J. (2015). Selective looking at natural scenes: Hedonic content and gender. *International Journal of Psychophysiology*, *98*, 54–58.

Britton, G. I., & Davey, G. C. L. (2014). Interrelationships between negative mood and clinical constructs: A motivational systems approach. *Frontiers in Psychology*, *5*, 393.

Brown, S. D., & Harris, G. (2012). Disliked food acting as a contaminant during infancy. A disgust based motivation for rejection. *Appetite*, *58*(2), 535–538.

Cisler, J. M., Olatunji B. O., & Lohr, J. M. (2009a). Disgust sensitivity and emotion regulation potentiate the effect of disgust propensity on spider fear, blood-injection-injury fear, and contamination fear. *Journal of Behaviour Therapy & Experimental Psychiatry*, *40*, 219-229.

Cisler, J. M., Olatunji, B. O., Lohr, J. M., & Williams, N. L. (2009b). Attentional bias differences between fear and disgust: implications for the role of disgust in disgust-related anxiety disorders. *Cognition and Emotion*, *23*(4), 675–687.

Clarke, A., Simpson, J., & Varese, F. (2018). A systematic review of the clinical utility of the concept of self-disgust. *Clinical Psychology & Psychotherapy*, *26*, 110-134.

Curtis, V., & de Barra, M. (2018). The structure and function of pathogen disgust. *Philosophical Transactions of the Royal Society B-Biological Sciences*, *373*, 20170208.

Davey, G. C. L. (1994a). Self-reported fears to common indigenous animals in a UK population: the role of disgust sensitivity. *British Journal of Psychology*, *85*, 541-554.

Davey, G. C. L. (1994b). Disgust. In V.S. Ramachandran (Ed.), *Encyclopedia of human behavior*. San Diego Press.

Davey, G. C. L. (2011). Disgust: the disease-avoidance emotion and its dysfunctions. *Philosophical Transactions of the Royal Society B: Biological Sciences*, *366*(1583), 3453-3465.

Davey, G. C. L. (2018). *The anxiety epidemic*. Little Brown Books.

Davey, G. C. L., & Bond, N. (2006). Using controlled comparisons in disgust psychopathology research: The case of disgust, hypochondriasis and health anxiety. *Journal of Behavior Therapy & Experimental Psychiatry*, *37*, 4-15.

Davey, G. C. L., Buckland, G., Tantow, B., & Dallos, R. (1998). Disgust and eating disorders. *European Eating Disorders Review*, *6*, 201-211.

Davey, G. C. L., Bickerstaffe, S., & MacDonald, B.A. (2006). Experienced disgust causes a negative interpretation bias: A causal role for disgust in anxiety psychopathology. *Behaviour Research & Therapy*, *44*, 1375-1384.

Davey, G. C. L., MacDonald, B., & Brierley, L. (2008). The effect of disgust on anxiety ratings to fear-relevant, disgust-relevant and fear-irrelevant stimuli. *Journal of Anxiety Disorders*, *22*, 1347-1354.

Davey, G. C. L., Dash, S., & Meeten, F. (2014). *Obsessive compulsive disorder*. Palgrave.

De Houwer, J., Thomas, S., & Baeyens, F. (2001). Associative learning of likes and dislikes: A review of 25 years of research on human evaluative conditioning. *Psychological Bulletin*, *127*, 853-869.

de Jong, P.J., & Merckelbach, H. (1998). Blood-injection-injury phobia and fear of spiders. Domain specific individual differences in disgust sensitivity. *Personality & Individual Differences*, *24*, 153-158.

DeJesus, J. M., Shutts, K., & Kinzler, K. D. (2015). Eww she sneezed! Contamination context affects children's food preferences and consumption. *Appetite, 87*, 303–309.

Everaert, J., Tierens, M., Uzieblo, K., & Koster, E. H. W. (2013). The indirect effect of attention bias on memory via interpretation bias: evidence for the combined cognitive bias hypothesis in subclinical depression. *Cognition & Emotion, 27*, 1450–1459.

Feder, Y. (2015). Contamination appraisals, pollution beliefs, and the role of cultural inheritance in shaping disease avoidance behavior. *Cognitive Science, 40*(6), 1561–1585.

Field, A. F., & Davey, G. C. L. (1997). Conceptual conditioning: Evidence for an artifactual account of evaluative learning. *Learning & Motivation, 28*, 446-464.

Field, A. F., & Davey, G. C. L. (1998). Evaluative conditioning: Arti-fact or -fiction?—A reply to Baeyens, De Houwer, Vansteenwegen, and Eelen (1998) *Learning & Motivation, 29*, 475-491.

Fleischman, D.S., Hamilton, L.D., Fessler, D.M.T., & Meston, C.M. (2015). Disgust versus lust: Exploring the interactions of disgust and fear with sexual arousal in women. *PLoS ONE, 10*, e0118151.

Gilbert, P., Clarke, M., Hempel, S., Miles, J. N. V., & Irons, C. (2004). Criticizing and reassuring oneself: An exploration of forms, styles and reasons in female students. *British Journal of Clinical Psychology, 43*(1), 31–50.

Giner-Sorolla, R., Kupfer, T., & Sabo, J. (2018). What makes moral disgust special? An integrative functional review. *Advances in Experimental Social Psychology, 57*, 223-289.

Graham, J., & Haidt, J. (2010). Beyond beliefs: Religions bind individuals into moral communities. *Personality and Social Psychology Review, 14*(1), 140-150.

Haidt, J. (2012). *The righteous mind: Why good people are divided by politics and religion*. Pantheon/Random House.

Haidt, J., McCauley, C., & Rozin, P. (1994). Individual differences in sensitivity to disgust: A scale sampling seven domains of disgust elicitors. *Personality & Individual Differences, 16*, 701-713.

Harris, A. A., Romer, A. L., Hanna, E. K., Keeling, L. A., LaBar, K. S. Sinnott-Armstrong, W., Strauman, T. J., Wagner, H. R., Marcus, M. D., & Zucker, N. L. (2019). The central role of disgust in disorders of food avoidance. *International Journal of Eating Disorders, 52*, 543-553.

Harvey, T., Troop, N. A., Treasure, J. L., & Murphy, T. (2002). Fear, disgust, and abnormal eating attitudes: A preliminary study. *International Journal of Eating Disorders, 32*, 213-218.

Herz, R. S., & Hinds, A. (2013). Stealing is not gross: Language distinguishes visceral disgust from moral violations. *The American Journal of Psychology, 126*, 275–286.

Hirsch, C. R., Clark, D. M., & Mathews, A. (2006). Imagery and interpretations in social phobia: Support for the combined cognitive biases hypothesis. *Behavior Therapy, 37*, 223–236.

Horberg, E. J., Oveis, C., Keltner, D., & Cohen, A. B. (2009). Disgust and the moralization of purity. *Journal of Personality and Social Psychology, 97*(6), 963-976.

Huppert J. D., Siev J., & Kushner E. S. (2007) When religion and obsessive-compulsive disorder collide: Treating scrupulosity in Ultra-Orthodox Jews. *Journal of Clinical Psychology, 63*, 925-941.

Ille, R., Schöggl, H., Kapfhammer, H.P., Arendasy, M., Sommer, M., & Schienle, A. (2014). Self-disgust in mental disorders—symptom-related or disorder specific. *Comprehensive Psychiatry, 55*, 938-943.

Inozu, M., Ulukut, F.O., Ergun, G., & Alcolado, G.M. (2014). The mediating role of disgust sensitivity and thought-action fusion between religiosity and obsessive compulsive symptoms. *International Journal of Psychology, 49*, 334-341.

Izard, C. E. (1993). Four systems for emotion activation: Cognitive and noncognitive processes. *Psychological Review, 100*, 68-90.

Izard, C. E. (2007). Basic emotions, natural kinds, emotion schemas, and a new paradigm. *Perspectives on Psychological Science, 2*, 260-269.

Knowles, K. A., Jessup, S. C., & Olatunji, B. O. (2018). Disgust in anxiety and obsessive-compulsive disorders: Recent findings and future directions. *Current Psychiatry Reports, 20*, 68.

Knowles, K. A., Cox, R. C., Armstrong, T., & Olatunji, B.O. (2019). Cognitive mechanisms of disgust in the development and maintenance of psychopathology: A qualitative review and synthesis. *Clinical Psychology Review*, *69*, 30-50.

Kollareth, D., & Russell, J. A. (2019). Disgust and the sacred: Do people react to violations of the sacred with the same emotion they react to something putrid. *Emotion*, *19*, 37-52.

Leathers-Smith, E., & Davey, G. C. L. (2011). The disgust threat interpretation bias is not moderated by anxiety and disgust sensitivity. *Journal of Experimental Psychopathology*, *2*, 63-76.

Lieberman, D., Tooby, J., & Cosmides, L. (2003). Does morality have a biological basis? An empirical test of factors governing moral statements relating to incest. *Proc. R. Soc. Lond.*, *B270*, 819–826.

Mancini, F., Grannani, A., & D'Olimpio, F. (2001). The connection between disgust and obsessions and compulsions in a non-clinical sample. *Personality & Individual Differences*, *31*, 1173-1180.

Marzillier, S. L. & Davey, G. C. L. (2004). The emotional profiling of disgust-eliciting stimuli: Evidence for primary and complex disgusts. *Cognition & Emotion*, *18*, 313-336.

Marzillier, S. L. & Davey, G. C. L. (2005). Anxiety and disgust: Evidence for a unidirectional relationship. *Cognition & Emotion*, *19*, 729-750.

Mason, E. C., & Richardson, R. (2010). Looking beyond fear: The extinction of other emotions implicated in anxiety disorders. *Journal of Anxiety Disorders*, *24*, 63-70.

Matchett, G., & Davey, G. C. L. (1991). A test of a disease-avoidance model of animal phobias. *Behaviour, Research and Therapy*, *29*, 91-94.

Mathews, A., & Mackintosh, B. (2000). Induced emotional interpretation bias and anxiety. *Journal of Abnormal Psychology*, *109*, 602-615.

Mayer, B., Muris, P., Busser, K., & Bergamin, J. (2009). A disgust mood state causes a negative interpretation bias, but not in the specific domain of body-related concerns. *Behaviour Research and Therapy*, *47*(10), 876–881.

Melli, G., Poli, A., Chiorri, C., & Olatunji, B.O. (2019). Is heightened disgust propensity truly a risk factor for contamination-related obsessive-compulsive disorder? *Behavior Therapy*, *50*, 621-629.

Mulkens, S.A., de Jong, P.J., & Merckelbach, H. (1996). Disgust and spider phobia. *Journal of Abnormal Psychology*, *105*, 464-468.

Muris, P., Merckelbach, H., Schmidt, H., & Tierny, S. (1999). Disgust sensitivity, trait anxiety and anxiety disorders symptoms in normal children. *Behaviour Research and Therapy*, *37*, 953-961.

Muris, P., Merckelbach, H., Nederkoorn, S., Rassin, E., Candel, I., & Horselenberg, R. (2000). Disgust and psychopathological symptoms in a nonclinical sample. *Personality & Individual Differences*, *29*, 1163-1167.

Neuberg, S. L., Kenrick, D. T., & Schaller, M. (2011). Human threat management systems: self-protection and disease-avoidance. *Neurosci. Biobehav. Rev.*, *35*, 1042-1051.

Ólafsson, R. P., Friðriksdóttir, A. E., Sveinsdóttir, S. þ., & Kristjánsson, A. (2019). Evidence for an attention bias toward disgust in contamination fear. *Journal of Experimental Psychopathology*. https://doi.org/10.1177/2043808719870043.

Olatunji, B. O. (2010). Changes in disgust correspond with changes in symptoms of contamination-based OCD: A prospective examination of specificity. *Journal of Anxiety Disorders*, *24*(3), 313-317.

Olatunji, B.O., & Armstrong, T. (2009). Contamination fear and effects of disgust on distress in a public restroom. *Emotion*, *9*, 592-597.

Olatunji B. O., Sawchuk C. N., Lohr J. M. & de Jong P. J. (2004). Disgust domains in the prediction of contamination fear. *Behaviour Research & Therapy*, *42*, 93-104.

Olatunji, B. O., Tolin, D. F., Huppert, J. D., & Lohr, J. M. (2005). The relation between fearfulness, disgust sensitivity and religious obsessions in a non-clinical sample. *Personality and Individual Differences*, *38*, 891–902.

Olatunji, B. O., Lohr, J. M., Sawchuk, C. N., & Tolin, D. F. (2007a). Multimodal assessment of disgust in contamination-related obsessive-compulsive disorder. *Behaviour Research & Therapy, 45*, 263-276.

Olatunji, B. O., Smits, J. A. J., Connolly, K., Willems, J., & Lohr, J. M. (2007b). Examination of the decline in fear and disgust during exposure to threat-relevant stimuli in blood-injection-injury phobia. *Journal of Anxiety Disorders, 21*, 445-455.

Olatunji, B. O., Wolitzky-Taylor, K. B., Willems, J., Lohr, J. M., & Armstrong, T. (2009). Differential habituation of fear and disgust during exposure to threat-relevant stimuli in contamination-based OCD: an analogue study. *Journal of Anxiety Disorders, 23*, 118–123.

Olatunji, B. O., Cisler, J., McKay, D., & Phillips, M. L. (2010). Is disgust associated with psychopathology? Emerging research in the anxiety disorders. *Psychiatry Research, 175*, 1-10.

Ouimet, A. J., Gawronski, B., & Dozois, D. J. A. (2009). Cognitive vulnerability to anxiety: A review and integrative model. *Clinical Psychology Review, 29*, 459-470.

Overton, P. G., Markland, F. E., Taggart, H. S., Bagshaw, G. L., & Simpson, J. (2008). Self-disgust mediates the relationship between dysfunctional cognitions and depressive symptomatology. *Emotion, 8*(3), 379.

Phillips, M. L., Senior, C., Fahy, T., & David, A. S. (1998). Disgust—the forgotten emotion of psychiatry. *British Journal of Psychiatry, 172*, 373-375.

Plakias, A. (2018). The response model of moral disgust. *Synthese, 195*, 5453-5472.

Poli, A., Melli, G., & Radomsky, A. S. (2019). Different disgust domains specifically relate to mental and contact contamination fear in obsessive-compulsive disorder: Evidence from a path analytic model in an Italian clinical sample. *Behavior Therapy, 50*, 380-394.

Powell, P. A., Simpson, J., & Overton, P. G. (2013). When disgust leads to dysphoria: A three-wave longitudinal study assessing the temporal relationship between self-disgust and depressive symptoms. *Cognition and Emotion, 27*(5), 900–913.

Powell, P. A., Simpson, J., & Overton, P. G. (2015a). An introduction to the revolting self: Self-disgust as an emotion schema. In P. A. Powell, P. G. Overton, & J. Simpson (Eds.), *The revolting self: Perspectives on the psychological, social, and clinical implications of self-directed disgust* (pp. 1–24). Karnac.

Powell, P. A., Simpson, J., & Overton, P. G. (2015b). Self-affirming trait kindness regulates disgust toward one's physical appearance. *Body Image, 12*, 98-107.

Ritter, R. S., & Preston, J. L. (2011). Gross gods and icky atheism: Disgust responses to rejected religious beliefs. *Journal of Experimental Social Psychology, 47*, 1225–1230.

Rottman, J., DeJesus, J., & Greenebaum, H. (2019). Developing disgust: Theory, measurement, and application. In V. LoBue et al. (Eds.) *Handbook of emotional development* (pp. 283-309). Springer.

Royzman, E., & Kurzban, R. (2011). Minding the metaphor: The elusive character of moral disgust. *Emotion Review, 3*, 269–271.

Royzman, E. B., Atanasov, P., Landy, J. F., Parks, A., & Gepty, A. (2014). CAD or MAD? Anger (not disgust) as the predominant response to pathogen-free violations of the divinity code. *Emotion, 14*, 892–907.

Rozin, P., & Fallon, A. E. (1987). A perspective on disgust. *Psychological Review, 94*, 23-41.

Rozin, P., Haidt, J., & McCauley, C. R. (2009). Disgust: The body and soul emotion in the 21st century. In B. O. Olatunji & D. McKay (Eds.), *Disgust and its disorders: Theory, assessment, and treatment implications* (pp. 9–29). American Psychological Association.

Rozin, P., Haidt, J., & McCauley, C. R. (2016). Disgust. In L. F. Barrett, M. Lewis, & J. M. Haviland-Jones (Eds.), *Handbook of emotions* (4th ed., pp. 815–834). Guilford Press.

Russell, P. A. (1979). Fear-evoking stimuli. In W. Sluckin (Ed.), *Fear in Animals and Man*. Van Nostrand Reinhold.

Salemink, E., van den Hout, M., & Kindt, M. (2007). Trained interpretive bias: Validity and effects on anxiety. *Journal of Behavior Therapy and Experimental Psychiatry, 38*, 212-224.

Salter-Pedneault, K., Tull, M. T., & Roemer, L. (2004). The role of avoidance of emotional material in the anxiety disorders. *Applied & Preventive Psychology, 11*, 95-114.

Schienle, A., Schafer, A., Stark, R., Walter, B., Franz, M., & Vaitl, D. (2003). Disgust sensitivity in psychiatric disorders: A questionnaire study. *Journal of Nervous & Mental Disease, 191*, 831-834.

Schienle, A., Ille, R., Sommer, M., & Arendasy, M. (2014). Diagnosis of self-disgust in the context of depression. *Verhaltenstherapie, 24*(1), 15–20.

Sica, C., Novara, C., & Sanavio, E. (2002). Religiousness and obsessive–compulsive cognitions and symptoms in an Italian population. *Behaviour Research and Therapy, 40*(7), 813-823.

Simpson, J., Hillman, R., Crawford, T., & Overton, P. G. (2010). Self-esteem and self-disgust both mediate the relationship between dysfunctional cognitions and depressive symptoms. *Motivation and Emotion, 34*(4), 399–406.

Smith, N. B., Steele, A. M., Weitzman, M. L., Trueba, A. F., & Meuret, A. E. (2015). Investigating the role of self-disgust in nonsuicidal self-injury. *Archives of Suicide Research, 19*(1), 60–74.

Sparks, A. M., Fessler, D. M. T., Chan, K. Q., Ashokkumar, A., & Holbrook, C. (2018). Disgust as a mechanism for decision making under risk: Illuminating sex differences and individual risk-taking correlates of disgust propensity. *Emotion, 18*, 942-958.

Stevenson, R. J., Oaten, M. J., Case, T. I., Repacholi, B. M., & Wagland, P. (2010). Children's response to adult disgust elicitors: Development and acquisition. *Developmental Psychology, 46* (1), 165–177.

Stevenson, R. J., Oaten, M. J., Case, T. I., & Repacholi, B. (2014). Is disgust prepared? A preliminary examination in young children. *The Journal of General Psychology, 141*(4), 326–347.

Stevenson, R. J., Case, T. I., Oaten, M. J., Stafford, L., & Saluja, S. (2019). A proximal perspective on disgust. *Emotion Review, 11*(3), 209-225.

Summerfeldt, L., Antony, M. M., Downie, F., Richter, M. A., & Swinson, R. P. (1998). Prevalence of particular obsessions and compulsions in a clinic sample. Unpublished manuscript.

Terrizzi, Jr., J. A., Shook, N. J., & Ventis, L. (2012). Religious conservatism: An evolutionary evoked disease-avoidance strategy. *Religions, Brains, & Behavior, 2*, 105-120

Thorpe, S. J. and Salkovskis, P. M. (1998). Studies on the role of disgust in the acquisition and maintenance of specific phobias. *Behaviour Research and Therapy, 36*, 877-893.

Tybur, J. M., Lieberman, D., Kurzban, R., & DeScioli, P. (2013). Disgust: Evolved function and structure. *Psychological Review, 120*, 65-84.

Vogt, J., Lozo, L., Koster, E. H. W., & de Houwer, J. (2011). On the role of goal relevance in emotional attention: disgust evokes early attention to cleanliness. *Cognition and Emotion, 25*(3), 466–478.

Ware J., Jain K., Burgess I. & Davey G. C. L. (1994). Disease-avoidance model: Factor analysis of common animal fears. *Behaviour Research & Therapy, 32*, 57-63.

Webb, K., & Davey, G. C. L. (1993). Disgust sensitivity and fear of animals: Effect of exposure to violent or revulsive material. *Anxiety, Stress and Coping, 5*, 329-335.

Wheatley, T., & Haidt, J. (2005). Hypnotic disgust makes moral judgments more severe. *Psychological Science, 16*, 780-784.

Woody, S. R., & Teachman, B. A. (2000). Intersection of disgust and fear: Normative and pathological views. *Clinical Psychology: Science & Practice, 7*, 291-311.

Zanjani, Z., Yaghubi, H., Shaeiri, M. R., & Fata, L. (2018). Relationship between disgust propensity and contamination obsessive-compulsive symptoms: The mediating role of information processing bias. *Iranian Red Crescent Medical Journal, 20*, e67133.

Chapter 12
Disgust in Neurological Disorders

Paul G. Overton, Ana B. Vivas, and Jane Simpson

A Systems Neuroscience View of Disgust

To begin to understand how a complex system such as disgust can be affected by the changes that underlie neurological disorders, it is necessary to consider firstly the basic requirements of a "disgust system" in the computational sense, and then see how those requirements might map onto the brain. When considering disgust in a neuroscientific context, the disgust system must consist of a sensory front end, presumably tuned to detect appropriate stimuli. The system needs to respond to a range of stimulus classes—visual disgust elicitors (e.g., faeces), auditory disgust elicitors (e.g., sounds of vomiting), and also the visual and auditory cues emitted by others indicating that they have encountered something that is disgusting. The latter are most likely communicated via facial expressions (visual) and/or prosody (auditory). The disgust system is also likely to have access to other classes of stimuli, for example odours and touch sensations (when things feel squishy, moist, scabid, wriggling). It also requires an output through which the detection and evaluation of disgust-relevant stimuli can influence bodily processes, including those associated with behavioural avoidance. In-between input and output an evaluation mechanism is required to assess the threat level and orchestrate the appropriate response

P. G. Overton (✉)
Department of Psychology, University of Sheffield, Sheffield, UK
e-mail: p.g.overton@sheffield.ac.uk

A. B. Vivas
Department of Psychology, CITY College, University of York Europe Campus, Thessaloniki, Greece
e-mail: vivas@york.citycollege.eu

J. Simpson
Division of Health Research, Lancaster University, Lancaster, UK
e-mail: j.simpson2@lancaster.ac.uk

P. A. Powell, N. S. Consedine (eds.), *The Handbook of Disgust Research*,
https://doi.org/10.1007/978-3-030-84486-8_12

(and perhaps label it on the basis of sensory feedback), or at least orchestrate the appropriate request or "bid" for a response to be made (more on that later). The result of that evaluation and the subsequent changes to bodily processes presumably underlies the experience of disgust. Extra complexity is added by the fact that each stage of the tripartite input-processer-output model can be influenced by other brain systems which are able to modulate the disgust system based on internal (hunger, thirst etc.) and external (stimuli eliciting approach or avoidance) contextual influences (see Fig. 12.1 for a schematic of the proposed disgust system).

The Biological Implementation of the Disgust System

In functional magnetic resonance imaging (fMRI) studies, in which participants viewed images of mutilation and contamination, the insular cortex appears to be one of the primary brain regions that exhibited elevated activity (Wright et al. 2004). The insular cortex is further implicated in disgust by evidence that electrical stimulation of the ventral sector of the monkey anterior insular cortex elicits behavioural responses consisting of typical disgust grimaces (such as the curling of the upper lip and the wrinkling of the nose), occasionally followed by retching (Caruana et al. 2011; Jezzini et al. 2012). Sometimes, more complex patterns have been observed, including food spitting or even refusal of food intake (Caruana et al. 2011; Jezzini et al. 2012). In humans, anterior insular cortex stimulation performed to localise an epileptic focus results in disgust-like sensations in the throat and mouth (Krolak-Salmon et al. 2003). Furthermore, fMRI studies show a role for the insular cortex (especially the anterior insular) in the perception of disgust in the facial expressions of others, in the direct experience of disgust, and in disgust-related imaginary experiences (Jabbi et al. 2008; Jehna et al. 2011; Krolak-Salmon et al. 2003; Phillips et al. 1997; Sambataro et al. 2006; Wicker et al. 2003). Likewise, selective impairment of disgust recognition has been reported in association with lesions of the insular cortex (Calder et al. 2000), and loss of insular volume is associated with deficits in identifying disgust (Adolphs et al. 2003).

In terms of the sensory inputs to the insular cortex carrying disgust-related information, prime candidates appear to be areas of the superior temporal cortex—the superior temporal sulcus and the superior temporal gyrus. The insular cortex receives a direct input from both areas (Mufson and Mesulam 1982). Evidence suggests that facial and vocal expressions of disgust both activate the superior temporal gyrus (Chen et al. 2014; Phillips et al. 1998; von dem Hagen et al. 2009), and facial expressions of disgust activate the superior temporal sulcus (von dem Hagen et al. 2009). These areas may play a more general role in conveying disgust-related information, in that the superior temporal gyrus is activated by non-vocal disgusting sounds (e.g., vomiting, Köchel et al. 2013), and the superior temporal area is activated by disgusting visual stimuli, such as rotting food, eating worms and food with insects (Pujol et al. 2018).

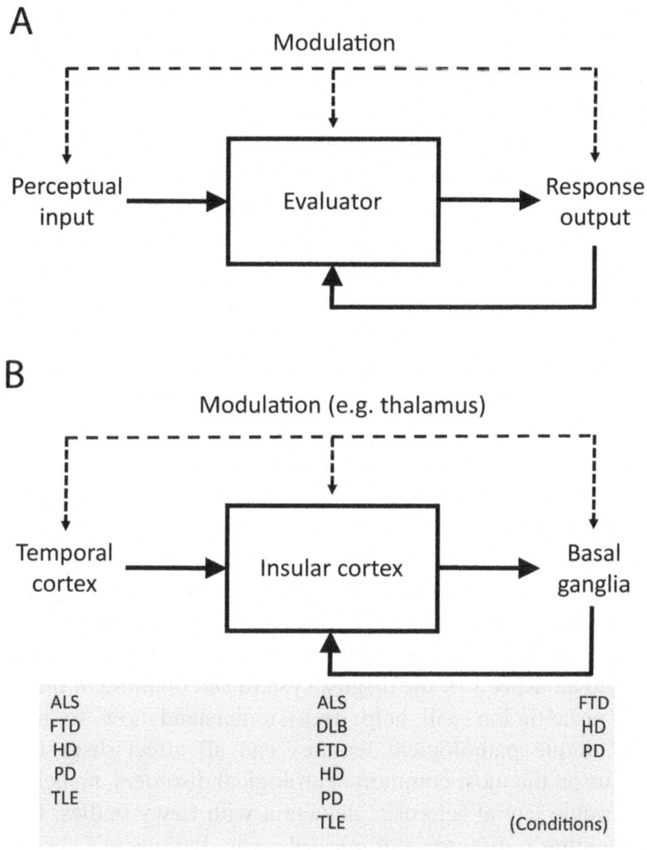

Fig. 12.1 Model of the putative "disgust system" in the brain. (**a**) At a computational level, the disgust system is considered to consist of a sensory front end, presumably tuned to detect appropriate stimuli. The disgust system also requires an output through which the detection and evaluation of disgust-relevant stimuli can influence bodily processes. In between input and output an evaluation mechanism is required to assess the threat level and orchestrate the appropriate response. The result of that evaluation and the subsequent changes to bodily processes presumably underlies the experience of disgust. Extra complexity is added by the fact that each stage of the tripartite input-processer-output model can be influenced by other brain systems on the basis of internal and external contextual influences. (**b**) The abstract model in "A" is mapped onto neural systems that are hypothesised to undertake the relevant processing steps. In addition, examples of neurological disorders ("conditions") considered in the text where evidence suggests aspects of disgust processing at various levels are affected. *ALS* amyotrophic lateral sclerosis; *DLB* dementia with Lewy bodies; *PD* Parkinson's disease; *FTD* frontotemporal dementia; *HD* Huntington's disease; *TLE* temporal lobe epilepsy

According to our earlier systems-level decomposition of the disgust system, the insular cortex appears to be the central node, which receives evaluated sensory input from the superior temporal area. The insular cortex in turn outputs directly to various areas of the brain, including the dorsal and ventral striatum (Reep and Winans 1982;

Shimizu and Norita 1991). In terms of more recent models of how the brain selects appropriate responses, inputs to the striatum are considered to be the substrate by which processing modules in the brain, in this case the insular cortex, bid for expression in terms of motor, autonomic and/or endocrine output (Gurney et al. 2001; Redgrave et al. 1999). Activity in insular outputs can be seen as putting a bid into the central selection device (thought to be the striatum and the associated basal ganglia); hence, insular outputs are hypothesised to constitute the output side of the disgust system. The process of orchestrating an output presumably involves an evaluation of the disgust elicitor and hence it can be assumed that the insular cortex is (at least partly) responsible for the experience of disgust, with the superior temporal cortex providing an appropriately tuned sensory front end.

The fundamental tripartite decomposition of the disgust system into input-processor-output is further complicated by the fact that each component can be modulated by afferent inputs from other brain areas. For example, disgust-related cortical areas (and the striatum) are modulated by the thalamus, as part of the widely documented basal ganglia loop system (Alexander et al. 1986). Although clearly complex, in terms of experimental approaches to exploring the issue of disgust in neurological disorders, the vast majority of studies have focused on the ability to recognise disgust facial expressions (i.e., the input side of the system). A much smaller set of studies have looked at the ability to generate expressions of disgust (i.e., the output side of the system), or the experience of disgust. That said, there are studies relevant to all aspects of the disgust system and thinking of the wider system and its neural instantiation will help us to understand how so many different disorders with unique pathological features can all affect disgust. The present chapter will focus on the most common neurological disorders, namely Alzheimer's disease, amyotrophic lateral sclerosis, dementia with Lewy bodies, frontotemporal dementia, Huntington's disease, multiple sclerosis, Parkinson's disease, temporal lobe epilepsy, and Tourette's syndrome. Some other common neurological disorders, namely migraine and progressive supranuclear palsy, are missing from that list because as far as we are aware, disgust processing has not been examined at all in migraine and only minimally in progressive supranuclear palsy (Ghosh et al. 2009).

Recognition of Disgust-Related Stimuli

Recognition of disgust-related stimuli in neurological disorders has received considerable attention, the vast majority of literature focusing on the recognition of facial expressions of disgust. Paradigmatically, facial expressions of disgust are usually tested in two (related) ways. First, via a Facial Emotion Recognition Task based on pictures from the Ekman and Friesen (1976) database (or some other validated facial expression dataset). In this task, hit rate (percent correct) for each photograph is measured, where subjects select from a choice of six different verbal labels, corresponding to the six basic emotions: anger, disgust, fear, happiness, sadness, and surprise. Second, facial expressions are morphed with neutral faces to obtain varying degrees of the emotional expressions (20–100%), and patients are exposed to faces exhibiting emotions at varying levels of intensity, again hit rate (at each intensity) is used as the measure.

Using these paradigms, deficits in the recognition of facial disgust have been identified in temporal lobe epilepsy (Bonora et al. 2011; Broicher et al. 2012; Bujarski et al. 2016 [using The Awareness of Social Inference Test]; Gomez-Ibanez et al. 2014; Hennion et al. 2015; Meletti et al. 2003, 2009; Sedda et al. 2013); amyotrophic lateral sclerosis (Aho-Özhan et al. 2016; Crespi et al. 2014; Oh et al. 2016; Zimmerman et al. 2007); Parkinson's disease (Dujardin et al. 2004; Kan et al. 2002; Sprengelmeyer et al. 2003; Suzuki et al. 2006); frontotemporal dementia (Bediou et al. 2009; Lavenu and Pasquier 2005; Lough et al. 2006); and Huntington's disease (Calder et al. 2010; Hayes et al. 2007, 2009a; Henley et al. 2008; Ille et al. 2011a, b; Labuschagne et al. 2013; Milders et al. 2003; Rees et al. 2014; Robotham et al. 2011; Snowden et al. 2008; Sprengelmeyer et al. 1996; Trinkler et al. 2013; Wang et al. 2003; Zarotti et al. 2019). In the case of frontotemporal dementia, the deficit is particularly apparent in the behavioural variant (Kumfor et al. 2011, 2013; however, see Oliver et al. 2014), and temporal lobe epilepsy patients appear to only exhibit the deficit when the medial part of the temporal lobe is involved (Meletti et al. 2009). In many cases deficits in the recognition of facial disgust have been shown to exist in the absence of a general breakdown in the ability to process facial stimuli. Perhaps even more importantly, they are not accounted for by elevated depression levels (in temporal lobe epilepsy, Bujarski et al. 2016; amyotrophic lateral sclerosis, Zimmerman et al. 2007; Parkinson's disease, Gray and Tickle-Degnen 2010; frontotemporal dementia, Bediou et al. 2009; or Huntington's disease, Ille et al. 2011a), which can affect the recognition of disgust (Douglas and Porter 2010).

From the point of view of the tripartite model discussed above, deficits in the recognition of disgust may not be too unexpected in these disorders. There is evidence for insular dysfunction in medial temporal lobe epilepsy (Chassoux et al. 2004), amyotrophic lateral sclerosis (Habert et al. 2007), behavioural variant frontotemporal dementia (Whitwell et al. 2009), Parkinson's disease (Christopher et al. 2014), and Huntington's disease (Ille et al. 2011b). In relation to faces specifically, all five disorders show evidence of dysfunction of the superior temporal sulcus (temporal lobe epilepsy, Moran et al. 2001; amyotrophic lateral sclerosis, Agosta et al. 2012; behavioural variant frontotemporal dementia, Zamboni et al. 2010; Parkinson's disease, Lotze et al. 2009; Huntington's disease, Nopoulos et al. 2010), the sensory input side of the disgust system for faces. Interestingly, some other common neurological disorders—Alzheimer's disease (Bediou et al. 2009; Henry et al. 2008; Park et al. 2017; Philippi et al. 2017, but see Sapey-Triomphe et al. 2015); dementia with Lewy bodies (Philippi et al. 2017); multiple sclerosis (Henry et al. 2009a; Prochnow et al. 2011); Tourette's syndrome (Drury et al. 2012; Mermillod et al. 2013; Sprengelmeyer et al. 1997)—do not exhibit a deficit in the recognition of facial expressions of disgust. That is significant for a number of reasons, principally amongst them is the fact that there is evidence for insular dysfunction in dementia with Lewy bodies (Blanc et al. 2015), multiple sclerosis

(Orbach et al. 2018), and Tourette's syndrome (Draper et al. 2016). We will consider why insular dysfunction in these disorders may not affect the recognition of facial disgust later on. Alzheimer's disease seems to be the odd one out, since there is little evidence of primary insular damage. In fact, there is more thinning in the insular cortex in dementia with Lewy bodies than Alzheimer's disease (Lebedev et al. 2013). That said, there is evidence for superior temporal sulcus damage in Alzheimer's disease (Eskildsen et al. 2015).

In addition to a deficit in the recognition of facial disgust, there is also evidence that Parkinson's disease patients are less able to decode disgust-related emotional features. This has been evidenced using stimuli delivered through prosodic elements of the human voice, when actors read semantically neutral sentences (such as "good morning") and nonsense sentences, using tone to convey the six basic emotions (Kan et al. 2002). Deficits in the processing of disgust-related prosody have also been found in medial temporal lobe epilepsy (Bonora et al. 2011), amyotrophic lateral sclerosis (Zimmerman et al. 2007), and Huntington's disease (Calder et al. 2010; Hayes et al. 2007)—other disorders have not been evaluated. Such deficits may not be too surprising, given that Parkinson's disease (Martin et al. 2009), medial temporal lobe epilepsy (Moran et al. 2001), amyotrophic lateral sclerosis (Agosta et al. 2012), and Huntington's disease (Rosas et al. 2008) are all associated with anatomical changes in the superior temporal gyrus, which supplies the disgust system with disgust-related auditory information.

Disgust-Related Emotional Responses

Parkinson's patients are impaired when it comes to making posed disgust-related facial expressions (Madeley et al. 1995; Simons et al. 2004), and also facial movements in response to a disgust-inducing (surgical) film, even though subjective disgust ratings are not reduced, suggesting that the experience of disgust is not affected (Smith et al. 1996). Again, changes in emotional expression, at least in terms of the face, are independent of depression (Simons et al. 2004). A deficit in the production of facial expressions is consistent with damage to the basal ganglia (affecting behavioural output) that is characteristic of the disorder (Obeso et al. 2000). A loss of the capacity to demonstrate disgust-related facial expressions has also been reported in frontotemporal dementia (Bathgate et al. 2001), including the behavioural variant (Eckart et al. 2012). Behavioural variant frontotemporal dementia is also associated with a reduced tendency to avoid (e.g., turn away from) disgust-eliciting film clips in the absence of a change in self-reported disgust (Otero and Levenson 2019).

These changes in disgust-related responses (i.e., the output side of the disgust system) in frontotemporal dementia are perhaps to be expected given that the disorder is associated with changes at the level of the striatum. Indeed, a recent review has argued that it should be considered a "frontostriatal disorder" (Looi et al. 2012). Huntington's disease is also a striatal disorder (Henley et al. 2008; Wang et al.

2003), and hence deficits in producing disgust-related facial expressions might be predicted in patients with the disorder (in the absence of a deficit in the ability to identify and describe feelings of disgust in themselves and others; Trinkler et al. 2013). Patients are particularly impaired when expressions are elicited rather than copied (Hayes et al. 2009b).

Finally, reduced physiological responses to disgust-inducing film clips has been reported in Alzheimer's disease (Verstaen et al. 2016; which also affects the striatum, de Jong et al. 2011), again in the absence of a change in self-reported disgust. However, facial expressions of disgust are not affected in either Alzheimer's disease (Verstaen et al. 2016) or behavioural variant frontotemporal dementia (Otero and Levenson 2019). The reason why only selective aspects of output are affected in these disorders, in the face of striatal dysfunction, is uncertain. It may be the case that the disorders selectively affect some aspects of the cortico-basal ganglia loop systems more than others as they pass through the striatum, leading in turn to the selective preservation of some aspects of output functionality.

Disgust-Related Emotional Experience

As mentioned above, Alzheimer's disease is not associated with a change in self-reported disgust, at least in response to disgust-inducing film clips (Verstaen et al. 2016). The same is also true of medial temporal lobe epilepsy (Kotwas et al. 2019). Similarly, in contrast to the widely reported impairments in emotion recognition and emotional expression in Parkinson's disease (see above), experience of core disgust seems to be relatively unimpaired (Smith et al. 1996). Our own work in Parkinson's disease suggests that for disgust directed at the self ("self-disgust"), experienced levels are actually higher (vs. healthy controls) both at baseline and in response to self-disgust inducing manipulations (Tsatali et al. 2019). In contrast, dementia with Lewy bodies (Philippi et al. 2017) and behavioural variant frontotemporal dementia (Eckart et al. 2012; but see Verstaen et al. 2016) have both been associated with a reduction in experienced disgust in response to disgust-inducing visual stimuli. Likewise, although Huntington's disease patients have been reported to retain the ability to identify and describe feelings of disgust in themselves and others (Trinkler et al. 2013), patients can recall less disgust-related scenarios than controls (but not those pertaining to other emotions) and the intensity of the recalled scenarios are considerably less disgusting (Hayes et al. 2007). This suggests that disgust-related emotional experience may not be completely intact. This is further demonstrated by the fact that judgements concerning disgust-related scenes are actually more intense in Huntington patients than in controls (Ille et al. 2011a).

Conclusions and Future Work

In many of the disorders considered in this chapter, the pattern of deficits can be mapped onto the known dysfunction or anatomical changes in various components of the disgust system we outlined earlier. However, as aforementioned, there are some disorders in which the insular cortex is compromised—in dementia with Lewy bodies (Blanc et al. 2015), multiple sclerosis (Orbach et al. 2018) and Tourette's syndrome (Draper et al. 2016)—but patients do not display a deficit in the recognition of facial disgust (other aspects of disgust processing have not been evaluated and represent an approach for further research). Likewise, the insular cortex is compromised in medial temporal lobe epilepsy (Chassoux et al. 2004) and Parkinson's disease (Christopher et al. 2014), yet disgust-related emotional experiences are not impaired. Clearly, this is a challenge to the model, certainly in its simplest form, and other factors have to be taken into account. There are at least three possible explanations for preserved (facial) disgust processing and emotional experience in the presence of a compromised insular cortex.

Firstly, dysfunction or anatomical changes do not necessarily mean that all aspects of processing are affected. In common with most pieces of neural tissue, we know very little about the computational processes in the insular cortex and hence cannot ascertain which changes are likely to result in specific deficits. This is compounded by the fact that only part of the insular cortex (the ventral anterior sector) is involved in disgust processing (Caruana et al. 2011; Jezzini et al. 2012). Elucidating the computational problem that's solved by the insular cortex and its algorithmic implementation will have to await further research, potentially bringing together computational neuroscience and bench neuroscience techniques.

Secondly, there is evidence that the disgust system has a degree of plasticity in adulthood, so it may be possible to recover from damage. Straube et al. (2010) reported a lack of impairment in disgust recognition and experience in a patient with a stroke involving the right insular cortex. Similarly, Couto et al. (2013) report the case of a woman with an ischemic focal lesion comprising the complete right anterior, mid, and posterior insular cortex. There was no significant difference from controls in the accuracy of her facial disgust recognition and she was actually better than controls at recognising prosodic signs of disgust. The substrate for recovery of function that these case studies suggest is unknown. The natural assumption based on the adult plasticity literature is that other areas of the brain "take over" the functions of the damaged region, however the basis for that process in the case of the disgust system awaits further investigation, probably using functional neuroimaging techniques in patients with lesions of the insular cortex.

Finally, a third possibility for preserved (facial) disgust processing and emotional experience in the presence of a compromised insular cortex is that the patients develop cognitive strategies to overcome the deficit. That may be easier in the case of Alzheimer's disease where the insular cortex is relatively preserved (Lebedev et al. 2013). For example, it might be the case that emotion regulation processes come into play, and these could be examined. In that regard it is interesting that emotional suppression appears to be preserved in Alzheimer's disease (Henry et al. 2009b).

Alzheimer's disease with its preserved insular cortex and unimpaired ability to recognise disgust in facial expressions is rather the odd one out. Most neurological disorders are associated with a dysfunction of the insular cortex, and where disgust is affected, in the majority of cases (self-disgust in Parkinson's disease aside), the tendency is for it to be reduced—for the recognition of disgust-related stimuli, disgust-related emotional responses, and disgust-related emotional experience. In the case of Alzheimer's disease, disgust-related deficits may be related more closely to the input (Eskildsen et al. 2015) and output (de Jong et al. 2011) sides of the disgust system.

Other chapters in this volume have identified a number of ways in which increased levels of disgust can have deleterious effects on both mental (e.g., through depression, see Davey, Chap. 11, this volume) and physical (e.g., through the avoidance of cancer screening behaviours, see Reynolds and Dev, Chap. 13, this volume) health, and hence the reduction of disgust-related emotional experience in dementia with Lewy bodies and behavioural variant frontotemporal dementia (Eckart et al. 2012; Philippi et al. 2017) may have some positive health-related benefits. However, the concomitant reduction in the inclination to avoid the ingestion of harmful substances may counterbalance that. In extreme cases, this may account for the coprophagia sometimes seen in frontotemporal dementia (Josephs et al. 2016). For those disorders in which the recognition of disgust-related stimuli and disgust-related emotional responses are impaired, those impairments are likely to lead to some difficulties in social interactions, which might suggest that disgust-related deficits should be a target for treatment. Unfortunately, disgust-related deficits are largely untouched by current pharmacotherapies aimed at the core features of many neurological disorders (including Parkinson's disease, Gray and Tickle-Degnen 2010; temporal lobe epilepsy, Meletti et al. 2009; and frontotemporal dementia, Lavenu and Pasquier 2005), and hence targeting those deficits would require an additional treatment strand to be added to the treatment regime for the disorders concerned.

Acknowledgements The authors would like to acknowledge the contribution of Dr. Nicolo Zarotti for his input relating to the data on Huntington's disease, and Dr. Marianna Tsatali for her input relating to the data on Parkinson's disease.

References

Adolphs, R., Tranel, D., & Damasio, A. R. (2003). Dissociable neural systems for recognizing emotions. *Brain and Cognition, 52*, 61-9.

Agosta, F., Valsasina, P., Riva, N., Copetti, M., Messina, M. J., Prelle, A., Comi, G., & Filippi, M. (2012). The Cortical Signature of Amyotrophic Lateral Sclerosis. *PLoS One, 7*, e42816.

Aho-Özhan, H. E., Keller, J., Heimrath, J., Uttner, I., Kassubek, J., Birbaumer, N., Ludolph, A. C., & Lulé, D. (2016). Perception of emotional facial expressions in amyotrophic lateral sclerosis (als) at behavioural and brain metabolic level. *PLoS One*, *11*, e0164655.

Alexander, G. E., DeLong, M. R., & Strick, P. L. (1986). Parallel organization of functionally segregated circuits linking basal ganglia and cortex. *Annual Review of Neuroscience*, *9*, 357-81.

Bathgate, D, Snowden, J S., Varma, A., Blackshaw, A., & Neary, D. (2001). Behaviour in frontotemporal dementia, Alzheimer's disease and vascular dementia. *Acta Neurologica Scandinavica*, *103*, 367-78.

Bediou, B., Ryff, I., Mercier, B., Milliery, M., Hénaff, M-A., D'Amato, T, Bonnefoy, M., Vighetto, A., & Krolak-Salmon, P. (2009). Impaired social cognition in mild Alzheimer disease. *Journal of Geriatric Psychiatry and Neurology*, *22*, 130–140.

Blanc, F., Colloby, S. J., Philippi, N., de Pétigny, X., Jung, B., Demuynck, C., Phillipps, C., Anthony, P., Thomas, A., Bing, F., Lamy, J., Martin-Hunyadi, C., O'Brien, J. T., Cretin, B., McKeith, I., Armspach, J.-P., & Taylor, J.-P. (2015). Cortical thickness in dementia with lewy bodies and alzheimer's disease: A comparison of prodromal and dementia stages. *PLoS One*, *10*, e0127396.

Bonora, A.., Benuzzi, F., Monti, G., Mirandola, L., Pugnaghi, M., Nichelli, P., & Meletti, S. (2011). Recognition of emotions from faces and voices in medial temporal lobe epilepsy. *Epilepsy & Behavior*, *20*, 648–654.

Broicher, S. D., Kuchukhidze, G., Grunwald, T, Krämer, G., Kurthen, M., & Jokeit, H. (2012). "Tell me how do I feel"—emotion recognition and theory of mind in symptomatic mesial temporal lobe epilepsy. *Neuropsychologia*, *50*, 118-28.

Bujarski, K. A., Flashman, L., Li Z., Tosteson, T. D., Jobst, B. C., Thadani, V. M., Kobylarz, E. J., Roberts, D. W., & Roth R. M. (2016). Investigating social cognition in epilepsy using a naturalistic task. *Epilepsia*, *57*, 1515-20.

Calder, A. J., Keane, J., Manes, F., Antoun, N., & Young, A. W. (2000). Impaired recognition and experience of disgust following brain injury. *Nature Neuroscience*, *3*, 1077-8.

Calder, A. J., Keane, J., Young, A. W., Lawrence, A. D., Mason, S., & Barker, R. A. (2010). The relation between anger and different forms of disgust: implications for emotion recognition impairments in Huntington's disease. *Neuropsychologia*, *48*, 2719–2729.

Caruana, F., Jezzini, A., Sbriscia-Fioretti, B., Rizzolatti, G., & Gallese, V. (2011). Emotional and social behaviors elicited by electrical stimulation of the insula in the macaque monkey. *Current Biology*, *21*, 195-9.

Chassoux, F., Semah, F., Bouilleret, V., Landre, E., Devaux, B., Turak, B., Nataf, F., & Roux, F.-X. (2004). Metabolic changes and electro-clinical patterns in mesio-temporal lobe epilepsy: a correlative study. *Brain*, *127*, 164-74.

Chen, C., Lee, Y. H., & Cheng, Y. (2014). Anterior insular cortex activity to emotional salience of voices in a passive oddball paradigm. *Frontiers in Human Neuroscience*, *8*, 743

Christopher, L., Koshimori, Y., Lang, A. E., Criaud, M., & Strafella, A. P. (2014). Uncovering the role of the insula in non-motor symptoms of Parkinson's disease. *Brain*, *137*, 2143-54.

Couto, B., Sedeño, L., Sposato, L. A., Sigman, M., Riccio, P M., Salles, A., Lopez, V., Schroeder, J., Manes, F., & Ibanez, A. (2013). Insular networks for emotional processing and social cognition: comparison of two case reports with either cortical or subcortical involvement. *Cortex*, *49*, 1420-34.

Crespi, C., Cerami, C., Dodich, A., Canessa, N., Arpone, M., Iannaccone, S., Corbo, M., Lunetta, C., Scola, E., Falini, A., & Cappa, S. F. (2014). Microstructural white matter correlates of emotion recognition impairment in Amyotrophic Lateral Sclerosis. *Cortex*, *53*, 1-8.

de Jong, L. W., Ferrarini, L., van der Grond, J., Milles, J. R., Reiber, J. H., Westendorp, R. G. J., Bollen, E. L. E. M., Middelkoop, H. A. M., & van Buchem, M. A. (2011). Shape abnormalities of the striatum in Alzheimer's disease. *Journal of Alzheimer's Disease*, *23*, 49-59.

Douglas, K. M., & Porter, R. J. (2010). Recognition of disgusted facial expressions in severe depression. *British Journal of Psychiatry*, *197*, 156-7.

Draper, A., Jackson, G. M., Morgan, P. S., & Jackson, S. R. (2016). Premonitory urges are associated with decreased grey matter thickness within the insula and sensorimotor cortex in young people with Tourette syndrome. *Journal of Neuropsychology*, *10*, 143-53.

Drury, H., Channon, S., Barrett, R., Young, M. B., Stern, J. S., Simmons, H., & Crawford, S. (2012). Emotional processing and executive functioning in children and adults with Tourette's syndrome. *Child Neuropsychology*, *18*, 281-298.

Dujardin, K., Blairy, S., Defebvre, L., Duhem, S., Noël, Y., Hess, U., & Destée, A. (2004). Deficits in decoding emotional facial expressions in Parkinson's disease. *Neuropsychologia*, *42*, 239-250.

Eckart, J. A., Sturm, V. E., Miller, B. L., & Levenson, R. W. (2012). Diminished disgust reactivity in behavioral variant frontotemporal dementia. *Neuropsychologia*, *50*, 786-90.

Ekman, P., & Friesen, E. V. (1976). *Pictures of facial affect*. Consulting Psychologists Press, Palo Alto.

Eskildsen, S. F., Coupé, P, Fonov, V. S., Pruessner, J. C., Collins, D. L., & Alzheimer's Disease Neuroimaging Initiative. (2015). Structural imaging biomarkers of Alzheimer's disease: predicting disease progression. *Neurobiology of Aging*, 36 *Suppl 1*, S23-31.

Ghosh, B. C. P., Rowe, J. B., Calder, A. J., Hodges, J. R., & Bak, T. H. (2009). Emotion recognition in Progressive Supranuclear Palsy. *Journal of Neurology, Neurosurgery and Psychiatry*, *80*, 1143–1145.

Gomez-Ibanez, A., Urrestarazu, E., & Viteri, C. (2014). Recognition of facial emotions and identity in patients with mesial temporal lobe and idiopathic generalized epilepsy: An eye-tracking study. *Seizure-European Journal of Epilepsy*, *23*, 892-898.

Gray, H. M., & Tickle-Degnen, L. (2010). A meta-analysis of performance on emotion recognition tasks in Parkinson's disease. *Neuropsychology*, *24*, 176–91.

Gurney, K., Prescott, T. J., & Redgrave, P. (2001). A computational model of action selection in the basal ganglia. I. A new functional anatomy. *Biological Cybernetics*, *84*, 401-10.

Habert, M. O., Lacomblez, L., Maksud, P, El Fakhri, G., Pradat, J. F., & Meininger, V. (2007). Brain perfusion imaging in amyotrophic lateral sclerosis: extent of cortical changes according to the severity and topography of motor impairment. *Amyotrophic Lateral Sclerosis*, *8*, 9-15.

Hayes, C. J., Stevenson, R. J., & Coltheart, M. (2007). Disgust and Huntington's disease. *Neuropsychologia*, *45*, 1135–1151.

Hayes, C. J., Stevenson, R. J., & Coltheart, M. (2009a). The processing of emotion in patients with Huntington's disease: variability and differential deficits in disgust. *Cognitive and Behavioral Neurology*, *22*, 249–257.

Hayes, C. J., Stevenson, R. J., & Coltheart, M. (2009b). Production of spontaneous and posed facial expressions in patients with Huntington's disease: Impaired communication of disgust. *Cognition & Emotion*, *23*, 118–134.

Henley, S. M. D., Wild, E. J., Hobbs, N. Z., Warren, J. D., Frost, C., Scahill, R. I., Ridgway, G. R., MacManus, D. G., Barker, R. A., Fox, N. C., & Tabrizi, S. J. (2008). Defective emotion recognition in early HD is neuropsychologically and anatomically generic. *Neuropsychologia*, *46*, 2152–2160.

Hennion, S., Szurhaj, W., Duhamel, A., Lopes, R., Tyvaert, L., Derambure, P., & Delbeuck, X. (2015). Characterization and prediction of the recognition of emotional faces and emotional bursts in temporal lobe epilepsy. *Journal Clinical and Experimental Neuropsychology*, *37*, 931-945.

Henry, J. D., Ruffman, T., McDonald, S., O'Leary, M. A., Phillips, L. H., Brodaty, H., & Rendell P. G. (2008). Recognition of disgust is selectively preserved in Alzheimer's disease. *Neuropsychologia*, *46*, 1363-70.

Henry, J. D., Phillips, L. H., Beatty, W. W., McDonald, S., Longley, W. A., Joscelyne, A., & Rendell, P. G. (2009a). Evidence for deficits in facial affect recognition and theory of mind in multiple sclerosis. *Journal of the International Neuropsychological Society*, *15*, 277-85.

Henry, J. D. P., Rendell, P. G., Scicluna, A., Jackson, M., & Phillips, L. H. (2009b). Emotion Experience, Expression, and Regulation in Alzheimer's Disease. *Psychology and Aging*, *24*, 252-7.

Ille, R., Holl, A K., Kapfhammer, H.-P., Reisinger, K., Schäfer, A., & Schienle, A. (2011a). Emotion recognition and experience in Huntington's disease: is there a differential impairment? *Psychiatry Research*, *188*, 377–82.

Ille, R., Schäfer, A., Scharmüller, W., Enzinger, C., Schöggl, H., Kapfhammer, H.-P., & Schienle, A. (2011b). Emotion recognition and experience in Huntington disease: a voxel-based morphometry study. *Journal of Psychiatry & Neuroscience*, *36*, 383–390.

Jabbi, M., Bastiaansen, J., & Keysers, C. (2008). A common anterior insula representation of disgust observation, experience and imagination shows divergent functional connectivity pathways. *PLoS One*, *3*, e2939.

Jehna, M., Neuper, C., Ischebeck, A., Loitfelder, M., Ropele, S., Langkammer, C., Ebner, F., Fuchs, S., Schmidt, R., Fazekas, F., & Enzinger, C. (2011). The functional correlates of face perception and recognition of emotional facial expressions as evidenced by fMRI. *Brain Research*, *1393*, 73-83.

Jezzini, A., Caruana, F., Stoianov, I., Gallese, V., & Rizzolatti, G. (2012). Functional organization of the insula and inner perisylvian regions. *Proceedings of the National Academy of Science of the United States of America*, *109*, 10077-82.

Josephs, K. A., Whitwell, J. L., Parisi, J. E., & Lapid, M. I. (2016). Coprophagia in neurologic disorders. *Journal of Neurology*, *263*, 1008-1014.

Kan, Y., Kawamura, M., Hasegawa, Y., Mochizuki, S., & Nakamura, K. (2002). Recognition of emotion from facial, prosodic and written verbal stimuli in Parkinson's disease. *Cortex*, *38*, 623-630.

Köchel, A., Schöngassner, F., & Schienle, A. (2013). Cortical activation during auditory elicitation of fear and disgust: a near-infrared spectroscopy (NIRS) study. *Neuroscience Letters*, *549*, 197-200.

Kotwas, I., Micoulaud-Franchi, J. A., Khalfa, S., McGonigal, A., Bastien-Toniazzo, M., & Bartolomei, F. (2019). Subjective and physiological response to emotions in temporal lobe epilepsy and psychogenic non-epileptic seizures. *Journal Affective Disorders*, *244*, 46-53.

Krolak-Salmon, P, Hénaff, M. A., Isnard, J., Tallon-Baudry, C., Guénot, M., Vighetto, A., Bertrand, O., & Mauguière, F. (2003). An attention modulated response to disgust in human ventral anterior insula. *Annals of Neurology*, *53*, 446-53.

Kumfor, F., Miller, L., Lah, S., Hsieh, S., Savage, S., Hodges, J. R., & Piguet, O. (2011). Are you really angry? The effect of intensity on facial emotion recognition in frontotemporal dementia. *Social Neuroscience*, *6*, 502-14.

Kumfor, F., Irish, M., Hodges J. R., & Piguet O. (2013). Discrete neural correlates for the recognition of negative emotions: Insights from Frontotemporal Dementia. *PLoS One*, *8*, e67457.

Labuschagne, I., Jones, R., Callaghan, J., Whitehead, D., Dumas, E. M., Say, M. J., Hart, E. P., Justo, D., Coleman, A., Santos, R. C. D., Frost, C., Craufurd, D., Tabrizi, S. J., Stout, J. C., & TRACK-HD Investigators (2013). Emotional face recognition deficits and medication effects in pre-manifest through stage-II Huntington's disease. *Psychiatry Research*, *207*, 118–126.

Lavenu, I., & Pasquier, F. (2005). Perception of emotion on faces in frontotemporal dementia and Alzheimer's disease: a longitudinal study. *Dementia and Geriatric Cognitive Disorders*, *19*, 37-41.

Lebedev, A. V., Westman, E., Beyer, M. K., Kramberger, M. G., Aguilar, C., Pirtosek, Z., & Aarsland, D. (2013). Multivariate classification of patients with Alzheimer's and dementia with Lewy bodies using high-dimensional cortical thickness measurements: an MRI surface-based morphometric study. *Journal of Neurology*, *260*, 1104-15.

Looi, J. C., Walterfang, M., Velakoulis, D., Macfarlane, M. D., Svensson, L. A., & Wahlund, L. O. (2012). Frontotemporal dementia as a frontostriatal disorder: neostriatal morphology as a biomarker and structural basis for an endophenotype. *Australian and New Zealand Journal of Psychiatry*, *46*, 422-34.

Lotze, M., Reimold, M., Heymans, U., Laihinen, A., Patt, M., & Halsband, U. (2009). Reduced ventrolateral fMRI response during observation of emotional gestures related to the degree of dopaminergic impairment in Parkinson disease. *Journal of Cognitive Neuroscience*, *21*, 1321-31.

Lough, S., Kipps, C. M., Treise, C., Watson, P., Blair, J. R., & Hodges, J. R. (2006). Social reasoning, emotion and empathy in frontotemporal dementia. *Neuropsychologia*, *44*, 950-8.

Madeley, P., Ellis, A. W., & Mindham, R. H. S. (1995). Facial expression and Parkinson's disease. *Behavioral Neurology*, *8*, 115–119.

Martin, W. R., Wieler, M., Gee, M., & Camicioli, R. (2009). Temporal lobe changes in early, untreated Parkinson's disease. *Movement Disorders*, *20024*, 1949-54.

Meletti, S., Benuzzi, F., Rubboli, G., Cantalupo, G., Stanzani Maserati, M., Nichelli, P., & Tassinari, C. A. (2003). Impaired facial emotion recognition in early-onset right mesial temporal lobe epilepsy. *Neurology*, *60*, 426-31.

Meletti S., Benuzzi F., Cantalupo G., Rubboli G., Tassinari C. A., & Nichelli P. (2009). Facial emotion recognition impairment in chronic temporal lobe epilepsy. *Epilepsia*, *50*, 1547-59.

Mermillod, M., Devaux, D., Derost, P., Rieu, I., Chambres, P., Auxiette, C., Legrand, G., Galland, F., Dalens, H., Coulangeon, L. M., Broussolle, E., Durif, F., & Jalenques, I. (2013). Rapid presentation of emotional expressions reveals new emotional impairments in Tourette's syndrome. *Frontiers in Human Neuroscience*, *7*, 149.

Milders, M., Crawford, J. R., Lamb, A., & Simpson, S. A. (2003). Differential deficits in expression recognition in gene-carriers and patients with Huntington's disease. *Neuropsychologia*, *41*, 1484–1492.

Moran, N. F., Lemieux, L., Kitchen, N. D., Fish, D. R., & Shorvon, S. D. (2001). Extrahippocampal temporal lobe atrophy in temporal lobe epilepsy and mesial temporal sclerosis. *Brain*, *124*, 167-75.

Mufson, E. J., & Mesulam, M. M. (1982). Insula of the old world monkey. II: Afferent cortical input and comments on the claustrum. *Journal of Comparative Neurology*, *212*, 23-37.

Nopoulos, P. C., Aylward, E. H., Ross, C. A., Johnson, H. J., Magnotta, V. A., Juhl, A. R., Pierson, R. K., Mills, J., Langbehn, D. R., Paulsen, J. S., & the PREDICT-HD Investigators and Coordinators of the Huntington Study Group (HSG) (2010). Cerebral cortex structure in prodromal Huntington Disease. *Neurobiology of Disease*, *40*, 544–554.

Obeso, J. A., Rodriguez-Oroz, M. C., Rodriguez, M., Lanciego, J. L., Artieda, J., Gonzalo, N., & Olanow, C. W. (2000). Pathophysiology of the basal ganglia in Parkinson's disease. *Trends in Neurosciences*, *23*, S8–S19.

Oh, S. I., Oh, K. W., Kim, H. J., Park, J. S., & Kim, S. H. (2016). Impaired Perception of Emotional Expression in Amyotrophic Lateral Sclerosis. *Journal of Clinical Neurology*, *12*, 295-300.

Oliver, L. D., Virani, K., Finger, E. C., & Mitchell, D. G. (2014). Is the emotion recognition deficit associated with frontotemporal dementia caused by selective inattention to diagnostic facial features? *Neuropsychologia*, *60*, 84-92.

Orbach, L., Menascu, S., Hoffmann, C., Miron, S., & Achiron, A. (2018). Focal cortical thinning in patients with stable relapsing-remitting multiple sclerosis: cross-sectional-based novel estimation of gray matter kinetics. *Neuroradiology*, *60*, 179-187.

Otero, M. C., & Levenson, R. W. (2019). Emotion regulation via visual avoidance: Insights from neurological patients. *Neuropsychologia*, *131*, 91-101.

Park, S., Kim, T, Shin, S. A., Kim, Y. K., Sohn, B. K., Park, H. J., Youn, J. H., & Lee, J. Y. (2017). Behavioral and Neuroimaging Evidence for Facial Emotion Recognition in Elderly Korean Adults with Mild Cognitive Impairment, Alzheimer's Disease, and Frontotemporal Dementia. *Frontiers in Aging Neuroscience*, *9*, 389.

Philippi, N., Kemp, J., Constans-Erbs, M., Hamdaoui, M., Monjoin, L., Ehrhard, E., Albasser, T., Botzung, A., Demuynck, C., Heim, G., Martin-Hunyadi, C., Bilger, M., Berly, L., Soulier, D., Cretin, B., Despres, O., & Blanc, F. (2017). Insular cognitive impairment at the early stage of dementia with Lewy bodies: a preliminary study. *Gériatrie et Psychologie Neuropsychiatrie du Vieillissement*, *15*, 329-338.

Phillips, M. L., Young, A. W., Senior, C., Brammer, M., Andrew, C., Calder, A. J., Bullmore, E. T., Perrett, D. I., Rowland, D., Williams, S. C., Gray, J. A., & David, A. S. (1997). A specific neural substrate for perceiving facial expressions of disgust. *Nature*, *389*, 495-8.

Phillips, M. L., Young, A. W., Scott, S. K., Calder, A. J., Andrew, C., Giampietro, V., Williams, S. C., Bullmore, E. T., Brammer, M., & Gray, J. A. (1998). Neural responses to facial and vocal expressions of fear and disgust. *Proceedings of the Royal Society B*, *265*, 1809-17.

Prochnow, D, Donell, J., Schaefer, R., Jorgens, S., Hartung, H. P., Franz, M., & Seitz, R. J. (2011). Alexithymia and impaired facial affect recognition in multiple sclerosis. *Journal of Neurology*, *258*, 1683-1688.

Pujol, J., Blanco-Hinojo, L., Coronas, R., Esteba-Castillo, S., Rigla, M., Martínez-Vilavella, G., Deus, J., Novell, R., & Caixàs, A. (2018). Mapping the sequence of brain events in response to disgusting food. *Human Brain Mapping*, *2039*, 369-380.

Redgrave, P., Prescott, T. J., & Gurney, K. (1999). The basal ganglia: a vertebrate solution to the selection problem? *Neuroscience*, *89*, 1009-23.

Reep, R. L., & Winans, S. S. (1982). Afferent connections of dorsal and ventral agranular insular cortex in the hamster Mesocricetus auratus. *Neuroscience*, *7*, 1265-88.

Rees, E. M., Farmer, R., Cole, J. H., Henley, S. M. D., Sprengelmeyer, R., Frost, C., Scahill, R. I., Hobbs, N. Z., & Tabrizi, S. J. (2014). Inconsistent emotion recognition deficits across stimulus modalities in Huntington′s disease. *Neuropsychologia*, *64*, 99–104.

Robotham, L., Sauter, D. A., Bachoud-Levi, A. C., Trinkler, I., Bachoud-Lévi, A.-C., & Trinkler, I. (2011). The impairment of emotion recognition in Huntington's disease extends to positive emotions. *Cortex*, *47*, 880–884.

Rosas, H. D., Salat, D. H., Lee, S. Y., Zaleta, A K., Pappu, V., Fischl, B, Greve, D., Hevelone, N., & Hersch, S. M. (2008). Cerebral cortex and the clinical expression of Huntington's disease: complexity and heterogeneity. *Brain*, *131*, 1057–1068.

Sambataro, F., Dimalta, S., Di Giorgio, A., Taurisano, P., Blasi, G., Scarabino, T., Giannatempo, G., Nardini, M., & Bertolino, A. (2006). Preferential responses in amygdala and insula during presentation of facial contempt and disgust. *European Journal of Neuroscience*, *24*, 2355-62.

Sapey-Triomphe, L. A., Heckemann, R. A., Boublay, N., Dorey, J. M., Hénaff, M. A., Rouch, I., Padovan, C., Hammers, A., Krolak-Salmon, P., & Alzheimer's Disease Neuroimaging Initiative. (2015). Neuroanatomical correlates of recognizing face expressions in mild stages of Alzheimer's disease. *PLoS One*, *10*, e0143586.

Sedda, A., Rivolta, D., Scarpa, P., Burt, M., Frigerio, E., Zanardi, G., Piazzini, A., Turner, K., Canevini, M. P., Francione, S., Russo, G. L., & Bottini, G. (2013). Ambiguous emotion recognition in temporal lobe epilepsy: the role of expression intensity. *Cognitive, Affective, and Behavioral Neuroscience*, *13*, 452-63.

Shimizu, H., & Norita, M. (1991). Connections of the insular cortex in kittens: an anatomical demonstration with wheatgerm agglutinin conjugated to horseradish peroxidase technique. *International Journal of Developmental Neuroscience*, *9*, 479-91.

Simons, G., Pasqualini, M. C., Reddy, V., & Wood, J. (2004). Emotional and non emotional facial expressions in people with Parkinson's disease. *Journal of the International Neuropsychology Society*, *10*, 521–555.

Smith, M. C., Smith, M. K., & Ellgring, H. (1996). Spontaneous and posed facial expression in Parkinson's disease. *Journal of the International Neuropsychology Society*, *2*, 383–391.

Snowden, J. S., Austin, N. A., Sembi, S., Thompson, J. C., Craufurd, D., & Neary, D. (2008). Emotion recognition in Huntington's disease and frontotemporal dementia. *Neuropsychologia*, *46*, 2638–2649.

Sprengelmeyer, R., Young, A. W., Calder, A. J., Karnat, A., Lange, H., Homberg, V., Perrett, D. I., & Rowland, D. (1996). Loss of disgust. Perception of faces and emotions in Huntington's disease. *Brain*, *119*, 1647–1665.

Sprengelmeyer, R., Young, A. W., Pundt, I., Sprengelmeyer, A., Calder, A. J., Berrios, G., Winkel, R., Vollmoeller, W., Kuhn, W., Sartory, G., & Przuntek, H. (1997). Disgust implicated in Obsessive-Compulsive Disorder. *Proceedings of the Royal Society B, 264*, 1767–1773.

Sprengelmeyer, R., Young, A. W., Mahn, K., Schroeder, U., Woitalla, D., Büttner, T, Kuhn, W., & Przuntek, H. (2003). Facial expression recognition in people with medicated and unmedicated Parkinson's disease. *Neuropsychologia, 41*, 1047-1057.

Straube, T, Weisbrod, A., Schmidt, S., Raschdorf, C., Preul, C., Mentzel, H. J., & Miltner, W. H. (2010). No impairment of recognition and experience of disgust in a patient with a right-hemispheric lesion of the insula and basal ganglia. *Neuropsychologia, 48*, 1735-41.

Suzuki, A., Hoshino, T., Shigemasu, K., & Kawamura, M. (2006). Disgust-specific impairment of facial expression recognition in Parkinson's disease. *Brain, 129*, 707–717.

Trinkler, I., de Langavant, L. C., & Bachoud-Levi, A. C. (2013). Joint recognition-expression impairment of facial emotions in Huntington's disease despite intact understanding of feelings. *Cortex, 49*, 549–558.

Tsatali, M., Overton, P. G., & Vivas, A. B. (2019). Self-reported and experimentally induced self-disgust is heightened in Parkinson's disease: Contribution of behavioural symptoms. *PLoS One, 14*, e0223663.

Verstaen, A., Eckart, J. A., Muhtadie, L., Otero, M. C., Sturm, V. E., Haase, C. M., Miller, B. L., & Levenson, R. W. (2016). Insular atrophy and diminished disgust reactivity. *Emotion, 16*, 903-12.

von dem Hagen, E. A., Beaver, J. D., Ewbank, M. P., Keane, J., Passamonti, L., Lawrence, A. D., & Calder, A. J. (2009). Leaving a bad taste in your mouth but not in my insula. *Social Cognitive and Affective Neuroscience, 4*, 379-86.

Wang K., Hoosain, R., Yang, R.-M. M., Meng, Y., & Wang, C.-Q. Q. (2003). Impairment of recognition of disgust in Chinese with Huntington's or Wilson's disease. *Neuropsychologia, 41*, 527–537.

Whitwell, J. L., Przybelski, S. A., Weigand, S. D., Ivnik, R J., Vemuri, P., Gunter, J. L., Senjem, M. L., Shiung, M. M, Boeve, B. F., Knopman, D. S., Parisi, J. E., Dickson, D. W., Petersen, R. C., Jack Jr, C. R., & Josephs, K. A. (2009). Distinct anatomical subtypes of the behavioural variant of frontotemporal dementia: a cluster analysis study. *Brain, 132*, 2932-46.

Wicker, B., Keysers, C., Plailly, J., Royet, J. P, Gallese, V., & Rizzolatti, G. (2003). Both of us disgusted in My insula: the common neural basis of seeing and feeling disgust. *Neuron, 40*, 655-64.

Wright, P., He, G., Shapira, N. A., Goodman, W. K., & Liu, Y. (2004). Disgust and the insula: fMRI responses to pictures of mutilation and contamination. *Neuroreport, 15*, 2347-51.

Zamboni, G., Grafman, J., Krueger, F., Knutson, K. M., & Huey, E. D. (2010). Anosognosia for behavioral disturbances in frontotemporal dementia and corticobasal syndrome: A voxel-based morphometry study. *Dementia and Geriatric Cognitive Disorders, 29*, 88-96.

Zarotti, N., Fletcher, I., & Simpson, J. (2019). New perspectives on emotional processing in people with symptomatic Huntington's disease: impaired emotion regulation and recognition of emotional body language. *Archives of Clinical Neuropsychology, 34*, 610-624.

Zimmerman, E. K., Eslinger, P. J., Simmons, Z., & Barrett, A. M. (2007). Emotional perception deficits in amyotrophic lateral sclerosis. *Cognitive and Behavioral Neurology, 20*, 79-82.

Chapter 13
How the "Ick Factor" Matters in Chronic Health Conditions

Lisa M. Reynolds and Vinayak Dev

Given its evolved purpose to protect against health threats (Curtis et al. 2011), disgust has particular relevance to chronic health conditions. In this chapter, we consider the primary reasons why disgust (or the "ick factor") is pertinent in the context of chronic conditions and discuss how, alongside some benefits, this emotion can be problematic. More specifically, this chapter highlights the research that has been conducted in the last decade on disgust and chronic conditions. Screening and diagnosis, treatment and decision-making, and long term adaptation to chronic health problems are used as a framework. The chapter closes with a discussion about clinical implications, gaps in the literature, and presents a proposed agenda for future research.

Disgust Evolved to Protect

All human emotions can be seen as having evolutionary roots related to facilitating adaptation to recurrent challenges or opportunities for survival and thus reproduction (Consedine et al. 2002; Nesse and Ellsworth 2009; Plutchik 2001; Tooby and Cosmides 2008). In contrast to the more general construct of "emotions", discrete emotions can be defined as, "episodic, relatively short-term, biologically-based patterns of perception, experience, physiology, action, and communication that occur in response to specific physical and social challenges and opportunities" (Keltner and Gross 1999, p. 468). Arguably, every discrete emotion has a different purpose in response to such challenges or opportunities. Some emotions, such as happiness, facilitate progress towards a goal; some, including fear, promote

L. M. Reynolds (✉) · V. Dev
Department of Psychological Medicine, University of Auckland, Auckland, New Zealand
e-mail: l.reynolds@auckland.ac.nz; vdev514@aucklanduni.ac.nz

P. A. Powell, N. S. Consedine (eds.), *The Handbook of Disgust Research*,
https://doi.org/10.1007/978-3-030-84486-8_13

self-preservation (Curtis and Biran 2001); and others, such as anger, support the removal of goal blockages (Johnson-Laird and Oatley 1992). This is especially true of discrete, as opposed to complex self-conscious, emotions. However, disgust—which functions as a protective, avoidance-promoting mechanism to minimise health threats—may be the only emotion whose purpose specifically relates to physical health (Curtis et al. 2004, 2011; Oaten et al. 2011).

As covered elsewhere in this volume (see Bradshaw and Gassen, Chap. 3, this volume), arguably, evolution has designed disgust to probabilistically reduce exposure to pathogens, especially through the body's entry points: mouth, skin, anus, and genitals (Tybur et al. 2013). The disgust experience is one of revulsion that occurs at the thought of incorporating a potentially contaminating object (Rozin and Fallon 1987), and is recognised by a cluster of established responses. Physiologically, disgust is characterised by lowered blood pressure, lowered galvanic skin response, and nausea (Curtis and Biran 2001). Behaviourally, the disgust response involves avoidance, withdrawal, spitting or ejection of noxious substances, and the emotion is universally recognised through a cluster of facial expressions, including constriction of nostrils, gape and tongue extension, furrowing of the eyebrows, raising of upper lip, and wrinkling of the nose (Consedine et al. 2007; Olatunji and Sawchuk 2005; Rozin et al. 2000, 2009). This cluster of avoidance tendencies and behavioural, physiological, and cognitive responses has important implications in the context of physical health and is the basis for our discussion below.

Given the potential costs of missing a health threat (i.e., getting sick), the disgust response errs on the side of conservativism (Oaten et al. 2011; Rozin et al. 1986). This conservative process means that the avoidance-promoting response can be activated in the absence of objective threat via laws of similarity and contagion beliefs (i.e., simply because aspects of a context seem similar to one associated with a genuine threat; Eskine et al. 2013; Stavrova et al. 2016). For example, disgust-generated withdrawal effects have been demonstrated in experimental studies where people stand at a greater distance to a confederate who displays an entirely non-infectious skin condition (birthmark) compared to when they do not (Newell 1999; Rumsey et al. 2004) and having a greater propensity to disgust has been causally implicated in stigmatisation of people with a non-contageous condition such as cancer (Azlan et al. 2020). This is an important consideration in the context of chronic health conditions, where individuals who might be non-infectious can become socially isolated despite posing no actual health threat.

Also of relevance in the context of chronic conditions, is the distinction between the *propensity* to experience disgust (i.e., the tendency/frequency in responding with disgust to a given situation) and disgust *sensitivity* (i.e., the extent that disgust is experienced as aversive or unpleasant; van Overveld et al. 2006). Although related, these two constructs are distinct from one another and independently predict behavioural avoidance in health contexts (van Overveld et al. 2010). People with smelling distortions exhibit an increased propensity (but not sensitivity) in experiencing disgust towards poor hygiene (Ille et al. 2016) and disgust sensitivity, but not propensity, predicts decisions to delay medical help seeking for bowel symptoms (Reynolds et al. 2018a). In a recent study, disgust propensity and sensitivity were actually shown to predict chemotherapy-related consequences in different and opposing ways (Dev et al. 2020). Thus, it is useful to make a distinction between a person's propensity to experience disgust and how sensitive (or aversive) their experience is.

Although the primary purpose of disgust has generalised to domains other than those that pose a threat to our physical health (Chapman and Anderson 2012), including moral judgements (Pizarro et al. 2011; see also Giner-Sorolla, Chap. 8, this volume), sexual function (de Jong et al. 2010; see also Borg and de Jong, Chap. 9, this volume), and career choices (Consedine et al. 2013), given the focus of our current discussion, below, we concentrate primarily on the domains most relevant to chronic health conditions: (1) *Pathogen disgust*—pathogen disgust is specific to health threats and is elicited in response to cues that indicate the potential presence of such a threat (Tybur et al. 2013). Pathogen cues are commonly found in chronic health contexts and include poor hygiene (e.g., pungent body odour); bodily fluids and products (e.g., blood, faeces, urine); violations of the body envelope (e.g., medical procedures, surgery, wounds); and death (Consedine 2008; Tybur et al. 2013). (2) *Self-disgust*—self-disgust is manifest in response to appraising aspects of oneself as "disgusting" (Powell et al. 2014). Much of the self-disgust literature has focused on mental health and psychopathology (e.g., depression; see Davey, Chap. 11, this volume); however, there is a growing body of work that argues that chronic health conditions which, by definition, often require ongoing, long term exposure to disgust elicitors, such as bodily products, disfigurement and the like, is also of relevance to the manifestation of self-disgust (Reynolds et al. 2018b).

Relevance to Chronic Health Conditions

Empirical work implicates pathogen and self-disgust in a range of chronic health conditions. Of particular relevance to this discussion are health conditions that have regular and/or sustained exposure to established disgust elicitors such as cancers (Dev et al. 2020), bowel-related conditions (Reynolds et al. 2014a), obesity (Lieberman et al. 2012), and skin-related conditions (Pereira et al. 2019). Most cancers and their treatments involve exposure to various elicitors including medical procedures and surgery (i.e., invasion of the body envelope); side-effects (e.g., vomit, diarrhoea); bodily changes (e.g., amputation, hair loss, radiotherapy burns); and existential elicitors, such as being confronted by the prospect of one's own death. Similarly, bowel conditions like inflammatory bowel disease or faecal incontinence inevitably involve close proximity to faecal matter (Reynolds et al. 2014a); excess body fat associated with obesity can signal the presence of pathogens (Lieberman et al. 2012); and chronic skin conditions can involve exposure to well-established elicitors, such as peeling skin and raw lesions, which may signal infiltration of the body envelope by pathogens (Pereira et al. 2019). In addition to considering the types of conditions where disgust is relevant, it is also useful to categorise the typical points along an illness trajectory where disgust has particular influence. These include screening and diagnosis, treatment and decision-making, and long term adaptation to a chronic health problem.

Screening and Diagnosis

Disgust has been strongly implicated in the screening and diagnosis of chronic health conditions. Given that many screening and diagnostic procedures involve exposure to elicitors such as testing bio-markers for disease (e.g., in blood, urine, and faeces) or inserting medical devices into body cavities that are ordinarily meant as exit rather than entry points (e.g., colonoscopies through the anus), it is unsurprising that dispositional disgust (both sensitivity and propensity) and state disgust (i.e., a persons current emotional state) might lead to screening avoidance and delays in attending healthcare services (Davis et al. 2017; Klasko-Foster et al. 2020; Reynolds et al. 2013, 2018a).

Avoidance of healthcare has particular significance in contexts such as screening, where early identification of disease enables timely management and prevention of disease progression (Garcia-Vidal et al. 2009; Provincial Health Services Authority 2019; World Health Organisation 2019). An emerging body of work has linked disgust with screening-related avoidance. Bowel cancer screening, which typically involves exposure to faeces and invasive rectal procedures, is of particular relevance here (e.g., Chambers et al. 2016b; Kotzur et al. 2016). Some recent work investigating bowel cancer screening behaviours and intentions has shown that both state disgust and the dispositional tendency to experience disgust is associated with less engagement with screening (Chambers et al. 2016a; Klasko-Foster et al. 2020). Likewise, another study revealed that greater insertion-related disgust predicted a lower likelihood of invasive bowel screening (such as colonoscopies) in the previous 5 years (Reynolds et al. 2018a). Other studies have also implicated disgust in this context with greater faecal disgust predicting greater screening-related avoidance behaviours over and above screening intentions (Davis et al. 2017; O'Carroll et al. 2015). The fact that disgust can predict screening behaviours above stated intentions demonstrates the additive potency of affect over cognitions in predicting health behaviours. Given the importance of screening in facilitating early intervention in a range of diseases, the robust association between greater disgust sensitivities and lower screening behaviours is one of considerable clinical importance.

Whilst screening programmes facilitate early detection of non-symptomatic disease, seeking timely medical advice is generally recommended in the presence of suspicious symptoms. However, as with screening avoidance, experimental studies have shown that people may delay seeking healthcare when symptoms are disgusting. For instance, disgust has been found to predict anticipated delay to sexual health services where disgust elicitors are involved (i.e., collecting genital discharge; McCambridge and Consedine 2014), and other experimental work has shown that people will delay seeking medical consultation in response to a scenario where they have suspicious bowel symptoms (diarrhoea and blood in stool; Reynolds et al.

2014b). Whilst these experimental studies provide an early indication that disgust is implicated in delayed presentation to healthcare, there is scant research that investigates this possibility with patient populations and further investigation of this possibility in clinical contexts is warranted.

Treatment and Decision-Making

Disgust can also negatively impact various aspects of treatment, including treatment uptake (Oppfeldt et al. 2016; Reynolds et al. 2013; Turner et al. 2018). Contemporary healthcare systems emphasise the importance of patient informed decision-making and, whilst the provision of detailed information can benefit patient understanding, it can also generate emotion. Where emotions are avoidance-promoting (as with disgust), treatment uptake has the potential to be disrupted. As noted above, symptoms that elicit disgust can influence decisions to delay help-seeking (McCambridge and Consedine 2014; Reynolds et al. 2014b). Similarly, experimental work indicates that disgust can impact intentions to undertake surgery. In a recent study, participants randomised to a condition where they received detailed information and pictures about construction and self-management of a stoma (a surgically created hole in the abdomen that excretes faeces) reported significantly higher disgust and lower intentions to undergo colorectal cancer surgery than controls (Turner et al. 2018). Furthermore, state disgust mediated memory recall such that disgusted participants were less able to remember pre-operative instructions. It is unclear why this might have been the case; however, one possibility is that the experience of state disgust uses valuable cognitive resources that are required in the process of laying down memories. Thus, overall, the experience of disgust appears to influence intentions and cognitive processes critical to rational decision-making.

The influence of disgust on treatment decision-making is also reflected in qualitative literature. For instance, people who have Chronic Obstructive Pulmonary Disease (COPD)—an inflammatory lung disease that causes obstructed airflow from the lungs and involves production and expulsion of excessive mucous (Miravitlles et al. 2014)—report concern about unpalatable symptoms in discussing intentions to undergo rehabilitation (Harrison et al. 2015). Other qualitative research implicates disgust in lower intentions of patients with Crohn's disease (an inflammatory bowel condition) to undergo recommended faecal microbiota transplantation (Oppfeldt et al. 2016). The act of transporting faeces, which harbours trillions of bacteria (Khoruts and Sadowsky 2016), from one person's colon to another is directly relevant to disgust in that it almost certainly triggers concerns about contamination and invasion of the body envelope (i.e., inserting foreign matter into the rectum). The barrier to uptake this innovative, potentially life changing (sometimes life-saving) procedure due to the "ick factor" has been a regular feature in media reporting (Chuong et al. 2015). Collectively, these studies indicate that disgust has the potential to influence important aspects of decision-making processes and treatment uptake. In the context of contemporary healthcare systems that increasingly emphasise patient informed decision-making, the influence of this avoidance-promoting emotion warrants further investigation.

Other research shows that heightened disgust sensitivities are associated with symptoms and wellbeing in patients as they undertake their treatment. For example, a recent unpublished study has shown that greater disgust sensitivity in a heterogenous sample of cancer patients predicted greater taste and smell-related changes during chemotherapy (Dev et al. 2020). Maintaining weight during treatment is an important factor in the health status of cancer patients (Deans et al. 2009), and changes to the taste and smell of food can negatively impact this process (Bernhardson et al. 2009). Similarly, other work has investigated cancer patients undergoing chemotherapy and found that greater disgust predicted increases in social, cognitive, and emotional avoidance (Reynolds et al. 2016) and that greater disgust-related side-effects, including bowel or bladder problems and nausea, were associated with greater depression and anxiety (Powell et al. 2016). The latter finding was mediated by self-disgust and partially moderated by disgust propensity (Powell et al. 2016). However, more research is required to determine the extent to which the anticipated impact of treatment might be influenced by either pathogen disgust or self-disgust. Whilst early indication suggests an association between disgust and lower psychological wellbeing in the context of cancer treatment, there is very little research that considers its role in other health conditions where disgust elicitors are prevalent.

Long Term Adaptation to a Chronic Health Problem

Beyond screening, diagnosis and treatment, disgust has also been implicated in the experience of long-term adaptation to various chronic health conditions. It has been noted in particular regarding health conditions which exhibit visible signs of poor health (Palmeira et al. 2019; Rosman 2004), involve exposure to established disgust elicitors (Dibley et al. 2019; Hunt 2019), and in certain neurological conditions (Ille et al. 2015; Trinkler et al. 2017; Verstaen et al. 2016; see also Overton et al., Chap. 12, this volume).

Recent research indicates that individuals living with visible signs of poor health can have high levels of disgust, including feelings of aversion and repugnance towards the self (Palmeira et al. 2019; Rosman 2004). Greater levels of disgust are reported by individuals with visible conditions such as alopecia (Rosman 2004; van Beugen et al. 2016), chronic leg ulcers (Nagaratnam et al. 2018), skin conditions (Narayanan et al. 2014), and obesity (Palmeira et al. 2019). Patients with psoriasis—dry, itchy, red raw, and scaly skin patches (Menter et al. 2008)—commonly describe their appearance with disgust and self-loathing (Narayanan et al. 2014). Even the brain scans of patients with psoriasis reveal pronounced reactions to criticism and facial signs of disgust (Schmidt 2015). In a study comparing psoriasis patients with healthy controls, skin-related disgust and shame were significantly higher in the psoriasis patients (Lahousen et al. 2016).

In such situations, individuals commonly engage in avoidance behaviours, including avoiding touching themselves, looking in mirrors, and masking the offending aspects of themselves (Burden et al. 2018; Palmeira et al. 2019; Powell et al. 2014). Women in particular appear to experience greater self-disgust than men (Palmeira et al. 2019; Rosman 2004) and are more likely to use disguises (e.g., wigs) to mask their bodily difference (Rosman 2004). Recent research reinforces the impact of self-disgust on psychological well-being; higher levels of self-disgust are implicated in depression (Powell et al. 2013), eating psychopathology in people who are overweight and obese (Palmeira et al. 2019), non-suicidal self-harm (Smith et al. 2015), and psychoticism (Ille et al. 2014). As such, the link between self-disgust and adaptation to chronic physical conditions is clinically important, with the construct itself being psychometrically validated and qualitatively consistent across the literature (Clarke et al. 2019).

Disgust is also reported where elicitors are less visible to others (Dibley et al. 2019; Hunt 2019). Bowel disease, which commonly involves increased exposure to one's own faeces, is of particular relevance and qualitative work notes that people with bowel disease are often disgusted by their condition (Brooks et al. 2015; Dibley et al. 2018, 2019; Woodward et al. 2016). In one study, a patient reported, "*. . .it's disgusting and I think it's horrible and smelly and going to the toilet all the time and seeing all this gunk and blood and mucus*", whilst another said, "*I don't think I'll ever change my mind about that, I've got it and it is disgusting*" (Woodward et al. 2016, p. 654). These observations are consistent with results of another qualitative study where patients with Irritable Bowel Syndrome identified that disgust was a primary feature of unfavourable body image and feeling different from others (Mohebbi et al. 2017). Importantly, adaptation to a chronic bowel condition has been shown to be worse in people who have greater sensitivity to disgust. For instance, stoma patients with higher disgust sensitivity (i.e., "range and intensity" as opposed to frequency) report lower life satisfaction, poorer adjustment to their colostomy, and are more likely to feel stigmatised (Smith et al. 2007). Of note, women are more likely to report disgust towards their stoma than their male counterparts (Juan et al. 2017). More broadly, this probably reflects the underlying gender differences in disgust (sensitivity and propensity; Clarke et al. 2016; Giel et al. 2016) *and* self-disgust (Palmeira et al. 2019).

As noted earlier, much of the literature investigating disgust in the context of chronic physical health conditions focuses on cancer. Not only do cancer patients tend to have greater levels of disgust sensitivity and self-disgust in comparison to matched controls, but disgust also tends to be a significant predictor of depressive symptoms in such individuals (Azlan et al. 2017a). In another study, greater disgust sensitivity amongst the partners of cancer patients was found to be associated with greater self-disgust, disgust propensity, and depression in those patients (Azlan et al. 2017b). Furthermore, results in this study indicated that patients' self-disgust had a mediating role in allowing disgust sensitivity of their partners to negatively influence their own psychological wellbeing (Azlan et al. 2017b). The association between patients' self-disgust and their partners' disgust may be explained by the partners' emotional (in this case, disgust-based) responses contributing to the emerging self-disgust schema in the patients (see Powell et al. 2014).

The experiences of patients with health conditions that comprise aversive symptoms or features are presumably made more debilitating when disgust-based reactions are observed in others. The following quote from a patient illustrates the detrimental impact of noting disgust in another person: *"I got a feeling of disgust from her. . . . It was hurtful and caused me a lot of anxiety. It probably knocked my confidence for a very long time. Because, as it got worse, it became a much bigger part of my life that was disgusting and secret and hidden* [*sic*]" (Dibley et al. 2019, p. 1201). Other work has observed that disgust sensitivity is a predictor of wanting less contact with colostomy patients (Smith et al. 2007). Similarly, this effect is demonstrated in a qualitative study that investigated social media responses to a selfie depicting a young woman in a bikini with an ostomy bag: *"EWWW EWWWW EWWWWW. . . .my mother wore one when she had cancer, and nobody wants to look at S**t. . .I'm sorry, but this is something that should be covered up. If you don't tell me you got one, I'll never know [sic]"* (Rademacher 2018, p. 3871). Similar themes are also observed in the context of amputees (Burden et al. 2018); however, the literature is limited in this context.

Disgust also appears to be an important component of bias towards people who have visible signs of poor health, including being overweight (Vartanian 2010) and having skin conditions (Green-Armytage et al. 2019; Halioua et al. 2016; Pereira et al. 2019; see also Vartanian et al., Chap. 10, this volume). Images of obese individuals have reliably elicited disgust responses in others (Lieberman et al. 2012) and, specifically, increased pathogen disgust proneness predicts more negative attitudes towards such individuals, ostensibly via implicit signals that excess body fat is an indicator of the potential presence of pathogens (Lieberman et al. 2012). Attitudes that obese individuals are lazy, unintelligent, and lacking in control are common (De Brún et al. 2014) and may provide a rationalisation for withdrawal and avoidance of such persons. Similarly, disgust is reported in response to people who have obvious signs of psoriasis and eczema (Green-Armytage et al. 2019; Halioua et al. 2016; Pereira et al. 2019), and laypersons and medical students both report not wanting to date, shake hands, or share a car with people who have enlarged psoriatic hand lesions (Pearl et al. 2019).

Clearly, disgust responses such as these have the capacity to impact on the psychological well-being of people with chronic conditions and can lead to people feeling isolated and stigmatised. However, there may also be situations where a person, who has feelings of self-directed disgust, may incorrectly assume that others feel similarly. In such situations, withdrawal from social connections may occur as a protective mechanism (Powell et al. 2014; Reynolds et al. 2018b). Although withdrawal from others appears likely in this context, apart from one study that demonstrated that disgust proneness in chemotherapy patients predicted social avoidance (Reynolds et al. 2016), there is little research investigating the extent of this possibility in chronic health and more work in the area is required.

Of note, disgust has also been implicated as having a causal role in certain health problems (Hildebrandt et al. 2015; Watkins et al. 2016). It has been argued that defects in the propensity to experience disgust can contribute to obesity by allowing over-consumption of food (Watkins et al. 2016). This effect has been shown to be especially true for females (Giel et al. 2016). Conversely, greater sensitivity to disgust may lead to under-consumption of food, which may be reflective of, or contribute towards, eating disorders (Anderson et al. 2018; Egolf et al. 2018; Hildebrandt et al. 2015). This is consistent with research that shows greater disgust (especially self-disgust) is associated with greater insulin restriction and more rigid, punitive approaches to diabetes management amongst patients with Type 1 diabetes (Merwin et al. 2015). It might be the case that feelings of disgust trigger vulnerable individuals to restrict insulin in the hope of losing weight and promoting feelings of well-being (Merwin et al. 2015).

Importantly, there is evidence to suggest that people might habituate to disgust elicitors over time (Olatunji et al. 2011; Rozin 2008), suggesting that problematic responses related to health conditions may lessen as time passes. However, one study found that while disgust lessened over time in response to one stimuli (i.e., a cold dead body) it did not reduce in response to another similar elicitor (i.e., a *warm* dead body; Rozin 2008) and other work has found disgust relatively rigid to change compared to other emotions (Olatunji et al. 2009). Thus, the impact of disgust may lessen over time in particular situations, but this may not translate to new situations. It is also possible that self-disgust elicitors might operate differently to pathogen elicitors, perhaps leading to sensitisation over time. Further research is required in this area to understand the extent to which characteristics of disgust are static versus dynamic.

Moving Forward

An escalating number of studies in the past decade have focussed on disgust in the context of chronic health conditions. Despite this recent activity, work investigating disgust in physical health has tended to cluster around the conditions that have the most obvious link to disgust elicitors including cancer, obesity, bowel conditions, and skin problems. However, there are other conditions that appear to be obvious candidates for future focus. Conditions where disfigurement is clearly visible to others or has altered bodily function (speaking, eating, drinking, excreting, etc.) seem like particularly strong candidates for further work. Such research could include faecal and urinary stoma patients, people who have experienced invasive burns, those who suffer dental problems, and/or those with obvious scarring or wounds. Considering the ways in which the avoidance-promoting disgust response might impact on such populations are worthy avenues for future research.

Arguably, however, disgust is relevant to *all* health conditions given its evolutionary purpose. Wherever a person exhibits signs of illness, blood is drawn, invasive medical procedures are conducted, the body is medicalised or cut open, bodily function is altered, exposure to body product occurs (etc.), a disgust response is possible. Given the likely relevance of this response across all aspects of health, clinicians could be routinely screening their patients for disgust sensitivity and propensity. In doing so, patients (and their partners) who have a greater propensity or sensitivity to disgust can be pre-emptively identified and targeted for early interventions. Correspondingly, it is also important for researchers and clinicians to consider developing better disgust screening tools that can identify the kinds of elicitors likely to be problematic, and the behavioural, physiological, and emotional responses that might occur in response.

Importantly, research is required that informs interventions that might mitigate the deleterious impacts of disgust on patients, their close associates, and healthcare professionals. There remains a need for interventions to be developed and tested that specifically target clinical contexts and the self-management tasks that chronic health conditions require. Whilst research investigating disgust-focused interventions in the context of chronic health is scant, emerging experimental work offers promise (Reynolds et al. 2013). A recent study has demonstrated that disgust negatively impacts clinical engagement with patients who present with disgusting symptoms and suggests that compassion might offer potential in mitigating this impact (Reynolds et al. 2019). Other strategies found to be helpful include acknowledging the presence of disgust (Kircanski et al. 2012); challenging thoughts, re-framing, and re-appraisal (Feinberg et al. 2014; Fink et al. 2018; Goldin et al. 2008; Wilson et al. 2018); using a detached stance (Shiota and Levenson 2012); exposure-based habituation (Adams et al. 2011; Olatunji et al. 2009); and acceptance-based therapies (Wolgast et al. 2011). However, while there has been an escalating interest in this area, the majority of this work has been experimental research either conducted with healthy volunteers or clinical populations who have psychopathologies (OCD, spider phobias, etc.). Investigation of the utility of clinical interventions with physical health populations has been almost completely overlooked and strategies almost certainly need to be tailored to the specific elicitor and type of disgust (i.e., pathogen or self-disgust), as different contexts are likely to require different interventional approaches.

Where disgust is self-directed or associated with critical self-evaluation it may be that self-compassion is of benefit. Self-compassion is the ability to direct compassion to ones' self and encapsulates components of self-kindness, common humanity, and mindfulness (Neff 2003). Cross-sectional work has found that higher levels of self-disgust are related to lower self-compassion (Palmeira et al. 2019), and recent clinical trials have shown that self-compassion training can lead to improved psychological and physiological outcomes amongst populations with chronic health conditions (Friis et al. 2016; Sherman et al. 2018). A study investigating the utility of a brief online self-compassion writing exercise with breast cancer patients found significant improvements post-intervention in body image distress and body appreciation (Sherman et al. 2018); and a replication of this study with stoma patients found that the intervention was effective in reducing body image distress in stoma patients with low disgust sensitivity (Harris 2019). Thus, there appears potential for further investigation of this interventional approach in other chronic health contexts.

Concluding Remarks

In sum, although there has been an escalating interest into the links between disgust and physical health in recent years, there remain vast areas of unexplored territories in this domain for interested researchers. Disgust appears relevant to any condition where health is threatened, and it is important to fully consider the interventional strategies that might be helpful for patients, caregivers, associates, and health professionals in this context.

References

Adams, T. G., Willems, J. L., & Bridges, A. J. (2011). Contamination aversion and repeated exposure to disgusting stimuli. *Anxiety, Stress and Coping, 24*(2), 157–165. https://doi.org/10.1080/10615806.2010.506953

Anderson, L. M., Reilly, E. E., Thomas, J. J., Eddy, K. T., Franko, D. L., Hormes, J. M., & Anderson, D. A. (2018). Associations among fear, disgust, and eating pathology in undergraduate men and women. *Appetite, 125*, 445–453. https://doi.org/10.1016/j.appet.2018.02.017

Azlan, H. A., Overton, P. G., Simpson, J., & Powell, P. A. (2017a). Differential disgust responding in people with cancer and implications for psychological wellbeing. *Psychology and Health, 32* (1), 19–37. https://doi.org/10.1080/08870446.2016.1235165

Azlan, H. A., Overton, P. G., Simpson, J., & Powell, P. A. (2017b). Effect of partners' disgust responses on psychological wellbeing in cancer patients. *Journal of Clinical Psychology in Medical Settings, 24*(3–4), 355–364. https://doi.org/10.1007/s10880-017-9521-z

Azlan, H. A., Overton, P. G., Simpson, J., & Powell, P. A. (2020). Disgust propensity has a causal link to the stigmatization of people with cancer. *Journal of behavioral medicine, 43*(3), 377–390. https://doi.org/10.1007/s10865-019-00130-4

Bernhardson, B.-M., Tishelman, C., & Rutqvist, L. E. (2009). Olfactory changes among patients receiving chemotherapy. *European Journal of Oncology Nursing, 13*(1), 9–15. https://doi.org/10.1016/j.ejon.2008.10.002

Brooks, A., Rowse, G., Ryder, A., Narula, P., Corfe, B., Norman, P., & Lobo, A. (2015). PTU-059 "i can cope right now, because i know where i have come from"; a qualitative exploration of the lived experience of young adults with inflammatory bowel disease. *Gut, 64*(Suppl 1), A85.2-A86. https://doi.org/10.1136/gutjnl-2015-309861.174

Burden, N., Simpson, J., Murray, C., Overton, P. G., & Powell, P. A. (2018). Prosthesis use is associated with reduced physical self-disgust in limb amputees. *Body Image, 27*, 109–117. https://doi.org/10.1016/j.bodyim.2018.08.001

Chambers, J. A., Callander, A. S., Grangeret, R., & O'Carroll, R. E. (2016a). Attitudes towards the Faecal Occult Blood Test (FOBT) versus the Faecal Immunochemical Test (FIT) for colorectal cancer screening: Perceived ease of completion and disgust. *BMC Cancer, 16*(96). https://doi.org/10.1186/s12885-016-2133-4

Chambers, J., O'Carroll, R., Brownlee, L., Libby, G., & Steele, R. (2016b). Colorectal cancer screening and perceived disgust: The importance of the "ICK" factor in Faecal Occult Blood Test uptake. *Colorectal Cancer: Open Access, 2*(1), 13–21. https://doi.org/10.21767/2471-9943.100013

Chapman, H. A., & Anderson, A. K. (2012). Understanding disgust. *Annals of the New York Academy of Sciences, 1251*(1), 62–76. https://doi.org/0.1111/j.1749-6632.2011.06369.x

Chuong, K. H., O'Doherty, K. C., & Secko, D. M. (2015). Media discourse on the social acceptability of fecal transplants. *Qualitative Health Research*, *25*(10), 1359–1371. https://doi.org/10.1177/1049732314568199

Clarke, N., Gallagher, P., Kearney, P. M., McNamara, D., & Sharp, L. (2016). Impact of gender on decisions to participate in faecal immunochemical test-based colorectal cancer screening: A qualitative study. *Psycho-Oncology, 25*(12), 1456–1462. https://doi.org/10.1002/pon.4085

Clarke, A., Simpson, J., & Varese, F. (2019). A systematic review of the clinical utility of the concept of self-disgust. *Clinical Psychology & Psychotherapy, 26*(1), 110–134. https://doi.org/10.1002/cpp.2335

Consedine, N. S. (2008). The health-promoting and health-damaging effects of emotions: the view from developmental functionalism. In M. Lewis, J. Haviland-Jones, & L. F. Barrett (Eds.), *Handbook of emotions* (3rd ed., pp. 676-690). Guilford.

Consedine, N. S., Magai, C., & Bonanno, G. A. A. (2002). Moderators of the emotion inhibition-health relationship: A review and research agenda. *Review of General Psychology*, *6*(2), 204–228. https://doi.org/10.1037//1089-2680.6.2.204

Consedine, N. S., Moskowitz, J. T. T., & Moscowitz, J. T. (2007). The role of discrete emotions in health outcomes: A critical review. *Applied and Preventive Psychology*, *12*(2), 59–75. https://doi.org/10.1016/j.appsy.2007.09.001

Consedine, N. S., Tzu-Chieh, Y., & Windsor, J. A. (2013). Nursing, pharmacy, or medicine? Disgust sensitivity predicts career interest among trainee health professionals. *Advances in Health Sciences Education*, *18*(5), 997–1008. https://doi.org/10.1007/s10459-012-9439-z

Curtis, V., & Biran, A. (2001). Dirt, disgust, and disease: Is hygiene in our genes? *Perspectives in Biology and Medicine*, *44*(1), 17–31. https://doi.org/10.1353/pbm.2001.0001

Curtis, V, Aunger, R., & Rabie, T. (2004). Evidence that disgust evolved to protect from risk of disease. *Proceedings of the Royal Society Biological Sciences: Series B*, *271*, S131–S133. https://doi.org/10.1098/rsbl.2003.0144

Curtis, V., de Barra, M., & Aunger, R. (2011). Disgust as an adaptive system for disease avoidance behaviour. *Philosophical Transactions of the Royal Society B: Biological Sciences*, *366*(1563), 389. https://doi.org/10.1098/rstb.2010.0117

Davis, M., Oaten, M., Occhipinti, S., Chambers, S. K., & Stevenson, R. J. (2017). An investigation of the emotion of disgust as an affective barrier to intention to screen for colorectal cancer. *European Journal of Cancer Care*, *26*(4), e12582. https://doi.org/10.1111/ecc.12582

De Brún, A., McCarthy, M., McKenzie, K., & McGloin, A. (2014). Weight stigma and narrative resistance evident in online discussions of obesity. *Appetite, 72*, 73–81. https://doi.org/10.1016/j.appet.2013.09.022

de Jong, P. J., van Lankveld, J., Elgersma, H. J., Borg, C., De Jong, P. J., Van Lankveld, J., & Borg, C. (2010). Disgust and sexual problems: Theoretical conceptualization and case illustrations. *International Journal of Cognitive Therapy*, *3*(1), 23–39. https://psycnet.apa.org/doi/10.1521/ijct.2010.3.1.23

Deans, D. A. C., Tan, B. H., Wigmore, S. J., Ross, J. A., De Beaux, A. C., Paterson-Brown, S., & Fearon, K. C. H. (2009). The influence of systemic inflammation, dietary intake and stage of disease on rate of weight loss in patients with gastro-oesophageal cancer. *British Journal of Cancer*, *100*(1), 63–69. https://doi.org/10.1038/sj.bjc.6604828

Dev, V., Consedine, N. S., & Reynolds, L. M. (2020). The "ick" factor: An unrecognised affective predictor of physical symptoms during chemotherapy. *Annals of Behavioral Medicine*. Advance online publication. https://doi.org/10.1093/abm/kaaa055

Dibley, L., Norton, C., & Whitehead, E. (2018). The experience of stigma in inflammatory bowel disease: An interpretive (hermeneutic) phenomenological study. *Journal of Advanced Nursing*, *74*(4), 838–851. https://doi.org/10.1111/jan.13492

Dibley, L., Williams, E., & Young, P. (2019). When family don't acknowledge: A hermeneutic study of the experience of kinship stigma in community-dwelling people with Inflammatory Bowel Disease. *Qualitative Health Research*. https://doi.org/10.1177/1049732319831795

Egolf, A., Siegrist, M., & Hartmann, C. (2018). How people's food disgust sensitivity shapes their eating and food behaviour. *Appetite, 127*, 28–36. https://doi.org/10.1016/j.appet.2018.04.014

Eskine, K. J., Novreske, A., & Richards, M. (2013). Moral contagion effects in everyday interpersonal encounters. *Journal of Experimental Social Psychology, 49*(5), 947–950. https://doi.org/10.1016/j.jesp.2013.04.009

Feinberg, M., Antonenko, O., Willer, R., Horberg, E. J., & John, O. P. (2014). Gut check: Reappraisal of disgust helps explain liberal-conservative differences on issues of purity. *Emotion, 14*(3), 513–521. https://doi.org/10.1037/a0033727

Fink, J., Pflugradt, E., Stierle, C., & Exner, C. (2018). Changing disgust through imagery rescripting and cognitive reappraisal in contamination-based obsessive-compulsive disorder. *Journal of Anxiety Disorders, 54*, 36–48. https://doi.org/10.1016/j.janxdis.2018.01.002

Friis, A. M., Johnson, M. H., Cutfield, R. G., & Consedine, N. S. (2016). Kindness matters: A randomized controlled trial of a mindful self-compassion intervention improves depression, distress, and HbA1c among patients with diabetes. *Diabetes Care, 39*(11), 1963–1971. https://doi.org/10.2337/dc16-0416

Garcia-Vidal, C., Rodríguez-Fernández, S., Teijón, S., Esteve, M., Rodríguez-Carballeira, M., Lacasa, J. M., Salvador, G., & Garau, J. (2009). Risk factors for opportunistic infections in infliximab-treated patients: The importance of screening in prevention. *European Journal of Clinical Microbiology and Infectious Diseases, 28*(4), 331–337. https://doi.org/10.1007/s10096-008-0628-x

Giel, K. E., Hartmann, A., Zeeck, A., Jux, A., Vuck, A., Gierthmuehlen, P. C. G., Wetzler-Burmeister, E., Sandholz, A., Marjanovic, G., & Joos, A. (2016). Decreased emotional perception in obesity. *European Eating Disorders Review, 24*(4), 341–346. https://doi.org/10.1002/erv.2444

Goldin, P. R., McRae, K., Ramel, W., & Gross, J. J. (2008). The neural bases of emotion regulation: Reappraisal and suppression of negative emotion. *Biological Psychiatry, 63*(6), 577–586. https://doi.org/10.1016/j.biopsych.2007.05.031

Green-Armytage, M., Simonds, L. M., John, M., & Woodger, N. (2019). Depictions of acne and psoriasis influence interpersonal aversion. *Psychology, Health and Medicine, 24*(1), 94–100. https://doi.org/10.1080/13548506.2018.1497188

Halioua, B., Sid-Mohand, D., Roussel, M. E., Maury-Le-Breton, A., De Fontaubert, A., & Stalder, J. F. (2016). Extent of misconceptions, negative prejudices and discriminatory behaviour to psoriasis patients in France. *Journal of the European Academy of Dermatology and Venereology, 30*(4), 650–654. https://doi.org/10.1111/jdv.13095

Harris, L. (2019). *Does an Online Self-Compassion Expressive Writing Exercise Offer Benefit to Stoma Patients? A Randomised Controlled trial* [Unpublished master's thesis], University of Auckland. http://hdl.handle.net/2292/53139

Harrison, S. L., Robertson, N., Apps, L., C. Steiner, Morgan, M. D. L., & Singh, S. J. (2015). "We are not worthy"-understanding why patients decline pulmonary rehabilitation following an acute exacerbation of COPD. *Disability and Rehabilitation, 37*(9), 750–756. https://doi.org/10.3109/09638288.2014.939770

Hildebrandt, T., Grotzinger, A., Reddan, M., Greif, R., Levy, I., Goodman, W., & Schiller, D. (2015). Testing the disgust conditioning theory of food-avoidance in adolescents with recent onset anorexia nervosa. *Behaviour Research and Therapy, 71*, 131–138. https://doi.org/10.1016/j.brat.2015.06.008

Hunt, M. G. (2019). Cognitive-behavioral therapy for Irritable Bowel Syndrome. In W. H. Sobin (Ed.), *Using central neuromodulators and psychological therapies to manage patients with disorders of gut-brain interaction* (pp. 95–141). Springer International Publishing. https://doi.org/10.1007/978-3-030-18218-2_5

Ille, R., Schöggl, H., Kapfhammer, H.-P., Arendasy, M., Sommer, M., & Schienle, A. (2014). Self-disgust in mental disorders—symptom-related or disorder-specific? *Comprehensive Psychiatry, 55*(4), 938–943. https://doi.org/10.1016/j.comppsych.2013.12.020

Ille, R., Wabnegger, A., Schwingenschuh, P., Katschnig-Winter, P., Kögl-Wallner, M., Wenzel, K., & Schienle, A. (2015). Role of disgust proneness in Parkinson's disease: A voxel-based morphometry study. *Journal of the International Neuropsychological Society, 21*(4), 314–317. https://doi.org/10.1017/S135561771500017X

Ille, R., Wolf, A., Tomazic, P. V., & Schienle, A. (2016). Disgust-related personality traits in men with olfactory dysfunction. *Chemical Senses, 41*(5), 427–431. https://doi.org/10.1093/chemse/bjw017

Johnson-Laird, P. N., & Oatley, K. (1992). Basic emotions, rationality, and folk theory. *Cognition & Emotion, 6*, 201–223. https://doi.org/10.1080/02699939208411069

Juan, L., Recardo, D., Mayte, V., & Esther, T. (2017). Gender differences in stoma-related quality of life in Puerto Ricans with IBD. *The American Journal of Gastroenterology, 113*(S1), S14. https://doi.org/10.14309/00000434-201802001-00056

Keltner, D., & Gross, J. J. (1999). Functional accounts of emotions. *Cognition & Emotion, 13*, 467–480. https://doi.org/10.1080/026999399379140

Khoruts, A., & Sadowsky, M. J. (2016). Understanding the mechanisms of faecal microbiota transplantation. *Nature Reviews Gastroenterology and Hepatology*. Nature Publishing Group. https://doi.org/10.1038/nrgastro.2016.98

Kircanski, K., Lieberman, M. D., & Craske, M. G. (2012). Feelings into words. *Psychological Science, 23*(10), 1086–1091. https://doi.org/10.1177/0956797612443830

Klasko-Foster, L. B., Keller, M. M., & Kiviniemi, M. T. (2020). Is it disgusting or am I just easily disgusted? The relation between situational disgust, dispositional disgust, and colonoscopy intentions. *European Journal of Cancer Care, 29 June 20*. https://doi.org/10.1111/ecc.13244

Kotzur, M., Wyke, S., Macdonald, S., Steele, R., Weller, D., Campbell, C., Crighton, E., McCowan, C., & Robb, K. (2016). 'Not that one': Understanding why women screen for breast and cervical, but not bowel cancer. *The European Health Psychologist, 18*, 1–1. https://www.ehps.net/ehp/index.php/contents/article/view/1845

Lahousen, T., Kupfer, J., Gieler, U., Hofer, A., Linder, M. D., & Schut, C. (2016). Differences between psoriasis patients and skin-healthy controls concerning appraisal of touching, shame and disgust. *Acta Dermato-Venereologica, 96*, 78–82. https://doi.org/10.2340/00015555-2373

Lieberman, D., Tybur, J. M., & Latner, J. D. (2012). Disgust sensitivity, obesity stigma, and gender: Contamination psychology predicts weight bias for women, not men. *Obesity, 20*(9), 1803–1814. https://doi.org/10.1038/oby.2011.247

McCambridge, S. A., & Consedine, N. S. (2014). For whom the bell tolls: Experimentally-manipulated disgust and embarrassment may cause anticipated sexual healthcare avoidance among some people. *Emotion, 14*(2), 407–415. https://doi.org/10.1037/a0035209

Menter, A., Gottlieb, A., Feldman, S. R., Van Voorhees, A. S., Leonardi, C. L., Gordon, K. B., Lebwohl, M., Koo, J. Y. M., Elmets, C. A., Korman, N. J., Beutner, K. R., & Bhushan, R. (2008). Guidelines of care for the management of psoriasis and psoriatic arthritis: Section 1. Overview of psoriasis and guidelines of care for the treatment of psoriasis with biologics. *Journal of the American Academy of Dermatology, 58*(5), 826–850. https://doi.org/10.1016/j.jaad.2008.02.039

Merwin, R. M., Dmitrieva, N. O., Honeycutt, L. K., Moskovich, A. A., Lane, J. D., Zucker, N. L., Surwit, R. S., Feinglos, M., & Kuo, J. (2015). Momentary predictors of insulin restriction among adults with Type 1 diabetes and eating disorder symptomatology. *Diabetes Care, 38*(11), 2025–2032. https://doi.org/10.2337/dc15-0753

Miravitlles, M., Worth, H., Soler Cataluña, J. J., Price, D., De Benedetto, F., Roche, N., Godtfredsen, N. S., van der Molen, T., Löfdahl, C. G., Padullés, L., & Ribera, A. (2014). Observational study to characterise 24-hour COPD symptoms and their relationship with patient-reported outcomes: Results from the ASSESS study. *Respiratory Research, 15*(1), 122. https://doi.org/10.1186/s12931-014-0122-1

Mohebbi, Z., Sharif, F., Peyrovi, H., Rakhshan, M., Alizade Naini, M., & Zarshenas, L. (2017). Self-perception of Iranian patients during their life with Irritable Bowel Syndrome: A qualitative study. *Electronic Physician, 9*(12), 5885–5893. https://doi.org/10.19082/5885

Nagaratnam, N., Nagaratnam, K., & Cheuk, G. (2018). Chronic leg and foot ulcers in the elderly. In N. Nagaratnam, K. Nagaratnam, & G. Cheuk (Eds.), *Geriatric diseases* (pp. 799–804). Springer International Publishing. https://doi.org/10.1007/978-3-319-33434-9_87

Narayanan, S., Guyatt, V., Franceschetti, A., & Hautamaki, E. (2014). Disease burden and patient reported outcomes among patients with moderate to severe psoriasis: An ethnography study. *Psoriasis: Targets and Therapy*, 1–7. https://doi.org/10.2147/ptt.s74906

Neff, K. (2003). Self-compassion: An alternative conceptualization of a healthy attitude toward oneself. *Self and Identity*, *2*(2), 85–101. https://doi.org/10.1080/15298860309032

Nesse, R. M., & Ellsworth, P. C. (2009). Evolution, emotions, and emotional disorders. *American Psychologist*, *64*(2), 129–139. https://doi.org/10.1037/a0013503

Newell, R. J. (1999). Altered body image: A fear-avoidance model of psycho-social difficulties following disfigurement. *Journal of Advanced Nursing*, *30*(5), 1230–1238. https://doi.org/10.1046/j.1365-2648.1999.01185.x

O'Carroll, R. E., Chambers, J. A., Brownlee, L., Libby, G., & Steele, R. J. C. (2015). Anticipated regret to increase uptake of colorectal cancer screening (ARTICS): A randomised controlled trial. *Social Science and Medicine*, *142*, 118–127. https://doi.org/10.1016/j.socscimed.2015.07.026

Oaten, M., Stevenson, R. J. J., & Case, T. I. I. (2011). Disease avoidance as a functional basis for stigmatization. *Philosophical Transactions of the Royal Society B: Biological Sciences*, *366* (1583), 3433–3452. https://doi.org/10.1098/rstb.2011.0095

Olatunji, B. O. O., & Sawchuk, C. N. N. (2005). Disgust: characteristic features, social manifestations, and clinical implications. *Journal of Social & Clinical Psychology*, *24*(7), 932–962. https://doi.org/10.1521/jscp.2005.24.7.932

Olatunji, B. O. O., Wolitzky-Taylor, K. B. B., Willems, J., Lohr, J. M. M., & Armstrong, T. (2009). Differential habituation of fear and disgust during repeated exposure to threat-relevant stimuli in contamination-based OCD: An analogue study. *Journal of Anxiety Disorders*, *23*(1), 118–123. https://doi.org/10.1016/j.janxdis.2008.04.006

Olatunji, B. O., Tart, C. D., Ciesielski, B. G., McGrath, P. B., & Smits, J. A. J. (2011). Specificity of disgust vulnerability in the distinction and treatment of OCD. *Journal of Psychiatric Research*, *45*(9), 1236–1242. https://doi.org/10.1016/j.jpsychires.2011.01.018

Oppfeldt, A. M., Dahlerup, J. F., Christensen, L. A., & Hvas, C. L. (2016). Faecal microbiota transplantation for recurring Clostridium difficile infection in a patient with Crohn's disease and ileorectal anastomosis. *BMJ Case Reports*, bcr2016217209. https://doi.org/10.1136/bcr-2016-217209

Palmeira, L., Pinto-Gouveia, J., & Cunha, M. (2019). The role of self-disgust in eating psychopathology in overweight and obesity: Can self-compassion be useful? *Journal of Health Psychology, 24*(13), 1807–1816. https://doi.org/10.1177/1359105317702212

Pearl, R. L., Wan, M. T., Takeshita, J., & Gelfand, J. M. (2019). Stigmatizing attitudes toward persons with psoriasis among laypersons and medical students. *Journal of the American Academy of Dermatology*, *80*(6), 1556–1563. https://doi.org/10.1016/j.jaad.2018.08.014

Pereira, P. A., Miot, H. A., Coutinho, T. V., & Schmitt, J. V. (2019). Aversion caused by skin diseases—a crowdsourcing study. *International Journal of Dermatology*, *58*(1), 60–66. https://doi.org/10.1111/ijd.14205

Pizarro, D., Inbar, Y., & Helion, C. (2011). On Disgust and Moral Judgment. *Emotion Review*, *3*(3), 267–268. https://doi.org/10.1177/1754073911402394

Plutchik, R. (2001). The nature of emotions: Human emotions have deep evolutionary roots, a fact that may explain their complexity and provide tools for clinical practice. *American Scientist*, *89* (4), 344–350. https://www.jstor.org/stable/27857503

Powell, P. A, Simpson, J., & Overton, P. G. (2013). When disgust leads to dysphoria: A three-wave longitudinal study assessing the temporal relationship between self-disgust and depressive symptoms. *Cognition and Emotion*, *27*(5), 900–913. https://doi.org/10.1080/02699931.2013.767223

Powell, P., Overton, P., & Simpson, J. (2014). The revolting self: An interpretative phenomenological analysis of the experience of self-disgust in females with depressive symptoms. *Journal of Clinical Psychology*, *70*(6), 562–578. https://doi.org/10.1002/jclp.22049

Powell, P. A., Azlan, H. A., Simpson, J., & Overton, P. G. (2016). The effect of disgust-related side-effects on symptoms of depression and anxiety in people treated for cancer: A moderated mediation model. *Journal of Behavioral Medicine*, *39*(4), 560–573. https://doi.org/10.1007/s10865-016-9731-0

Provincial Health Services Authority. (2019). Why is screening important? Retrieved from http://www.bccancer.bc.ca/screening/colon/get-screened/why-is-screening-important

Rademacher, M. A. (2018). "The most inspiring bikini photos you'll see this summer": A thematic analysis of mass audiences' interpretations of ostomy selfies. *New Media and Society*, *20*(10), 3858–3878. https://doi.org/10.1177/1461444818761876

Reynolds, L. M., Consedine, N. S., Pizarro, D., & Bissett, I. (2013). Disgust and behavioral avoidance in colorectal cancer screening and treatment: A systematic review and research agenda. *Cancer Nursing*, *36*(2), 122–130. https://doi.org/10.1097/NCC.0b013e31826a4b1b

Reynolds, L. M., Bissett, I. P., & Consedine, N. S. (2014a). Predicting the patients who will struggle with anal incontinence: Sensitivity to disgust matters. *Colorectal Disease*, *17*(1), 73–80. https://doi.org/10.1111/codi.12781

Reynolds, L. M., McCambridge, S. A., Bissett, I. P., & Consedine, N. S. (2014b). Trait and state disgust: An experimental investigation of disgust and avoidance in colorectal cancer decision scenarios. *Health Psychology*, *33*(12), 1495–1506. https://doi.org/10.1037/hea0000023

Reynolds, L. M., Bissett, I. P., Porter, D., & Consedine, N. S. (2016). The "ick" factor matters: Disgust prospectively predicts avoidance in chemotherapy patients. *Annals of Behavioral Medicine*, *50*(6), 935–945. https://doi.org/10.1007/s12160-016-9820-x

Reynolds, L. M., Bissett, I. P., & Consedine, N. S. (2018a). Emotional predictors of bowel screening: The avoidance-promoting role of fear, embarrassment, and disgust. *BMC Cancer*, *18*(1), 518. https://doi.org/10.1186/s12885-018-4423-5

Reynolds, L. M., McCambridge, S. A., & Consedine, N. S. (2018b). Self-disgust and adaptation to chronic physical health conditions: Implications for avoidance and withdrawal. In P. A. Powell, P. G. Overton, & J. Simpson (Eds.), *The revolting self: Perspectives on the psychological, social, and clinical implications of self-directed disgust*. Routledge.

Reynolds, L. M., Powell, P., Lin, Y. S., Ravi, K., Chung, C. K., & Consedine, N. S. (2019). Fighting the flinch: Experimentally induced compassion makes a difference in health care providers. *British Journal of Health Psychology*, 1–17. https://doi.org/10.1111/bjhp.12390

Rosman, S. (2004). Cancer and stigma: Experience of patients with chemotherapy-induced alopecia. *Patient Education and Counseling*, *52*(3), 333–339. https://doi.org/10.1016/s0738-3991(03)00040-5

Rozin, P. (2008). Hedonic "adaptation": Specific habituation to disgust/death elicitors as a result of dissecting acadaver. *Judgment and Decision Making, 3*(2), 191–194.

Rozin, P., & Fallon, A. E. (1987). A perspective on disgust. *Psychological Review*, *94*(1), 23–41. https://doi.org/10.1037/0033-295X.94.1.23

Rozin, P., Millman, L., & Nemeroff, C. (1986). Operation of the laws of sympathetic magic in disgust and other domains. *Journal of Personality and Social Psychology, 50*(4), 703–712. https://psycnet.apa.org/doi/10.1037/0022-3514.50.4.703

Rozin, P., Haidt, J., & McCauley, C. R. (2000). Disgust. In M. Lewis & S. M. Haviland-Jones (Eds.), *Handbook of Emotions* (2nd ed., pp. 637–653). Guilford Press.

Rozin, P., Haidt, J., & McCauley, C. R. (2009). Disgust: The body and soul emotion in the 21st century. In B. O. Olatunji & D. McKay (Eds.), *Disgust and its disorders: Theory, assessment and treatment implications* (pp. 9–29). American Psychological Association.

Rumsey, N., Clarke, A., White, P., Wyn-Williams, M., & Garlick, W. (2004). Altered body image: Appearance-related concerns of people with visible disfigurement. *Journal of Advanced Nursing*, *48*(5), 443–453. https://doi.org/10.1111/j.1365-2648.2004.03227.x

Schmidt, C. (2015). Suicidal thoughts end Amgen's blockbuster aspirations for psoriasis drug. *Nature Biotechnology*, *33*(9), 894–895. https://doi.org/10.1038/nbt0915-894b

Sherman, K. A., Przezdziecki, A., Alcorso, J., Jon Kilby, C., Elder, E., Boyages, J., Koelmeyer, L., & Mackie, H. (2018). Reducing body image-related distress in women with breast cancer using a structured online writing exercise: Results from the my changed body randomized controlled trial. *Journal of Clinical Oncology, 36*, 1930–1940. https://doi.org/10.1200/JCO.2017.76.3318

Shiota, M. N., & Levenson, R. W. (2012). Turn down the volume or change the channel? Emotional effects of detached versus positive reappraisal. *Journal of Personality and Social Psychology*, *103*(3), 416–429. https://doi.org/10.1037/a0029208

Smith, D., Loewenstein, G., Rozin, P., Sheriff, R., & Ubel, P. (2007). Sensitivity to disgust, stigma and colostomy. *Journal of Research in Personality* *41*(4), 787–803. https://doi.org/10.1021/nn300902w.Release

Smith, N. B., Steele, A. M., Weitzman, M. L., Trueba, A. F., & Meuret, A. E. (2015). Investigating the role of self-disgust in nonsuicidal self-injury. *Archives of Suicide Research*, *19*(1), 60–74. https://doi.org/10.1080/13811118.2013.850135

Stavrova, O., Newman, G. E., Kulemann, A., & Fetchenhauer, D. (2016). Contamination without contact: An examination of intention-based contagion. *Judgment and Decision Making*, *11*(6), 554–571.

Tooby, J., & Cosmides, L. (2008). The evolutionary psychology of the emotions and their relationship to internal regulatory variables. In Michael Lewis, L. F. Barrett, & J. M. Haviland-Jones (Eds.), *Handbook of emotions* (3rd ed., pp. 114–137). Guilford Press.

Trinkler, I., Devignevielle, S., Achaibou, A., Ligneul, R. V, Brugières, P., Cleret de Langavant, L., De Gelder, B., Scahill, R., Schwartz, S., & Bachoud-Lévi, A. C. (2017). Embodied emotion impairment in Huntington's Disease. *Cortex*, *92*, 44–56. https://doi.org/10.1016/j.cortex.2017.02.019

Turner, M. C., O'Brien, J. D., Kahn, R. M., Mantyh, C. R., Migaly, J., & Ariely, D. (2018). Impact of disgust on intentions to undergo colorectal surgery. *Diseases of the Colon & Rectum*, *61*(12), 1386–1392. https://doi.org/10.1097/DCR.0000000000001254

Tybur, Joshua M., Lieberman, D., Kurzban, R., & DeScioli, P. (2013). Disgust: Evolved function and structure. *Psychological Review American Psychological Association*, *120*(1), 65–84. https://doi.org/10.1037/a0030778

van Beugen, S., Maas, J., van Laarhoven, A. I. M., Galesloot, T. E., Rinck, M., Becker, E. S., van de Kerkhof, P. C. M., van Middendorp, H., & Evers, A. W. M. (2016). Implicit stigmatization-related biases in individuals with skin conditions and their significant others. *Health Psychology*, *35*(8), 861–865. https://doi.org/10.1037/hea0000404

van Overveld, W., de Jong, P. J., Peters, M. L., Cavanagh, K., & Davey, G. C. L. (2006). Disgust propensity and disgust sensitivity: Separate constructs that are differentially related to specific fears. *Personality and Individual Differences*, *41*(7), 1241–1252. https://doi.org/10.1016/j.paid.2006.04.021

van Overveld, W. J. M., de Jong, P. J., Peters, M. L., van Overveld, M., de Jong, P. J., & Peters, M. L. (2010). The Disgust Propensity and Sensitivity Scale—Revised: Its predictive value for avoidance behavior. *Personality and Individual Differences*, *49*(7), 706–711. https://doi.org/10.1016/j.paid.2010.06.008

Vartanian, L. R. (2010). Disgust and perceived control in attitudes toward obese people. *International Journal of Obesity*, *34*(8), 1302–1307. https://doi.org/10.1038/ijo.2010.45

Verstaen, A., Eckart, J. A., Muhtadie, L., Otero, M. C., Sturm, V. E., Haase, C. M., Miller, B. L., & Levenson, R. W. (2016). Insular atrophy and diminished disgust reactivity. *Emotion*, *16*(6), 903–912. https://doi.org/10.1037/emo0000195

Watkins, T. J., Di Iorio, C. R., Olatunji, B. O., Benningfield, M. M., Blackford, J. U., Dietrich, M. S., Bhatia, M., Theiss, J. D., Salomon, R. M., Niswender, K., & Cowan, R. L. (2016). Disgust proneness and associated neural substrates in obesity. *Social Cognitive and Affective Neuroscience*, *11*(3), 458–465. https://doi.org/10.1093/scan/nsv129

Wilson, A. C., Aldao, A., & Cheavens, J. (2018). Appraising reappraisal: Exploring its role in the context of behavioral exposure for contamination fears. *Journal of Clinical Psychology*, *74*(9), 1583–1593. https://doi.org/10.1002/jclp.22613

Wolgast, M., Lundh, L.-G., & Viborg, G. (2011). Cognitive reappraisal and acceptance: An experimental comparison of two emotion regulation strategies. *Behaviour Research and Therapy*, *49*(12), 858–866. https://doi.org/10.1016/j.brat.2011.09.011

Woodward, S., Dibley, L., Coombes, S., Bellamy, A., Clark, C., Czuber-Dochan, W., Everelm, L., Kutner, S., Sturt, J., & Norton, C. (2016). Identifying disease-specific distress in patients with inflammatory bowel disease. *British Journal of Nursing*, *25*(12), 649–660. https://doi.org/10.12968/bjon.2016.25.12.649

World Health Organisation. (2019). WHO | Early diagnosis. Retrieved from https://www.who.int/cancer/prevention/diagnosis-screening/en/

Chapter 14
From Disease to Democracy: How Disgust Shapes Western Politics

Natalie J. Shook, Benjamin Oosterhoff, and Barış Sevi

Political ideology generally refers to a set of beliefs, values, and ethical principles that dictate how a society should function and be structured (Erikson and Tedin 2003). Stemming from the French Revolution, political ideology is often conceptualised as falling on a continuum from right-wing, or conservative, to left-wing, or liberal, views (Everett 2013). Conservative ideology prioritises tradition, adherence to norms, and submission to authority, whereas liberal ideology prioritises individual freedom and equality (e.g., Altemeyer 1998; Jost et al. 2003). These different ideological perspectives often lead to divergent opinions and attitudes regarding social and economic matters.

Over the past several decades, a substantial amount of research in psychology and political science has been dedicated to identifying personality, motivational, and cognitive factors that distinguish right-wing and left-wing ideology. For example, conservatives tend to be more fearful of threat or loss, have a greater need for closure, and be more intolerant of ambiguity, whereas liberals tend to be more open to new experiences, more accepting of change, and less dogmatic (see Jost et al. 2003, for a review). From a motivated social cognition perspective, Jost et al. (2003) have argued that uncertainty and threat are core features underlying these personality and cognitive factors, which subsequently motivate ideological viewpoints. Conservative ideology generally consists of values and beliefs that favour resistance to change and acceptance of inequality or maintenance of the status quo.

N. J. Shook (✉)
School of Nursing, University of Connecticut, Storrs, CT, USA
e-mail: natalie.shook@uconn.edu

B. Oosterhoff
Department of Psychology, Montana State University, Bozeman, MT, USA
e-mail: benjamin.oosterhoff@montana.edu

B. Sevi
Department of Psychological Sciences, University of Connecticut, Storrs, CT, USA
e-mail: bs0059@mix.wvu.edu

P. A. Powell, N. S. Consedine (eds.), *The Handbook of Disgust Research*,
https://doi.org/10.1007/978-3-030-84486-8_14

As such, those who are more fearful of threat or less tolerant of uncertainty may endorse conservative ideology as a means of achieving safety and security. In contrast, liberal ideology generally consists of values and beliefs that encourage openness to change and rejection of inequality or the status quo. Those who are less sensitive to threat or more tolerant of uncertainty may adopt the progressive beliefs of liberal ideology as a way of enhancing equality and social advancement. Thus, threat perception is proposed to be a key motivator that influences where individuals fall on the political continuum, and threat perception is central to disgust (Oaten et al. 2009).

Infectious Disease Avoidance

Humans have faced many threats to survival and reproduction, but one of the most impactful (both historically and contemporarily) is infectious disease. Despite advances in modern medicine, hygiene, and sanitation, infectious diseases remain top causes of human death worldwide (World Health Organization 2018). Scholars have proposed that various psychological processes (e.g., disgust) evolved in part to reduce infectious disease threat (Schaller 2006; see also Bradshaw and Gassen, Chap. 3, this volume), and that culture (e.g., group norms and values) in part emerged in such a way to facilitate the management of infectious disease threat within and between groups (Gangestad et al. 2006). Indeed, some scholars have suggested that parasites were more influential than other adaptive challenges in shaping the complexity of human psychology and social behaviour (Thornhill and Fincher 2014).

Traditionally, disgust has been conceptualised as an emotion that evolved to reduce oral contamination (Darwin 1872). If you smell sour milk or take a bite of a rotten apple, you generally will experience disgust, which motivates you to turn away from the spoiled milk and spit out the rotten food. The smell or taste may even cause you to gag or vomit. These prophylactic behaviours decrease the likelihood of ingesting substances that could lead to sickness. However, disgust is experienced in response to a number of stimuli (e.g., garbage, blood) not associated with oral contamination. Despite their differences, these stimuli are similar in that they are all potential sources of pathogens. Disgust encourages individuals to avoid stimuli that may be a source of pathogen threat, which could lead to infectious diseases. Thus, disgust is proposed to serve a broader disease-avoidance mechanism (Curtis and Biran 2001; Oaten et al. 2009). Further, the intensity and threshold of activation for this disease-avoidance emotion varies across people. Disgust sensitivity (or proneness to disgust) has been proposed as a stable individual difference (Haidt et al. 1994; Tybur et al. 2009; see also Tybur, Chap. 6, this volume). Greater disgust sensitivity is associated with avoidance tendencies, thereby reducing exposure to potential pathogens and infectious disease threat (Shook et al. 2019).

A primary means of transmitting pathogens and infectious disease is contact with other people. As such, cultural values and practices may have formed over time to, in part, manage and limit pathogen transmission within and between groups (Gangestad et al. 2006). Value systems that promote adherence to social norms, ingroup cohesion, and avoidance of outgroup members (i.e., social conservatism) may be adaptive strategies for avoiding potentially contaminated outgroup members, as well as maintaining group norms and customs (e.g., food preparation, hygiene, etc.) that limit disease transmission (Terrizzi Jr et al. 2013; Thornhill et al. 2009; see also Vartanian et al., Chap. 10, this volume). Indeed, historic and contemporary disease prevalence rates predict cultural endorsement of social conservatism, such as collectivism, traditionalism, and subordination of women (Fincher et al. 2008; Murray and Schaller 2010; Thornhill et al. 2009; Tybur et al. 2016). At an individual level, disgust sensitivity is associated with several different forms of social conservatism (Shook et al. 2015; Terrizzi Jr et al. 2013), such as collectivism (Clay et al. 2012), right-wing authoritarianism (Hodson and Costello 2007), and religious fundamentalism (Terrizzi Jr et al. 2012). Thus, disgust sensitivity may encourage individuals to endorse social conservatism as a means of infectious disease avoidance.

Disgust and Politics

Given its connections with socially conservative value systems, disgust has implications for politics, particularly when considered from the perspective of threat management. Disgust is a response to a potential pathogen or disease threat in one's environment. The experience of disgust, or the dispositional tendency to be more sensitive to disgust, may lead individuals to perceive their social world as more threatening or unsafe (i.e., encourage a more dangerous worldview; Shook et al. 2017a). This worldview may, in turn, motivate the adoption of conservative ideologies that promote security and control (Duckitt 2001; Duckitt et al. 2002; Perry et al. 2013). Indeed, individuals are thought to adopt politically conservative ideologies as a means of managing the perception of threat (Jost et al. 2003), and political conservatives tend to perceive the world to be more dangerous than political liberals (Altemeyer 1998; Duckitt 2001). Thus, disgust and disgust sensitivity may increase perceptions of threat, which has implications for political ideology. Based on this theoretical rational, considerable empirical evidence has linked disgust sensitivity with politics, including political ideology, party affiliation, specific political attitudes, political candidates, political advertising and rhetoric, and voting behaviour. We review and integrate this research below, followed by recommendations for future research in the field of disgust and politics.

Political Ideology and Party Affiliation

A large body of research has demonstrated that disgust sensitivity is associated with political ideology and party affiliation. Specifically, greater disgust sensitivity is related to greater political conservativism (e.g., Inbar et al. 2009, 2012b; Terrizzi Jr et al. 2010; Tybur et al. 2015) and conservative party affiliation (e.g., Brenner and Inbar 2015; Shook et al. 2017b; Tybur et al. 2015). Some have failed to find significant associations between disgust sensitivity and political ideology (Tybur et al. 2010), but this exception may stem from issues with the measurement of political ideology (see Terrizzi Jr et al. 2013). That is, ideology is not unidimensional; there are important distinctions between social versus economic conservatism (Claessens et al. 2020). Measuring political ideology with a single item that does not differentiate social and economic values (as in Tybur et al. 2010) may obscure the association between disgust and political ideology. The general pattern of findings across studies aligns with the broader literature connecting greater trait disgust sensitivity with multiple forms of social conservatism (see Terrizzi Jr et al. 2013, for a review).

The association between disgust sensitivity and political conservativism has been demonstrated using multiple methodologies beyond self-report, including neurobiological measures (e.g., fMRI; Ahn et al. 2014), physiological measures (e.g., skin conductance; Smith et al. 2011), and visual attention measures (e.g., eye tracking; Oosterhoff et al. 2018). These studies demonstrate that political conservatives exhibit stronger neural and physiological responses and greater visual attention biases to disgust-related stimuli relative to political liberals. Importantly, ideological differences in neurobiological and physiological response patterns appear to be specific to disgust and are not found with other negative (e.g., fear or anxiety inducing), neutral, or positive stimuli. Of note, a recent study did not find an association between political conservatism and physiological reactions to disgust images or threat sensitivity more generally (Bakker et al. 2020). However, political conservatives who are instructed to reappraise disgust endorse more liberal beliefs (Feinberg et al. 2014; Lee et al. 2013), highlighting that the modulation of disgust reactions may explain differences in political ideology. Associations between disgust sensitivity and political conservatism also appear to be consistent across cultures (Inbar et al. 2012b) and potentially stronger for social compared to economic political conservatism (e.g., Oosterhoff et al. 2018).

Specific Political Beliefs

Political ideology is a necessarily broad construct that covers a number of different issues from taxes to healthcare to international relations. This diversity of topics has raised speculation as to which dimension(s) of ideology are most strongly associated with disgust. A few studies have examined the extent to which disgust sensitivity is related to specific political issues. Disgust sensitivity is generally associated with social, not economic, issues and policies (e.g., Aarøe et al. 2020; Brenner and Inbar 2015; Inbar et al. 2009; Terrizzi Jr et al. 2010). That is, greater disgust sensitivity is associated with more conservative views on social issues, such as abortion or gay marriage, but is unrelated to economic issues, such as raising the minimum wage. Socially conservative political attitudes are characterised by more traditional, restrictive, exclusionary social structures and behaviour. Thus, they may serve as a means of maintaining group norms and limiting disease transmission.

To date, most research supports links between disgust and two distinct types of political attitudes: those related to purity and those related to social exclusion. Purity issues concern behaviours that may pose a contamination threat to the body or soul, and often include topics such as homosexuality, abortion, vaccines, food, and drug use. Several correlational studies have linked trait disgust sensitivity to greater opposition to gay marriage (Brenner and Inbar 2015; Inbar et al. 2009; Kam and Estes 2016), abortion (Inbar et al. 2009; Kam and Estes 2016), genetically modified organisms (Clifford and Wendell 2016; Scott et al. 2016; see also Powell, Chap. 15, this volume), organic foods (Clifford and Wendell 2016), vaccines (Clay 2017; Clifford and Wendell 2016), and drug use (Oosterhoff and Shook 2017). Some experimental evidence suggests that inducing disgust decreases liking of gay men in both political liberals and conservatives (Inbar et al. 2012a) and decreases support for gay marriage (Adams et al. 2014). However, experimental evidence linking disgust to changes in attitudes in other purity domains is limited (Clifford and Wendell 2016).

Disgust may also have important implications for political attitudes that favour social exclusion, particularly those regarding racial/ethnic outgroup members and the homeless. Historically, people from unfamiliar groups and regions may have carried different pathogens and had different traditions regarding food preparation or hygiene, which represented a potential disease threat. Those who are more sensitive to disgust may therefore endorse more negative attitudes towards out-groups as a means of distancing themselves from others that were once a source of possible infectious disease (e.g., Faulkner et al. 2004; but see van Leeuwen and Petersen 2018, for an opposing perspective of the association between disgust and outgroup attitudes). Similarly, homeless people often suffer from high rates of illness and lack access to hygiene products, which can serve as disease cues (Gelberg et al. 2000). Those higher in disgust sensitivity may therefore seek to distance themselves from, and support policies designed to socially exclude, the homeless to avoid contact with pathogens. Some research supports these propositions, as greater trait-level disgust sensitivity has been linked with stronger xenophobic and anti-immigration attitudes (Aarøe et al. 2017; Brenner and Inbar 2015; Faulkner et al. 2004; Hodson et al. 2013; Kam and Estes 2016; Karinen et al. 2019; Navarrete and Fessler 2006) and support for exclusionary policies towards the homeless (Clifford and Piston 2017). Experimental evidence further demonstrates that inducing disgust increases xenophobia (Faulkner et al. 2004) and support for exclusionary policies towards the homeless (Clifford and Piston 2017).

Political Candidates

Disgust may also affect perceptions and judgments about political candidates. Using an indirect measure of attitudes (i.e., the affect misattribution task; Payne et al. 2005), Shook et al. (2017b) found that greater disgust sensitivity was related to more negative attitudes towards Barack Obama, the Democratic candidate for the 2012 U.S. presidential election, and more positive attitudes towards Mitt Romney, the Republican candidate. Similarly, greater disgust sensitivity was associated with more negative self-reported emotional responses to President Barack Obama (Stewart et al. 2019). These findings are consistent with past research suggesting that disgust sensitivity may motivate greater political conservatism.

Regardless of political leanings, political candidates may prompt or alleviate feelings of disgust. Physical attractiveness is often used as a cue for physical health (e.g., Thornhill and Gangestad 1999) and potential disease threat (White et al. 2013). Prior research has shown that the physical attractiveness of non-incumbent parliamentary candidates is a significant predictor of getting more votes (e.g., Berggren et al. 2010). That is, more attractive candidates are more likely to be elected. This, in part, may be due to disease threat concerns. White et al. (2013) found that priming disease threat, which induced disgust, increased the preference for attractive political candidates, compared to preferences in a control priming condition. Physical attractiveness of a political candidate may signal healthiness and lower likelihood of disease threat.

In addition to disgust potentially influencing evaluations of candidates, party affiliation of the candidates may evoke feelings of disgust. Using fMRI, neural activation was recorded in response to images of the 2004 U.S. presidential candidates: George Bush (Republican), John Kerry (Democrat), and Ralph Nader (Independent) (Kaplan et al. 2007). Participants exhibited greater activation in the insula and putamen when viewing images of candidates from political parties other than their own, as compared to viewing their political party's candidate. Both the insula and putamen are associated with perception of disgust in faces (Phillips et al. 1998). These findings suggest that regardless of party affiliation and political beliefs, simply viewing images of opposing party candidates may increase neural activity in regions of the brain associated with disgust (for more on the neurology of disgust see Overton et al., Chap. 12, this volume).

Political Advertisement and Rhetoric

Persuasive communications are often designed to elicit emotional responses in order to change recipients' attitudes or motivate behaviour change (Petty et al. 2003). Although fear is the most commonly utilised negative emotion, disgust is also used to try to shape public attitudes and influence voting behaviour (Gadarian and van der Vort 2018; Hatemi and McDermott 2012). A primary area in which emotions, particularly disgust, have been used is negative messaging in political campaign advertisements. The goal of negative messaging is to attack the rival candidate and worsen their image. These types of advertisements may include disgust as a peripheral emotional cue (Newhagen and Reeves 1989) and induce short-term disgust in viewers when describing opposing candidates (Pinkleton et al. 2002). The objective being to create an association between the opposing candidate and feelings of disgust. However, such attempts are not always successful. During the 1993 Canadian federal elections, the Progressive Conservative party aired negative advertisements highlighting the facial paralysis of the Liberal leader Jean Chrétien. These advertisements were aimed to help the Conservative candidate Kim Campbell, potentially by prompting feelings of disgust that were affiliated with the opposition candidate. People are particularly sensitive to physical abnormalities or characteristics (e.g., deformities, lesions, obesity) that may be indicative of infectious disease (e.g., Oaten et al. 2011; Park et al. 2003, 2007) and often experience disgust in response to individuals with physical abnormalities (Park et al. 2003; Ryan et al. 2012; see also Vartanian et al., Chap. 10, this volume). Yet, these advertisements had the opposite effect, with most people feeling angry and disgusted towards Kim Campbell, not Jean Chrétien (Haddock and Zanna 1997). As with fear appeals, the utilisation of disgust can backfire. Potentially, the attempt to elicit disgust was too transparent or offensive, leading to the Campbell political campaign being viewed as disingenuous or having made a moral transgression.

For several policy issues, political rhetoric has incorporated disgust-related terms or imagery that evokes disgust (e.g., Hatemi and McDermott 2012; Nussbaum 2010). In particular, social issues that are associated with disgust sensitivity and may involve a potential disease threat tend to be the targets of disgust rhetoric (e.g., Clifford and Wendell 2016; Kam and Estes 2016). Generally, this language is used to encourage conservative political beliefs (Gadarian and van der Vort 2018). But, as noted above, the use of disgust can lead to mixed results. Gadarian and van der Vort (2018) found that the use of disgust rhetoric decreased support for gay rights in some individuals as intended. However, for other individuals, disgust rhetoric evoked anger towards the message, resulting in increased support for gay rights. Although sometimes effective, the use of disgust as a persuasive tool in political messages can lead to unintended outcomes.

Voting Behaviour

Disgust may also have implications for voting behaviour. At a basic level, disgust may influence whether one chooses to vote. Media coverage of politics and campaign advertising often highlights scandals, misdeeds, and norm violations by politicians, which can arouse feelings of disgust. Indeed, disgust is a common reaction towards politics, and increased feelings of disgust towards politics was related to lower voting turnout for the 2010 midterm elections (Vandenbroek 2011). As disgust motivates avoidance of objects, people, and situations that are perceived to have elicited the disgust-related feelings, linking disgust with politics may deter political engagement.

Disease threat and disgust may also motivate voting for particular parties. Data from the 2014 U.S. midterm elections and Canadian polling data indicated that as the Ebola outbreak in West Africa became more salient, intentions to vote for conservative candidates increased (Beall et al. 2016; Schaller et al. 2017; c.f., Tiokhin and Hruschka 2017). In the Netherlands, individuals who tended to vote for the conservative "Freedom Party" were higher in disgust sensitivity than individuals who tended to vote for other political groups (Brenner and Inbar 2015). For the 2008 U.S. presidential election, greater disgust sensitivity was correlated with greater intentions to vote for the Republican candidate (i.e., John McCain; Inbar et al. 2012b). Aggregate disgust sensitivity scores by state predicted Barack Obama's margin of victory in each state (i.e., for states with higher aggregate levels of disgust sensitivity, individuals were less likely to vote for Obama over McCain). Similarly, disgust sensitivity predicted voting behaviour in the 2012 U.S. presidential election (Shook et al. 2017b). Those higher in disgust sensitivity were more likely to vote for Mitt Romney than Barack Obama. A similar pattern was found for the 2016 U.S. presidential election (Billingsley et al. 2018). Those who were higher in disgust sensitivity were more likely to have had voted for the Republican candidate (Donald Trump) over the Democratic candidate (Hillary Clinton). Finally, a cross-national study utilising multiple large nationally representative samples from the U.S. and Denmark found that greater disgust sensitivity was associated with voting for conservative political candidates (Aarøe et al. 2020).

Summary

Most of the literature on disgust and politics has focused on the extent to which disgust, particularly disgust sensitivity, is related to political ideology and attitudes towards specific political issues. In general, there is clear evidence connecting greater disgust sensitivity with stronger support for politically conservative ideologies and attitudes. Disgust sensitivity has been repeatedly associated with stronger self-identification as a political conservative using multiple methodologies and across different cultural contexts. When examining specific political attitudes, disgust appears to be most strongly connected with social issues, particularly those that concern individual purity (e.g., abortion, gay marriage, vaccines, GMOs, drug use) and social exclusion (e.g., anti-immigration, anti-homeless). Experimental evidence generally supports correlational evidence; however, the findings from these studies are more mixed. Although sparse, a couple of studies have found that disgust sensitivity is related to attitudes towards political candidates and that exposure to opposing political candidates may evoke feelings of disgust. Furthermore, disgust may be used as a persuasive tool in campaign advertising and political rhetoric, with mixed results. All of this may then translate to political behaviour. Although few in number, studies have reliably found that disgust sensitivity is related to, and in some cases predicts, voting behaviour. Overall, the consistent pattern is that greater disgust sensitivity is related to conservative political ideology, attitudes, and behaviour.

Limitations

Although there is a growing body of research focusing on the role of disgust in politics, this is still a nascent area of work. As such, the number of studies is relatively small, and there are limitations. Methodologically, most of the existing research is correlational and cross-sectional. Thus, causation and temporal order cannot be determined. There may be third variables that explain the observed correlations, or there may be more complex relations. For example, there have been a few experimental studies testing the effect of disgust on specific political attitudes, but these studies have produced more mixed findings than the correlational studies. This could simply be due to differences in disgust manipulations, measures, or samples, or these experimental findings may indicate that the relation between disgust and politics is not as simple as portrayed by the correlational research. There are fundamental differences between state-level disgust and trait-level disgust sensitivity (Tybur et al. 2014). Experiencing disgust in the moment is not the same as individual differences in trait disgust sensitivity, and thus, these two variables may differentially influence political attitudes. More experimental research is required to determine the causal role of disgust in politics. However, this may be easier to operationalise for some aspects of politics than others. For instance, attitudes towards specific political issues may be more malleable than overriding political ideology (Shook and Oosterhoff 2020; Strandberg et al. 2018). Given the amorphous nature of political ideology, a state disgust induction may be insufficient to shift these general beliefs. Instead, for such variables, longitudinal research would be useful to demonstrate the prospective influence of disgust on political outcomes. Such efforts may be especially useful during late adolescence and emerging adulthood when political attitudes and values are thought to develop and be more malleable (Hess and Torney 1967).

Another shortcoming of the existing research is the reliance on self-report measures. Although some studies have incorporated physiological (e.g., fMRI) and indirect (e.g., affect misattribution procedure) measures, most have relied heavily on self-report measures of disgust and political attitudes. This raises concerns about social desirability and common method variance that may amplify effect sizes. Of particular concern is that disgust sensitivity is primarily assessed through self-report measures, such as the Disgust Scale (Haidt et al. 1994) or Disgust Scale-Revised (Olatunji et al. 2007), Three Domain Disgust Scale (Tybur et al. 2009), and Disgust Propensity and Sensitivity Scale Revised (van Overveld et al. 2006). Future research should incorporate more indirect or physiological measures to assess disgust sensitivity more objectively (e.g., facial expression; see also Consedine, Chap. 2, this volume).

Disgust sensitivity has been conceptualised in multiple ways. The Disgust Scale/Disgust Scale-Revised and Three Domain Disgust Scale assess feelings of disgust in response to specific categories of targets. The Disgust Propensity and Sensitivity Scale assesses how frequently individuals feel disgust in general (not specific to a target) and the extent to which feeling disgust concerns them. Although these measures are generally positively correlated, there is variability in the strength of these associations (e.g., Shook et al. 2017a; Terrizzi Jr et al. 2012). Moreover, the extent to which these measures are associated with political and social values varies (Terrizzi Jr et al. 2013). Criticisms have been raised about some of these measures and the extent to which they validly assess disgust, particularly the measures that assess disgust in response to specific categories (Elad-Strenger et al. 2020; Tybur et al. 2014; Wagemans et al. 2019).

Several of the subscales from the disgust sensitivity measures (e.g., sexual disgust, moral disgust) involve categories that are not directly related to disease threat, but assess ways that the disgust system has been "preadapted" or "co-opted" to respond to social or moral transgressions (Rozin and Haidt 2013; Tybur et al. 2009, 2013). In these domains, there is much greater cultural variability in what is considered a social or moral transgression, which is often dictated by or reflected in political ideology. As such, it is unclear to what extent these subscales assess disgust sensitivity per se versus sensitivity to social norms.

A final limitation is that most of the research on disgust and politics has been based on U.S. samples. A few studies have involved non-U.S. samples (e.g., Aarøe et al. 2020; Brenner and Inbar 2015; Inbar et al. 2012b), producing similar results. However, those studies are few and still involve Western, democratic societies. Political systems and ideology in non-Western or non-democratic societies are different in a number of important ways (e.g., individual freedom, influence of religion, etc.). Consequently, the role of disgust in politics may be quite different in non-Western, non-democratic societies. More research in other cultures is therefore necessary.

Future Directions

Two mechanisms have been proposed to explain links between disgust and political conservatism. As outlined earlier, sensitivity to disease threat (i.e., greater disgust sensitivity) may motivate the adoption of socially conservative attitudes as a means of upholding social norms that historically served a disease avoidance function (e.g., Terrizzi Jr et al. 2013). Furthermore, disgust may encourage a more dangerous worldview, which motivates seeking security through conservative ideologies (Shook et al. 2017a). Others have questioned this explanation and argued that the link between disgust sensitivity and political conservatism is completely explained by individual differences in sexual strategies—or differences in whether people prefer short-term versus long-term mating goals (Tybur et al. 2015). Indeed, some evidence indicates that sexual disgust sensitivity and attitudes about sex account for associations between pathogen disgust sensitivity and political conservatism (Billingsley et al. 2018; Tybur et al. 2015). However, other researchers have failed to replicate the sexual strategies model (Aarøe et al. 2020), and one study found support for both the disease avoidance and sexual strategies arguments (Shook et al. 2015). Based on existing evidence, links between disgust sensitivity and political ideologies are likely explained by individual differences in both disease threat and sexual strategies. Much more research is needed to test and compare these mechanisms and elucidate how disgust sensitivity may influence political ideology.

Future research should consider the association between disgust and politics from a developmental perspective. That is, there may be important differences in disgust processes across the lifespan. Political beliefs generally begin to form in adolescence (Hess and Torney 1967; Rekker et al. 2015), which is a period marked by changes in disgust sensitivity (e.g., Borg et al. 2019). It is unknown to what extent disgust or disgust sensitivity may be involved in this early shaping of political attitudes and behaviour. Older age is often associated with greater political conservatism (e.g., Cornelis et al. 2009). Theoretically, this change in political beliefs could align with a disease avoidance model. With age, immune function declines (Makinodan and Kay 1980). Older adults are generally more susceptible to infectious diseases, and it often takes older adults longer to clear infections, increasing the likelihood of secondary infections. With this increased threat of infectious disease, older adults may be more disgust sensitive, which could contribute to their greater political conservatism. On the whole, age differences or developmental changes in disgust sensitivity have not been examined. However, one study did report that disgust sensitivity may decrease with age, at least until middle age (Curtis et al. 2004). Most of the participants in this sample (75%) were between 17 and 45 years old; only 10% of the sample were over 45 years. Thus, disgust sensitivity in older adulthood has not been systematically assessed. Potentially, there may be developmental periods when individuals are more sensitive to the effects of disgust on political attitudes. More research using lifespan samples or longitudinal designs are necessary to understand the time course of disgust and how it affects the initial development of, and life course changes, in political ideology and behaviour.

Additionally, most studies examining the intersection between disgust and politics have either focused on links between disgust and political ideology and social values more broadly, or a narrow range of political attitudes (e.g., Inbar et al. 2012a, b; Terrizzi Jr et al. 2013). These efforts have provided valuable insight into how disgust is linked with various components of politics. However, few studies have comprehensively examined how disgust sensitivity is connected with a broad range of political attitudes. Such endeavours may advance theory by providing much needed specificity on links between disgust and political attitudes. Network analysis provides a promising avenue of this research given that it is capable of isolating unique associations among a large range of constructs simultaneously while managing threats of multicollinearity. These aspects of network analysis are especially advantageous given that specific political attitudes are often moderately correlated with one another (e.g., Shook and Clay 2011).

Conclusions

In the past two decades, there has been growing interest in the role disgust may play in politics. Although relatively small, the existing body of evidence has consistently linked the experience of disgust and greater disgust sensitivity with conservative political ideology, attitudes, and behaviour. Much work still needs to be done to demonstrate the causal role of disgust and elucidate the mechanism(s) underlying the association between disgust and politics. However, the extant literature provides a strong foundation upon which to build.

References

Aarøe, L., Petersen, M. B., & Arceneaux, K. (2017). The behavioral immune system shapes political intuitions: Why and how individual differences in disgust sensitivity underlie opposition to immigration. *American Political Science Review, 111*, 277-294.

Aarøe, L., Petersen, M. B., & Arceneaux, K. (2020). The behavioral immune system shapes partisan preferences in modern democracies: Disgust sensitivity predicts voting for socially conservative parties. *Political Psychology*. Advance online publication.

Adams, T. G., Stewart, P. A., & Blanchar, J. C. (2014). Disgust and the politics of sex: Exposure to a disgusting odorant increases politically conservative views on sex and decreases support for gay marriage. *PloS One*, *9*(5), e95572.

Ahn, W. Y., Kishida, K. T., Gu, X., Lohrenz, T., Harvey, A., Alford, J. R., Smith K. B., Yaffe, G., Hibbing, J. R., Dayan, P., & Montague, P. R. (2014). Nonpolitical images evoke neural predictors of political ideology. *Current Biology, 24*, 2693-2699.

Altemeyer, B. (1998). The other "authoritarian personality". In *Advances in Experimental Social Psychology* (Vol. 30, pp. 47-92). Academic Press.

Bakker, B. N., Schumacher, G., Gothreau, C., & Arceneaux, K. (2020). Conservatives and liberals have similar physiological responses to threats. *Nature Human Behaviour, 4*, 613-621.

Beall, A. T., Hofer, M. K., & Schaller, M. (2016). Infections and elections: Did an Ebola outbreak influence the 2014 US federal elections (and if so, how)?. *Psychological Science, 27*(5), 595-605.

Berggren, N., Jordahl, H., & Poutvaara, P. (2010). The looks of a winner: Beauty and electoral success. *Journal of Public Economics, 94*(1-2), 8-15.

Billingsley, J., Lieberman, D., & Tybur, J. M. (2018). Sexual disgust trumps pathogen disgust in predicting voter behavior during the 2016 US presidential election. *Evolutionary Psychology, 16*, 1-15

Borg, C., Oosterwijk, T. A., Lisy, D., Boesveldt, S., & De Jong, P. J. (2019). The influence of olfactory disgust on (genital) sexual arousal in men. *PloS one*, *14*(2), e0213059.

Brenner, C. J., & Inbar, Y. (2015). Disgust sensitivity predicts political ideology and policy attitudes in the Netherlands. *European Journal of Social Psychology, 45*, 27-38.

Claessens, S., Fischer, K., Chaudhuri, A., Sibley, C. G., & Atkinson, Q. D. (2020). The dual evolutionary foundations of political ideology. *Nature Human Behaviour, 4*, 336-345.

Clay, R. (2017). The behavioral immune system and attitudes about vaccines: contamination aversion predicts more negative vaccine attitudes. *Social Psychological and Personality Science, 8*, 162-172.

Clay, R., Terrizzi Jr, J. A., & Shook, N. J. (2012). Individual differences in the behavioral immune system and the emergence of cultural systems. *Social Psychology*, *43*(4), 174.

Clifford, S., & Piston, S. (2017). Explaining public support for counterproductive homelessness policy: The role of disgust. *Political Behavior, 39*, 503-525.

Clifford, S., & Wendell, D. G. (2016). How disgust influences health purity attitudes. *Political Behavior, 38*, 155-178.

Cornelis, I., Van Hiel, A., Roets, A., & Kossowska, M. (2009). Age differences in conservatism: Evidence on the mediating effects of personality and cognitive style. *Journal of Personality, 77* (1), 51-88.

Curtis, V., & Biran, A. (2001). Dirt, disgust, and disease: Is hygiene in our genes?. *Perspectives in Biology and Medicine, 44*(1), 17-31.

Curtis, V., Aunger, R., & Rabie, T. (2004). Evidence that disgust evolved to protect from risk of disease. *Proceedings of the Royal Society of London. Series B: Biological Sciences, 271*, S131-S133.

Darwin, C. (1872). *The Expression of the Emotions in Man and Animals*. John Murray.

Duckitt, J. (2001). A dual-process cognitive-motivational theory of ideology and prejudice. In *Advances in Experimental Social Psychology* (Vol. 33, pp. 41-113). Academic Press.

Duckitt, J., Wagner, C., Du Plessis, I., & Birum, I. (2002). The psychological bases of ideology and prejudice: Testing a dual process model. *Journal of Personality and Social Psychology, 83*(1), 75.

Elad-Strenger, J., Proch, J., & Kessler, T. (2020). Is disgust a "conservative" emotion? *Personality and Social Psychology Bulletin, 46*(6), 896-912.

Erikson, R. S., & Tedin, K. L. (2003). *American public opinion (6th ed.)*. New York: Longman.

Everett, J. A. (2013). The 12 item social and economic conservatism scale (SECS). *PloS One, 8*(12), e82131.

Faulkner, J., Schaller, M., Park, J. H., & Duncan, L. A. (2004). Evolved disease-avoidance mechanisms and contemporary xenophobic attitudes. *Group Processes & Intergroup Relations, 7*(4), 333-353.

Feinberg, M., Antonenko, O., Willer, R., Horberg, E. J., & John, O. P. (2014). Gut check: Reappraisal of disgust helps explain liberal–conservative differences on issues of purity. *Emotion, 14*, 513-521.

Fincher, C. L., Thornhill, R., Murray, D. R., & Schaller, M. (2008). Pathogen prevalence predicts human cross-cultural variability in individualism/collectivism. *Proceedings of the Royal Society B: Biological Sciences, 275*(1640), 1279-1285.

Gadarian, S. K., & van der Vort, E. (2018). The gag reflex: Disgust rhetoric and gay rights in American politics. *Political Behavior, 40*(2), 521-543.

Gangestad, S. W., Haselton, M. G., & Buss, D. M. (2006). Evolutionary foundations of cultural variation: Evoked culture and mate preferences. *Psychological Inquiry, 17*(2), 75-95.

Gelberg, L., Andersen, R. M., & Leake, B. D. (2000). The Behavioral Model for Vulnerable Populations: application to medical care use and outcomes for homeless people. *Health Services Research, 34*, 1273-1302.

Haddock, G., & Zanna, M. P. (1997). Impact of negative advertising on evaluations of political candidates: The 1993 Canadian federal election. *Basic and Applied Social Psychology, 19*(2), 205-223.

Haidt, J., McCauley, C., & Rozin, P. (1994). Individual differences in sensitivity to disgust: A scale sampling seven domains of disgust elicitors. *Personality and Individual Differences, 16*(5), 701-713.

Hatemi, P., & McDermott, R. (2012). Policing the perimeter: Disgust and purity in democratic debate. *PS. Political Science & Politics, 45*(4), 675–687

Hess R. D., & Torney J. V. (1967). *The Development of Political Attitudes in Children*. Aldine.

Hodson, G., & Costello, K. (2007). Interpersonal disgust, ideological orientations, and dehumanization as predictors of intergroup attitudes. *Psychological Science, 18*(8), 691-698.

Hodson, G., Choma, B. L., Boisvert, J., Hafer, C. L., MacInnis, C. C., & Costello, K. (2013). The role of intergroup disgust in predicting negative outgroup evaluations. *Journal of Experimental Social Psychology, 49*, 195-205.

Inbar, Y., Pizarro, D. A., & Bloom, P. (2009). Conservatives are more easily disgusted than liberals. *Cognition and Emotion, 23*, 714-725.

Inbar, Y., Pizarro, D. A., & Bloom, P. (2012a). Disgusting smells cause decreased liking of gay men. *Emotion, 12*, 23-27.

Inbar, Y., Pizarro, D., Iyer, R., & Haidt, J. (2012b). Disgust sensitivity, political conservatism, and voting. *Social Psychological and Personality Science, 3*, 537-544.

Jost, J. T., Glaser, J., Kruglanski, A. W., & Sulloway, F. J. (2003). Political conservatism as motivated social cognition. *Psychological Bulletin, 129*(3), 339.

Kam, C. D., & Estes, B. A. (2016). Disgust sensitivity and public demand for protection. *The Journal of Politics, 78*, 481-496.

Kaplan, J. T., Freedman, J., & Iacoboni, M. (2007). Us versus them: Political attitudes and party affiliation influence neural response to faces of presidential candidates. *Neuropsychologia, 45* (1), 55-64.

Karinen, A. K., Molho, C., Kupfer, T. R., & Tybur, J. M. (2019). Disgust sensitivity and opposition to immigration: Does contact avoidance or resistance to foreign norms explain the relationship?. *Journal of Experimental Social Psychology, 84*, 103817.

Lee, J. J., Sohn, Y., & Fowler, J. H. (2013). Emotion regulation as the foundation of political attitudes: does reappraisal decrease support for conservative policies?. *PloS One, 8*, e83143.

Makinodan, T., & Kay, M. M. (1980). Age influence on the immune system. In *Advances in immunology* (Vol. 29, pp. 287-330). Academic Press.

Murray, D. R., & Schaller, M. (2010). Historical prevalence of infectious diseases within 230 geopolitical regions: A tool for investigating origins of culture. *Journal of Cross-Cultural Psychology, 41*(1), 99-108.

Navarrete, C. D., & Fessler, D. M. (2006). Disease avoidance and ethnocentrism: The effects of disease vulnerability and disgust sensitivity on intergroup attitudes. *Evolution and Human Behavior, 27*, 270-282.

Newhagen, J., & Reeves, B. (1989). *Emotion and Memory Responses for Negative Political Advertising: A Study of Television Commercials Used in the 1988 Presidential Election.* Paper presented at the Annual Meeting of the Association for Education in Journalism and Mass Communication, Washington D.C., August 1989.

Nussbaum, M. C. (2010). *From disgust to humanity: Sexual orientation and constitutional law.* Oxford University Press.

Oaten, M., Stevenson, R. J., & Case, T. I. (2009). Disgust as a disease-avoidance mechanism. *Psychological Bulletin, 135*(2), 303-321.

Oaten, M., Stevenson, R. J., & Case, T. I. (2011). Disease avoidance as a functional basis for stigmatization. *Philosophical Transactions of the Royal Society B: Biological Sciences, 366*, 3433-3452.

Olatunji, B. O., Williams, N. L., Tolin, D. F., Abramowitz, J. S., Sawchuk, C. N., Lohr, J. M., & Elwood, L. S. (2007). The Disgust Scale: item analysis, factor structure, and suggestions for refinement. *Psychological Assessment, 19*(3), 281.

Oosterhoff, B., & Shook, N. J. (2017). From drug laws to recreational substance use: The adaptationist role of disgust sensitivity. *Personality and Individual Differences, 104*, 544-553.

Oosterhoff, B., Shook, N. J., & Ford, C. (2018). Is that disgust I see? Political ideology and biased visual attention. *Behavioural Brain Research, 336*, 227-235.

Park, J. H., Faulkner, J., & Schaller, M. (2003). Evolved disease-avoidance processes and contemporary anti-social behavior: Prejudicial attitudes and avoidance of people with physical disabilities. *Journal of Nonverbal Behavior, 27*(2), 65-87.

Park, J. H., Schaller, M., & Crandall, C. S. (2007). Pathogen-avoidance mechanisms and the stigmatization of obese people. *Evolution and Human Behavior, 28*(6), 410-414.

Payne, B. K., Cheng, C. M., Govorun, O., & Stewart, B. D. (2005). An inkblot for attitudes: affect misattribution as implicit measurement. *Journal of Personality and Social Psychology, 89*(3), 277.

Perry, R., Sibley, C. G., & Duckitt, J. (2013). Dangerous and competitive worldviews: A meta-analysis of their associations with social dominance orientation and right-wing authoritarianism. *Journal of Research in Personality, 47*(1), 116-127.

Petty, R. E., Fabrigar, L. R., & Wegener, D. T. (2003). *Emotional factors in attitudes and persuasion.* In R. J. Davidson, K. R. Scherer, & H. H. Goldsmith (Eds.), *Series in affective science. Handbook of affective sciences* (pp. 752–772). Oxford University Press.

Phillips, M. L., Young, A. W., Scott, S., Calder, A. J., Andrew, C., Giampietro, V., Williams, S. C., Bullmore, E. T., Brammer, M., & Gray, J. A. (1998). Neural responses to facial and vocal expressions of fear and disgust. *Proceedings of the Royal Society of London. Series B: Biological Sciences, 265*(1408), 1809-1817.

Pinkleton, B. E., Um, N. H., & Austin, E. W. (2002). An exploration of the effects of negative political advertising on political decision making. *Journal of Advertising, 31*(1), 13-25.

Rekker, R., Keijsers, L., Branje, S., & Meeus, W. (2015). Political attitudes in adolescence and emerging adulthood: Developmental changes in mean level, polarization, rank-order stability, and correlates. *Journal of Adolescence, 41*, 136–147.

Rozin, P., & Haidt, J. (2013). The domains of disgust and their origins: Contrasting biological and cultural evolutionary accounts. *Trends in Cognitive Sciences, 17*, 367-368.

Ryan, S., Oaten, M., Stevenson, R. J., & Case, T. I. (2012). Facial disfigurement is treated like an infectious disease. *Evolution and Human Behavior, 33*(6), 639-646.

Schaller, M. (2006). Parasites, behavioral defenses, and the social psychological mechanisms through which cultures are evoked. *Psychological Inquiry, 17*(2), 96-101.

Schaller, M., Hofer, M. K., & Beall, A. T. (2017). Evidence That an Ebola Outbreak Influenced Voting Preferences, Even After Controlling (Mindfully) for Autocorrelation: Reply to Tiokhin and Hruschka (2017). *Psychological Science, 28*(9), 1361-1363.

Scott, S. E., Inbar, Y., & Rozin, P. (2016). Evidence for absolute moral opposition to genetically modified food in the United States. *Perspectives on Psychological Science, 11*, 315-324.

Shook, N. J., & Clay, R. (2011). Valence asymmetry in attitude formation: A correlate of political ideology. *Social Psychological and Personality Science, 2*(6), 650-655.

Shook, N. J., & Oosterhoff, B. (2020). Testing the effects of pathogen threat and sexual strategies on political ideology. *Politics and the Life Sciences.* Advance online publication.

Shook, N. J., Terrizzi Jr, J. A., Clay, R., & Oosterhoff, B. (2015). In defense of pathogen disgust and disease avoidance: A response to Tybur et al.(2015). *Evolution and Human Behavior, 36*, 498-502.

Shook, N. J., Ford, C. G., & Boggs, S. T. (2017a). Dangerous worldview: A mediator of the relation between disgust sensitivity and social conservatism. *Personality and Individual Differences, 119*, 252-261.

Shook, N. J., Oosterhoff, B., Terrizzi Jr, J. A., & Brady, K. M. (2017b). "Dirty politics": The role of disgust sensitivity in voting. *Translational Issues in Psychological Science, 3*, 284-297.

Shook, N. J., Thomas, R., & Ford, C. G. (2019). Testing the relation between disgust and general avoidance behavior. *Personality and Individual Differences, 150*, 109457.

Smith, K. B., Oxley, D., Hibbing, M. V., Alford, J. R., & Hibbing, J. R. (2011). Disgust sensitivity and the neurophysiology of left-right political orientations. *PloS One, 6*, e25552.

Stewart, P. A., George, J. R., & Adams, T. (2019). Does predisposition toward disgust affect emotional response to political leaders? Evidence from the 2012 US Presidential Election. *Social Science Quarterly, 100*(6), 2033-2046.

Strandberg, T., Sivén, D., Hall, L., Johansson, P., & Pärnamets, P. (2018). False beliefs and confabulation can lead to lasting changes in political attitudes. *Journal of Experimental Psychology: General, 147*(9), 1382-1399.

Terrizzi Jr, J. A., Shook, N. J., & Ventis, W. L. (2010). Disgust: A predictor of social conservatism and prejudicial attitudes toward homosexuals. *Personality and Individual Differences, 49*(6), 587-592.

Terrizzi Jr, J. A., Shook, N. J., & Ventis, W. L. (2012). Religious conservatism: An evolutionarily evoked disease-avoidance strategy. *Religion, Brain & Behavior, 2*(2), 105-120.

Terrizzi Jr, J. A., Shook, N. J., & McDaniel, M. A. (2013). The behavioral immune system and social conservatism: A meta-analysis. *Evolution and Human Behavior, 34*, 99-108.

Thornhill, R., & Fincher, C. L. (2014). The parasite-stress theory of sociality, the behavioral immune system, and human social and cognitive uniqueness. *Evolutionary Behavioral Sciences, 8*(4), 257.

Thornhill, R., & Gangestad, S. W. (1999). Facial attractiveness. *Trends in Cognitive Sciences, 3* (12), 452-460.

Thornhill, R., Fincher, C. L., & Aran, D. (2009). Parasites, democratization, and the liberalization of values across contemporary countries. *Biological Reviews, 84*(1), 113-131.

Tiokhin, L., & Hruschka, D. (2017). No evidence that an Ebola outbreak influenced voting preferences in the 2014 elections after controlling for time-series autocorrelation: A Commentary on Beall, Hofer, and Schaller (2016). *Psychological Science, 28*(9), 1358-1360.

Tybur, J. M., Lieberman, D., & Griskevicius, V. (2009). Microbes, mating, and morality: individual differences in three functional domains of disgust. *Journal of Personality and Social Psychology, 97*(1), 103.

Tybur, J. M., Merriman, L. A., Hooper, A. E. C., McDonald, M. M., & Navarrete, C. D. (2010). Extending the behavioral immune system to political psychology: Are political conservatism and disgust sensitivity really related?. *Evolutionary Psychology, 8*(4), 599-616.

Tybur, J. M., Lieberman, D., Kurzban, R., & DeScioli, P. (2013). Disgust: Evolved function and structure. *Psychological Review, 120*(1), 65.

Tybur, J. M., Frankenhuis, W. E., & Pollet, T. V. (2014). Behavioral immune system methods: Surveying the present to shape the future. *Evolutionary Behavioral Sciences, 8*(4), 274.

Tybur, J. M., Inbar, Y., Güler, E., & Molho, C. (2015). Is the relationship between pathogen avoidance and ideological conservatism explained by sexual strategies?. *Evolution and Human Behavior, 36*, 489-497.

Tybur, J. M., Inbar, Y., Aarøe, L., Barclay, P., Barlow, F. K., de Barra, M., ... ŽeŽelj, I. (2016). Parasite stress and pathogen avoidance relate to distinct dimensions of political ideology across 30 nations. *Proceedings of the National Academy of Sciences, 113*, 12408–12413.

van Leeuwen, F., & Petersen, M. B. (2018). The behavioral immune system is designed to avoid infected individuals, not outgroups. *Evolution and Human Behavior, 39*(2), 226–234.

van Overveld, W. J. M., De Jong, P. J., Peters, M. L., Cavanagh, K., & Davey, G. C. L. (2006). Disgust propensity and disgust sensitivity: Separate constructs that are differentially related to specific fears. *Personality and Individual Differences, 41*(7), 1241-1252.

Vandenbroek, L. M. (2011). *Disentangling aversion: Experimentally testing the impact of disgust and anger on political participation*. Paper presented at the Annual Meeting of the American Political Science Association, Seattle, September 2011.

Wagemans, F. M., Brandt, M. J., & Zeelenberg, M. (2019). Weirdness of disgust sensitivity items predicts their relationship to purity moral judgments. *Personality and Individual Differences, 146*, 182-187.

White, A. E., Kenrick, D. T., & Neuberg, S. L. (2013). Beauty at the ballot box: Disease threats predict preferences for physically attractive leaders. *Psychological Science, 24*(12), 2429-2436.

World Health Organization. (2018). *World health statistics 2018: monitoring health for the SDGs*. World Health Organization.

Chapter 15
Disgust and Consumer Behaviour

Philip A. Powell

It is difficult to find an exact opposite to an emotional experience as visceral and embodied as disgust, but attraction comes close. In consumer marketing, attraction is king. Products, brands, and advertising campaigns are all designed to *attract* consumers (Hammond 2008); via the halo effect, *attractive* actors are used to demonstrate the benefits of purchasable goods (Baker and Churchill 1977); product design is centred around *attracting* potential customers away from the competition (Crilly et al. 2004). Attraction sells and, on-the-face-of-it, one may conclude that disgust does not. As an emotion characterised by avoidance and rejection, it is true that disgust constrains markets (Roth 2007). However, the role of revulsion in consumer behaviour is much more layered. After all, what better motivation could there be to blow your wages on bleach than revolting TV-advert personifications of germs swarming around your house (Morales et al. 2012)? What about the products that utilise mixtures of humour and the grotesque to *appeal* to consumers, such as UK children book series "Horrible Histories" (Scanlon 2011)? And what of the market for highbrow art (and YouTube zit-popping videos) designed to induce morbid fascination (Menninghaus et al. 2017)? In this chapter, I argue that disgust, like all other discrete emotions, shapes consumer decision making. However, it does so not only by constraining consumption, but also by motivating it, under certain conditions. I discuss the association between disgust and sustainability, a significant societal problem driven by consumption; and consider the extent to which (and why) certain consumption habits elicit repugnance. Finally, I identify some outstanding research questions for researchers interested in disgust and consumer behaviour, and how they may be approached. Let's start with the obvious.

P. A. Powell (✉)
School of Health and Related Research, University of Sheffield, Sheffield, UK
e-mail: p.a.powell@sheffield.ac.uk

P. A. Powell, N. S. Consedine (eds.), *The Handbook of Disgust Research*,
https://doi.org/10.1007/978-3-030-84486-8_15

Disgust Constrains Consumer Behaviour

As discussed in detail elsewhere (see Bradshaw and Gassen, Chap. 3, this volume), the emotion of disgust evolved to motivate avoidance and rejection, particularly in the context of health, but has also expanded to the domains of sexual and moral behaviour (Tybur et al. 2013). In 2006, The Harvard Economist, Alvin Roth, published a working paper entitled "Repugnance as a Constraint on Markets", which was published in the *Journal of Economic Perspectives* the following year, and has since been cited over 500 times (Roth 2007). The basic tenet of the piece was the observation that "repugnance"—loosely defined—is an irrational influence that limits the inclusion of particular kinds of transactions within markets, as they are considered too inappropriate or provoke societal repulsion. Roth notes a string of possible transactions that were once, or are presently, "repugnant" in the West. These range from the mundane charging interest on loans; to the contemporary disgust relevant marketing of horse and dog meat; to the fanciful dwarf-tossing (in the interests of avoiding ambiguity, that's dwarf *throwing*). While these and other examples serve to illustrate that disgust can act as a constraint on markets, they also support the idea that economic repugnance is socio-culturally-defined and specific to *time* and *place*.

Despite its influence on subsequent generations of economic thought, a key issue with Roth's approach is that his use of the term "repugnance" is imprecise. This matters because different emotional states have evolved specifically as responses to different classes of stimuli and have differing effects on cognitive and motivational systems, eliciting heterogeneous behavioural tendencies within economic exchanges (e.g., Lerner et al. 2004). Roth (2007) uses "repugnance" as a catch-all for negative affective reactions towards transactions perceived as socio-morally inappropriate, rather than an identification of the specific influence of disgust on markets *per se*. Indeed, in later works, Roth states that he and his colleagues "...use 'repugnant' in its economic sense [...] in a repugnant transaction the participants are willing to transact, but third parties disapprove and wish to prevent the transaction (rather than in its psychological sense of eliciting disgust among potential participants)" (Leider and Roth 2010). Other emotions, such as anger or contempt, have been shown to become confused with disgust in issues of socio-moral significance, with language blurring the distinction between the two (Herz and Hinds 2013; Nabi 2002; see also Giner-Sorolla, Chap. 8, this volume). Fear too has been argued by some to play as much, if not more, of a role in restricting consumption habits in certain areas, such as genetically-modified (GM) foods (Royzman et al. 2017). Further, different *kinds* of disgust (i.e., those elicited by moral versus pathogen-based elicitors) are likely to be effective in regulating different types of markets, ranging from the marketing of life insurance for children to atypically-shaped fruits and vegetables. Thus, while many examples used by Roth and his colleagues involve disgusting stimuli, their analysis is not specific to what we might actually term disgust.

Since Roth's exposition of market repugnance, a number of more recent empirical works have explored and confirmed the idea that disgust responses, specifically, act to constrain certain kinds of consumer behaviour. Much of this work has focussed on the food and drink market, which is unsurprising given disgust's origins as an emotion to reduce the health risks of oral ingestion (Rozin et al. 2009). Further, much of the evidence has been concentrated on new or emerging food technologies (Egolf et al. 2019). Regarding GM foods, a study by Scott et al. (2016) illustrated that "absolutist" opponents to GM (i.e., those that were opposed to GM regardless of the evidence on its risks and benefits) were more sensitive to disgust, and, further, that participants' disgust responses predicted their support for GM market restrictions. While the primacy of disgust in opposition to GM has been challenged by Edward Royzman and colleagues (Cusimano et al. 2018; Royzman et al. 2017), who argue for a core role of fear, a number of sources support disgust as a factor in reduced intentions to purchase or approve of GM foods (e.g., see Prokop et al. 2013; Townsend and Campbell 2004).

In other work, Siegrist et al. (2018) have provided experimental evidence for a role of perceived unnaturalness and evoked disgust in leading to lower acceptance of meat from in vitro cultivation (vs. traditionally slaughtered animals). Importantly, the provision of supplementary technical explanations about the production of cultured meat and its benefits was ineffective in reducing disgust reactions (Siegrist et al. 2018). Such data are consistent with the idea that disgust is often an unreasoned emotion (Russell and Giner-Sorolla 2011), and that people can exhibit absolutist (irrational) opposition to new technologies they find disgusting (Scott et al. 2016). Disgust responses have also been shown to restrict a willingness-to-pay for water recycled from wastewater (e.g., Powell et al. 2019; Rozin et al. 2015), with a perceived threat to health as a core, but not exhaustive, explanation (Powell et al. 2019). Taken together, the evidence outlined above suggests that disgust constrains novel food and drink markets, which may either represent an ostensible threat to health (e.g., in the case of recycled wastewater), and/or involve genetic or cellular processing and are perceived as unnatural or immoral (Siegrist et al. 2018).

In addition to novel food technologies, disgust has helped to shape the *types* or *variants* of otherwise accepted products consumers buy. Reminders of "animality" in Western meat products, including those that resemble animals and/or involve an emotional connection to humans, typically elicit increased disgust (Kubberød et al. 2008). Accordingly, cuts of meat are typically prepared to avoid such cues. Likewise, disgust sensitivity is a stronger predictor than pro-environmental attitudes and reported risk-taking behaviour of reduced willingness-to-pay for atypically shaped (non-prototypical) fruits and vegetables (Powell et al. 2019); an effect that is partially explained by perceptions of worse taste and unnaturalness.

A further line of research has shown that incidental disgust, for example in response to perceived contamination concerns, may reduce purchase intentions for otherwise desirable products (e.g., Guido et al. 2018). Faraji-Rad and Pham (2017) showed that induced disgust led to reduced willingness-to-pay for a carton of fruit drink (compared to a control condition), if participants had been primed with themes of uncertainty. Driven by the law of contagion (i.e., once in contact, always in contact) (Rozin and Nemeroff 1990), Argo et al. (2006) tested the effect of three tactile contamination cues (proximity to contact, time elapsed since contact, and number of contacts) on product evaluations and purchase intentions of a target t-shirt in a university bookstore. Participants received a cover story that involved having them try on the t-shirt before rating it. The contamination cues were manipulated across three studies by having a confederate Sales Associate tell the participant that: the target item was on the sales rack, return rack, or being tried on in the dressing room (proximity cues); had been tried on just now or a few days ago (temporal cues); and/or had had one or a lot of people try it on (frequency cues). The researchers found that participants had significantly more negative evaluations, lower purchase intentions, and reduced willingness-to-pay, as proximity, time, and the number of contamination sources increased. Disgust ratings mediated this effect. Similar results have been found in more recent studies, using both tangible and intangible contamination cues (i.e., with and without perceptible residue) (Gérard and Helme-Guizon 2018), and across a variety of access-based services (i.e., where money is paid for temporary access to physical goods, including car-sharing and utility tools) (Hazée et al. 2019).

Much of the abovementioned research has focused on contamination in the context of pathogen disgust, that is a perceived or ostensible threat to health from potential infection, but there has been work on socio-moral disgust too. Chan et al. (2014) ran three experiments where participants were exposed to different moral violations (including incest, theft, and fraud) in multiple modalities, while being provided with a beverage to drink. The authors confirmed their hypothesis that participants drank less of the beverage (i.e., exhibited less oral consumption) when exposed to the moral violations, than in a control condition. Amar et al. (2018) explored in a series of studies how counterfeiting may affect product usage, of both counterfeit and genuine items resembling the counterfeit, as mediated by ratings of moral disgust. Consistent with the above studies, students endorsed language suggesting they found counterfeits more morally disgusting (e.g., "morally repulsive") than genuine products (although still with an average rating on the lower end of a polarised Likert scale) (Amar et al. 2018). These studies are interesting and illustrate the deleterious effect counterfeiting may have on a market (although consumer behaviour was not assessed *per se*). Nevertheless, the extent to which these studies will replicate and extend to observable consumer behaviour and/or the extent to which they represent *genuine* feelings of disgust versus other affective reactions that are often conflated with disgust, such as moral anger (Herz and Hinds 2013), is unclear. On balance, it is likely that moral disgust (and the associated "contamination") are likely to inhibit consumer behaviour at least to some degree. This effect is illustrated vividly by the reduced endorsement of products when they become associated with disgust-relevant moral violations, such as the paedophilic accusations levelled against pop star Michael Jackson (Johnson 2005).

To this end, disgust and its associated elicitors are often the opposite emotion that marketers want associated with their products. Take, for example, the relatively recent introduction of insect-based foods into the UK market. Many, if not all, producers avoid physical images of insects, a reliable disgust elicitor (La Barbera et al. 2018), on their packaging. Instead, companies use ground up flours and processed insect protein to facilitate the marketing of these types of goods (an activity that I have called "masking"—see Powell 2017). In an attempt to promote product adoption and sales, elicitors of disgust are deliberately avoided.

More broadly, in more conventional markets, the presentation of products has become increasingly sterile. The selection, washing, and packaging of fresh produce has been designed to remove any elicitors of disgust (e.g., dirt and grime) (Curtis 2007) and so our disgust response has acted to shape consumer markets in the way products are presented. Disgust has also been used strategically to discourage consumer behaviour in certain target areas, where excess consumption is viewed as problematic. Examples include anti-obesity campaigns (i.e., products with high amounts of sugar and/or fat content) (Lupton 2015); smoking (Cameron and Williams 2015)—a behaviour which has become increasing moralised over time (Rozin and Singh 1999); and excess alcohol consumption (Collymore and McDermott 2016). Yet, given disgust's oral origins, most research into disgust and consumer behaviour has been restricted to the food and drink domain. Despite some initial probing investigations (e.g., Hazée et al. 2019), much less is known, for example, about how disgust affects consumer behaviour in real-time across broader retail environments. While it is clear that disgust acts as a constraint on consumption, there would appear to be some situations where disgust may promote consumer behaviour, an area to which we next turn.

Disgust Promotes Certain Consumer Behaviours

If you are inclined, on a cold, dark, soggy evening, you could kill a few minutes searching YouTube for "Domestos germ adverts". You will be rewarded with a selection of advertisements for the popular toilet cleaner that involve the cartoon personifications of microbes, designed to have the kind of characteristics that are intended to make you want to expunge them maniacally. These things are created to be ugly, slimy, spotty, and asymmetrical—think 50 shades of green. These are all characteristics that are known to activate our disgust response (Oum et al. 2011), and advertisers are counting on stimulating disgust to help motivate you to buy their products to expel these repugnant abominations from your home.

As the above example illustrates, as well as constraining markets, disgust can encourage consumer behaviour. A headline-grabbing paper by Di Muro and Noseworthy (2013) showed that people were more likely to spend more using, and take more chances with, worn and dirty bank notes than fresh and clean ones (except when in-front of social others), and that feelings of contamination and/or disgust (and pride) explained these effects. The researchers concluded that the *appearance* of money matters, and that consumption could be stimulated by people's disgust-fuelled desire to reject literal "dirty money" (see also Galoni and Noseworthy 2015). While the marketers' desire for promoting the circulation of "dirty money" has failed to materialise, disgust can stimulate consumption in at least three other, non-mutually-exclusive situations. First, when the consumer product has been designed to solve a problem that elicits disgust in the consumer (e.g., McAteer 2019). Second, disgust may be leveraged in advertising to attract attention and promote salience, in order to improve the advert's effects on sales, even when the product itself does not solve a disgusting problem (e.g., Hubbard 1993). Third, and perhaps the most interesting, is the possibility that a disgusting characteristic of a product is what the consumer finds appealing about it in the first place (e.g., Menninghaus et al. 2017). Let us explore these ideas in succession.

First, disgust is a motivator of consumption for products designed to address problems that elicit disgust in either the consumer or people around the consumer. Examples of these kinds of products include cleaning products (McAteer 2019), new and/or disposable (versus reusable and remanufactured) variants of consumer goods (Abbey et al. 2015), and cosmetics that mask physical features known to elicit disgust in others (such as deodorants to reduce body odour) (Ubel et al. 2017). In a Polish study, Hełka and Stefanowicz (2016) investigated the effects of disgust cues on attitudes towards, and willingness-to-buy, a new face cream. Exposure to a poster of a disgusting skin disease, versus a poster of a rat (non-associated disgust condition) or an advert for a private university, was associated with more positive attitudes towards the cream and a greater willingness-to-buy, whereas the irrelevant disgust poster was not. Elsewhere, Chan (2019) tested whether visualising causal sex (vs. visualising a romantic walk or yesterday's events) influenced consumer responses to hygiene products, including toothpaste, soap, and face scrub. In three studies, the authors found increased product liking and an increased willingness-to-pay for the hygiene products in the casual sex condition than comparator conditions (see also Tybur et al. 2011). The implications are that situations that make people feel dirty or disgusted may promote the consumption of products that are marketed as delivering cleanliness. The common factor in these works is that disgust promotes consumer behaviour when products are designed to *remedy* a disgusting problem.

Second, disgust is seemingly paradoxical, in that it promotes rejection and avoidance but also attracts attention, presumably in an adaptive manner to alert an organism to potential pathogen threats (van Hooff et al. 2013). Some elicitors of disgust, such as death and mutilation and sexual behaviour, also have a degree of shock value. Such attention-attracting and shocking qualities have been used in advertisements, to promote contemporary social issues, make a statement, and deliver the impression of brands having a social conscience, to target consumers. The classic example of this is the adverts of the clothing company, *Benetton*, in the 1980s, which featured, amongst other things: coloured condoms, a priest and nun kissing, immigration, people dying of AIDS, child labour, and pigs rooting in piles of rubbish (Hubbard 1993). These adverts did not show any products, yet were designed to position "the company as a concerned, socially-active, cutting edge and global fashion apparel company" using provocative (and disgust eliciting) images that "attract attention, make a statement, and create dialogue and action" (Hubbard 1993, p. 46).

While it is difficult to measure the success of such "shockvertising" campaigns, which is likely to differ depending on the target audience (Hubbard 1993), it is clear that when utilising disgust advertisers should proceed with caution. Empirical work has shown that disgust in adverts can simply lead to more negative attitudes towards adverts, and not necessarily greater brand recall, than non-disgust equivalents (Dens et al. 2008). A fine-line must be trod between (a) the extra attention and "buzz" afforded by controversial, shocking, and potentially disgusting adverts and (b) the role of revulsion in promoting product avoidance. For example, in a study testing two adverts for a beef sandwich from a hypothetical fast food chain (one featuring the sandwich, and one featuring it alongside raw meat), Shimp and Stuart (2004) found that the advert with raw meat was associated with more negative purchase intentions and that reported disgust fully mediated this effect. Marketers should thus ensure that any disgust elicited from an advert is not attributed to the product itself, but to the problem for which the product (or brand, or company itself) seeks to raise attention and potentially alleviate.

Third, experiences of disgust that are elicited by products themselves may lead to sales in certain circumstances. This includes consumer goods, such as popular media products, which often blur the line between humour and disgust, including the successful UK children's book series "Horrible Histories" and the US TV show "Jackass". In this context, revolting scenarios are used to engender blended emotions of disgust and humour to appeal to target consumers, particularly children and adolescents (Oppliger and Zillmann 1997). Related to this is the "buzz" that can be generated by viral media and advertising campaigns that feature elements of the grotesque (Rubenking 2019). Such media is likely successful in generating attention and sales because the disgust stimuli does not present a perceived health or moral threat to the consumer; observers are more likely to find someone else drinking a "sweatsuit cocktail" humorous than if they have to chug it down themselves (Hemenover and Schimmack 2007). Nevertheless, the appeal of disgust in these contexts is contrary to the idea that disgust solely promotes aversion; something that Nina Strohminger has called "hedonic disgust" (Strohminger 2013, 2014). The role of hedonic disgust, and how it may operate in consumer behaviour, for example in the taboo and fetish market, is under-researched. However, initial investigations into "morbid fascination" by Suzanne Oosterwijk and colleagues have shown that people prefer *social* negative images over decontextualised morbid, or neutral, images (Oosterwijk 2017), and that supraliminal morbid fascination may recruit different brain regions than disgust or fear states *per se* (Oosterwijk et al. 2016).

In the context of the art market, Wagner et al. (2014) found more positive ratings of the same photos when framed as art photographs than documentary photographs, while the level of disgust ratings were identical, suggesting context matters. A distancing-embracing model for the enjoyment of negative emotions in art, including disgust, has since been proposed by Menninghaus and colleagues, where the link between negative emotions and aesthetic pleasure is mediated by mixed affective states and emotion regulatory processes (Menninghaus et al. 2017). Although this possibility needs empirical work in traditional consumer contexts, under the right conditions, disgust sells.

Therefore, in at least the three ways outlined above, disgust (or Roth's "repugnance") is more than just a market constraint, but also a market facilitator under particular conditions. The ways that disgust can both inhibit and facilitate consumer behaviour are summarised in Table 15.1. Both of these influences have implications for significant social issues with consumerism at their core. None of these issues is as

Table 15.1 Examples associated with disgust that constrain or facilitate consumer behaviour

Factor	Example
Constraints	
1. Product elicits disgust as a perceived threat to health or viewed as physically atypical	Oddly-shaped fruit and vegetables (Powell et al. 2019)
2. Product elicits disgust as it is perceived to represent a violation of social-moral values or viewed as unnatural	Meat from in vitro cultivation (Siegrist et al. 2018)
3. Product elicits disgust because it is associated with a person, brand or company that elicits disgust	Music associated with accused paedophiles (Johnson 2005)
4. Obtaining the product *with money* elicits disgust, as a perceived socio-moral violation	Selling organs online (Roth 2007)
5. Consumer perceives that the product has been contaminated by other people (or counterfeits)	Used or remanufactured goods (Abbey et al. 2015)
6. Disgust is used strategically in advertising to discourage consumption of a product	Anti-smoking campaigns (Cameron and Williams 2015)
7. Consumer is incidentally in a disgusted state (i.e., not elicited by the product itself)	Lower ratings of products when disgusted (Motoki and Sugiura 2018)
Facilitators	
1. Product is designed to solve a problem that elicits disgust in the consumer or others connected to them	Cleaning products (McAteer 2019)
2. Disgust is used effectively in shock advertising campaigns, i.e., to promote a brand as one with a social conscience	Benetton advertising campaigns (Hubbard 1993)
3. Product elicits mixed (blended) emotions of disgust *and* positive states, like humour or atheistic pleasure (hedonic disgust)	Art with disgusting content (Menninghaus et al. 2017)
4. Advertisements induce morbid curiosity or fascination that promotes circulation and consumption	Viral YouTube videos (Rubenking 2019)
5. Money is physically dirty and encourages consumer to spend (i.e., to get rid of it)	Worn bank notes (Di Muro and Noseworthy 2013)

perhaps of as much contemporary significance as environmental sustainability, and the need to curb people's consumption of disposable goods in order to help reduce the associated greenhouse gas emissions driving climate change. While countries like China may be the most significant *producers* of greenhouse gases as manufacturers of consumer goods, the *consumer* carbon footprint is much higher in other rich Western countries, like Luxemburg, that consume many of these products (Caro et al. 2015). We explore the effect disgust may have on sustainability, via consumer behaviour, next.

Disgusted Consumers Are Bad for a Sustainability Agenda

Mixed affective states and forays into hedonistic disgust aside; disgust, above all, is an emotion of rejection. Perhaps more to the point, an emotion that promotes rejection and disposal presents a problem for movements towards *reduced* consumption and environmental sustainability. I wrote about this issue in a piece for *The Conversation* in 2017 (Powell 2017). Key features of disgust, including its conservativeness for triggering false alarms and ideational quality, mean the "yuck factor" can be a tricky customer when it comes to encouraging sustainable consumer behaviour. Indeed, there is evidence suggesting that disgust contributes to a disposable consumer culture.

A recent European survey study exploring the role of sensitivity to disgust on food behaviour found a small but significant positive association between level of food disgust sensitivity and a higher frequency of food waste behaviour (Egolf et al. 2018). Disgust has also been identified as a barrier to food-based recycling activities such as kitchen caddy food waste composting in Australia (Ames and Cook 2020). Evidence shows that the disgust response promotes selective consumption, including, for example, sanitised and prototypical versions of organic products, such as fresh produce (Jaeger et al. 2018). Further, a body of research exploring the effects of contamination concerns on consumer behaviour illustrates how consumers may be put off from engaging with the second-hand market and buying used goods (Argo et al. 2006; Gérard and Helme-Guizon 2018; Hazée et al. 2019). Consumer comments found on online message boards about charity shops illustrate this phenomenon: "Honestly, I think buying clothes and shoes 2nd hand is quite gross!" and "I've picked up some real bargains on good brands. But some of my friends think its (sic) absolutely disgusting and shameful".

What is more, revulsion can be a potential barrier to *new* sustainability initiatives that may inadvertently elicit disgust in consumers, including alternative, more sustainable sources of protein (e.g., insect-based proteins) (Gmuer et al. 2016; La Barbera et al. 2018), and the production of products from reclaimed ingredients (Herbes et al. 2018). This includes the use of biomethane, refined from biogas derived from organic waste, which can be re-introduced into the consumption cycle. An example of this is the fuel used in the UK "bio-bus" which was launched in the West Country in 2015 and has since been adopted elsewhere. Again, the disgust lexicon was apparent online: "I bet the exhaust fumes stink" and "I think this is kind of disgusting. . .a bus running on human waste". Herbes et al. (2018) studied consumer attitudes towards bio-based packaging and found evidence of consumer disgust as an obstacle: "Because I find it disgusting"; "The thought of using manure for producing food packaging arouses a feeling of disgust in me"; "Because it stinks"; and "Kind of disgusting—A little unsanitary to me". Meng and Leary (2021) investigated consumer purchase intentions for clothing made from recycled plastic bottles, finding that participants were less likely to purchase a t-shirt made from used plastic bottles (than unused plastic bottles or cotton) and this effect was moderated by disgust sensitivity. In a recent study, we showed that people who were more sensitive to disgust were willing to pay less for drinks and medicines that contained reclaimed ingredients from sewage, including fizzy drinks with recycled sources of carbon dioxide (Powell et al. 2019).

Thus, overcoming the "yuck factor" in certain novel and technologically viable sustainability solutions may be important for encouraging their widespread uptake. Powell (2017) outlines at least three ways to do this. First, the product (and associated brand or company) is altered or presented in such a way so that any disgust cues are masked. The problem with this approach is that you still have the psychological essence of, for example, knowing the carbon dioxide in your fizzy drink came from human waste. Nevertheless, masking disgust cues does matter. An experiment marketing "treated wastewater" as "recycled water" showed that people were more willing to use, and willing to pay more for, the latter than the former (Menegaki et al. 2009). Second, educational and marketing techniques can be used to encourage cognitive reappraisal in the consumer, so that they regulate (and overcome) automatic disgust reactions. While some work has shown promise for reappraisal in disgust regulation (e.g., Olatunji et al. 2017), we recently found little support for a moderating role of dispositional disgust reappraisal on the effect of disgust responses predicting willingness-to-pay for more sustainable alternatives (Powell et al. 2019). The jury is thus out on whether reappraisal may be helpful in this context. Finally, inducing antithetical emotional states to disgust, such as compassion for the product(s) and/or the environment, may be helpful in reducing disgust's influence. Such an approach has been shown to be beneficial in other areas (e.g., in healthcare) (Reynolds et al. 2019), and unpublished data from our research group suggests appeals to compassion may be beneficial in increasing willingness-to-pay for atypical fruit and vegetables (but not as beneficial as positive appeals to comparable taste and superior pro-environmental qualities). Regardless of the technique employed, there is some room for optimism, in that disgust is reduced in situations of scarcity, illustrated by the dampened food-related disgust to less-than-appetising dishes exhibited by people in food-deprived states (Hoefling et al. 2009).

While disgust presents certain challenges for consumer sustainability, it is important to note that it does not have to be entirely antagonistic towards this endeavour. In particular, if used strategically, analogous to public health campaigns, disgust can be used as a tool to help *increase* sustainable living. This could be achieved, for example, by presenting graphic depictions of disposable waste in a way that elicits disgust (e.g., large piles of rotting rubbish) or by associating the mass production of disposable consumer goods with, for example, disgusting pollutants. One significant contribution to global greenhouse gases is meat farming, so highlighting the disgusting elements of this practice, including both their living conditions as well as the death and butchery of animals, could be useful from an environmental perspective. Such an approach brings with it the threats of unappreciated paternalism and threats to the livelihoods of people relying on these markets, which would need to be balanced against any uses of disgust to promote sustainable consumption. In an interesting twist on the negative perceived contamination of clothing made from used plastic bottles, described above, Meng and Leary (2021) demonstrated that such an effect could be used *positively* and increase willingness-to-pay if the source of contamination was an attractive member of the opposite sex. Used responsibly, there are opportunities for disgust (and the associated principles of contamination) to be leveraged to help promote sustainable consumption habits—but what do we actually know about consumption patterns and their link to revulsion? Let us find out.

Certain Consumption Habits Are Disgusting

The economic psychologist Stephen Lea and his late colleague Paul Webley introduced the hypothesis that money functions both as a tool and a *drug*, with characteristics that mimic that of other known addictive substances (Lea and Webley 2006). While the drug metaphor is useful in understanding how people respond to money, it is, as acknowledged by the authors themselves, an imperfect analogy. One of the ways the drug metaphor breaks down is that addicts of other addictive stimuli are often viewed with disgust (Harris and Fiske 2007), while those who pursue money, and display markers of legitimatised wealth, tend to receive higher social status (Kraus et al. 2017). Part of this polarisation may be due to the behaviours that are associated with these respective addictions, with the former typically associated with actions that violate socio-moral purity norms and represent a potential health threat. Rich, successful people too, while often revered, can become the target of disgust-based criticism when associated with immorality, such as corruption or money earned through unfair means (Deigh 2006). Consider, for example, the ugly cartoon portrayals of "fat cat" bosses and bankers, who pursue money at the expense of others' welfare (see Fig. 15.1). Here, disgust is used to promote and reflect societal antipathy towards these actors, emphasising characteristics known to elicit the emotion, including fatness and ugliness (which are also physical

Fig. 15.1 "Fat cats" who pursue greed at the expense of others' welfare are viewed with disgust. (Image available for reuse, Wikimedia Commons, https://commons.wikimedia.org/wiki/File:The_Subsidised_Mineowner.jpg)

manifestations of characterological greed). While wealth per se is not a repulsive characteristic (it is often the opposite), the *greedy* pursuit or consumption of money (or any other desirable resource) in a manner which suggests illegitimacy appears to trigger disgust, and particularly so if it has deleterious effects on others (Kempen 2017). In this sense, the "dirty money" discussed above in relation to *physically dirty* currency takes on its *metaphorical* connotations as being representative of wealth acquired through corrupt or ill-gotten means, and is consequently less desirable (Stellar and Willer 2013; Tasimi and Gelman 2017).

One of the reasons that greedy consumption habits may trigger disgust is that they are often seen as reflecting an unequal and unfair distribution of resources, which violate socio-moral boundaries (Chapman et al. 2009). A number of empirical studies using economic games have shown that people experience disgust in response to unfair offers (e.g., Chapman et al. 2009) and that disgusted participants are more likely to reject offers perceived as unfair (e.g., Moretti and di Pellegrino 2010). Socio-moral disgust reactions then may affect consumption transactions when elements of unfairness are salient and help regulate in-group resource consumption. The extent to which such responses represent a manifestation of visceral disgust specifically (rather than shared variance between discrete emotional states, often unaccounted for in empirical studies), is a question for further research to untangle. Nonetheless, based on these and other complementary findings (including a tendency for people primed with disgust to cheat more on economic tasks) (e.g., Winterich et al. 2014), it has been posited that disgust might stimulate a mind-set related to resource scarcity (as an emotion that promotes self-protection) (Schnall 2017). Indeed, an emerging fMRI study indicated that stimuli depicting food *wasting* behaviour was associated with activation in brain regions previously identified as important in moral and physical disgust (Marczak et al. 2019). One can easily imagine the disgust reactions if food wasting or rich indulgences were juxtaposed with images of starving children. Note here the potential self-other hypocrisy, given that felt disgust is typically associated with increased fussiness around food (Egolf et al. 2018; Motoki and Sugiura 2018). Indeed, if the "resource scarcity" theory is supported empirically, this would mean that not only is disgust directed towards others with bad moral character (Giner-Sorolla and Chapman 2017; see also Giner-Sorolla, Chap. 8, this volume), but it may stimulate unethical and selfish consumption behaviour in the self.

One of the most interesting phenomena associated with money and transactional human behaviour is how certain trading activities appear to elicit disgust *only* when they involve monetary exchange. Consider, for example, the donation of organs (one of Alvin Roth's archetypal repugnant markets) (Roth 2007). With consent and when given freely, the practice of organ donation does not typically elicit disgust in others (except perhaps in a minority of absolutist moral-opposed individuals). However, when paid for, such an exchange becomes repugnant and is inhibited. The same distinction is made in many areas of the world between freely-given and bought sexual activity. Indeed, a number of social practices have arisen to *avoid* the exchange of payment in return for risqué goods, including the "charity girls" of the early twentieth century, who would receive goods (not money) and give sex as "a gift" (not a commodity) (Schilke and Rossman 2018). Schilke and Rossman (2018) use the term "obfuscated exchange" to describe this irrational human behaviour. Obfuscation may occur wherever resources are exchanged in a scenario where people would otherwise object to that exchange occurring within a market context, finding it morally reprehensible. Why money taints otherwise viable resource-sharing behaviour is not fully understood, but may involve issues of legality and social, cultural, and moral norms; power balances; and issues of unfairness and perceived exploitation and coercion. For example, offering to pay more for the same "repugnant" good than the seller suggested is judged as more ethical, while *incentivising* the sale of a greater amount of the good is judged as less ethical, especially when the seller is poorer and assumingly more vulnerable to exploitation (Ambuehl et al. 2015). While fascinating, the effects of money rendering an otherwise acceptable exchange of resources disgusting is under-researched. We turn now to explore some other unknowns between disgust and consumer behaviour.

The Unknowns on Disgust and Consumer Behaviour

The work reviewed above suggests disgust can be both a constrainer and facilitator of consumer behaviour, that it may be a barrier to sustainability, and that certain consumption patterns elicit revulsion. However, there is still much to learn on how disgust links to consumerism. First, we need to *mind the metaphor*: the extent to which disgust has been empirically demonstrated to influence, and be influenced by, consumer decision-making is often conflated with the effects of other correlated negative affective states, including fear and anger. For example, the extent to which the average person feels *genuine*, *visceral* moral disgust to counterfeit fountain pens (Amar et al. 2018), or finds used power tools physiologically disgusting due to potential contamination (Hazée et al. 2019), is questionable, and requires further investigation (which includes outcome measures beyond self-report ratings). This is analogous to the problem identified in moral psychology, whereby self-reports of disgust and disgust language can be used to describe situations that are best characterised by an emotional experience of anger (Herz and Hinds 2013; Nabi 2002). Other areas of research suggest a critical role for fear, over and above disgust, in consumer scenarios where disgust, if measured in isolation, has been shown to be an important factor (Cusimano et al. 2018; Royzman et al. 2017).

What is typically missing in disgust-based consumer research are the complementary measures of correlated discrete emotions (or even negative affective states), that would allow researchers to isolate a unique role for disgust (see also Consedine, Chap. 2, this volume). This problem is most apparent in the use of self-report measures of emotion. While disgust does not have to be measured by self-report alone (Chapman et al. 2009; Marczak et al. 2019), there is a notable absence of alternative modalities of assessment in this area of research. Additional studies employing alternative modes of assessment, which are reliably and uniquely associated with disgust, such as levator labii activation in facial EMG (Vrana 1993) and the electrogastrogram (Meissner et al. 2011), may help to strengthen the evidence that it is disgust, and not shared variance with other negative affective states, that is driving observed effects. It would be interesting to know, for example, the extent to which money or greed is *physiologically* disgusting. More advanced self-report measures, such as ecological momentary assessments (e.g., Vansteelandt et al. 2005), could also add another layer of knowledge about how disgust factors into consumer decision-making in real-time retail environments. Most work exploring consumer disgust has centred on oral consumption and it would be nice to see some novel research probing beyond this.

Second, we need to *explore effective regulation*. To the extent that disgust is deleterious to consumer behaviour and societal progress, there is much more work to be done into the effective regulation of the emotion. We are still unclear on the best ways to regulate and manage disgust, particularly in a consumer context. Indeed, evidence suggests disgust may be a particularly difficult emotion to regulate (Olatunji et al. 2007). In regulating disgust in consumers, framing may have some promise (e.g., Menegaki et al. 2009). Initial investigations on disgust regulation in other fields, such as mental health, show some potential for cognitive reappraisal (e.g., Fink et al. 2018; Olatunji et al. 2017), while other studies support good old-fashioned exposure and habituation (e.g., Rozin 2008). As work has shown that disgust can be a barrier to the consumption of pro-social, sustainable alternatives (Powell et al. 2019) and the circular economy (Argo et al. 2006; Gérard and Helme-Guizon 2018; Hazée et al. 2019; Meng and Leary 2021), the literature would benefit from additional experiments on disgust regulation in applied settings. Effective regulation also applies in situations where disgust should be divorced from market decision-making (i.e., in deciding what products are permissible to marketise). This is in itself a thorny issue with its own unknowns; in which situations is a repugnant market irrationally impeded by disgust? While some behaviour in repugnant markets, such as obfuscated exchanges, is clearly irrational, it is perhaps difficult to argue that people should not feel disgusted by, for example, a marketised forum for the selling of children (with the exception of exhausted parents).

Third, we need to do more work on *hedonic disgust*. Most research has focussed on the negative side of disgust. Less has considered how disgust, blended with other more appealing emotions, or as part of the fetish market, may actually attract consumers. Is the disgust elicited in a hedonic context the same kind of disgust that repels people from eating mouldy sandwiches, or is it fundamentally distinct, for example, when it is blended with humour or aesthetic pleasure? Work using fMRI suggests that morbid fascination is associated with differential brain activity than revulsion (Oosterwijk et al. 2016), and observed behaviour suggests that something different is going on phenomenologically. Mixed methods research further exploring hedonic disgust and why people are drawn to it, what they experiencing when they find products with elements of disgust appealing, and how such elements can be leveraged in the market to generate appeal, would be useful. Initial work in design has produced a framework for designing "rich" experiential states, including the grotesque, which include negative emotions like disgust, blended with positive states such as fascination (Fokkinga and Desmet 2013; Menninghaus et al. 2017). I expect such frameworks to be further enriched over time, with inputs from affective and consumer psychology.

Finally, we need to critically examine the *resource scarcity theory* (Schnall 2017) in consumer behaviour and assess how it holds up to further scrutiny. To recap, the basic tenets of this theory are: i) disgust evolved to stop us getting ill and has been exapted to function in response to broader socio-moral threats or violations; ii) as disgust operates to protect against pathogen threats and threats to one's moral character, we can consider it a "self-protective emotion"; iii) because disgust is a "self-protective emotion", it instils a mind-set of "resource scarcity" in resource transactions (Schnall 2017). While some evidence is consistent with this theory, including disgust elicited to unfair resource distribution by others (Chapman et al. 2009; Moretti and di Pellegrino 2010) and cheating behaviour benefiting the self (Winterich et al. 2014), a number of other observations are not. For example, experienced disgust is typically associated with a reduced flexibility in consumption, including atypical foods (Powell et al. 2019), which is not what we might expect if resources were thought to be scarce. Indeed, disgust has been shown to be more likely to be *reduced* during times of necessity (scarcity), such as when people are deprived of food or living in a harsher environment (Batres and Perrett 2020; Hoefling et al. 2009), which appears contrary to the idea that heightened disgust stimulates a resource scarcity mind-set.

Furthermore, the work suggesting people like to spend physically dirty money, because it disgusts them is not consistent within a resource scarcity framework (Di Muro and Noseworthy 2013; Galoni and Noseworthy 2015). There are a few ways these differences could be reconciled, including by considering that disgust may not be a unitary emotion, but consists of subcomponents that may have differing effects on behaviour (Simpson et al. 2006). Further, instead of resource scarcity, disgust may promote a self-focused attitude associated with elevated self-importance, or a "holier-than-thou" mind-set, which reflects the pattern of evidence reviewed above. Further work will be required to disentangle these different theoretical interpretations. It is worthy of note that this "unknown" is related to the first, to the extent that we could also think of other emotions, such as fear, as self-protective, in that they help us to avoid and escape from danger. So more theoretical development is required that delineates the unique role for disgust in resource consumption behaviour.

Conclusions

Over a decade ago, in 2007, Alvin Roth published a now highly cited paper arguing that repugnance constrains markets (Roth 2007). In this chapter we have explored how disgust, just like Roth's broader construct of economic repugnance, indeed does act as a constraint on consumer behaviour, but also how, and under which conditions, it serves to facilitate it. In doing so, we have learned that the relationship between disgust (or "repugnance") and consumer behaviour is much more complex than it may first appear. The disgust emotion is a barrier to certain sustainability initiatives in the consumer sector, but it is also a potential tool to be leveraged, strategically, to encourage pro-social and pro-environmental consumption behaviour. Further, the emotion may play a critical role in helping to ensure fairness within the intra-social distribution of resources, although there is a risk of disgust encouraging selfish economic behaviour, perhaps as a result of an extended form of a self-protection mind-set. Unknowns on disgust and consumer behaviour include the unique role for the emotion over and above other, correlated negative affective states, and the phenomenology of, and necessary conditions for, pleasurable, or hedonic consumer disgust. While initial models are being developed to provide guidance on how producers and marketers can "design for disgust" within the consumer industry, we still have much to learn. I look forward to being suitably grossed out in the future.

References

Abbey, J. D., Meloy, M. G., Blackburn, J., & Guide, V. D. R. (2015). Consumer markets for remanufactured and refurbished products. *California Management Review, 57*(4), 26-42. https://doi.org/10.1525/cmr.2015.57.4.26

Amar, M., Ariely, D., Carmon, Z., & Yang, H. Y. (2018). How counterfeits infect genuine products: The role of moral disgust. *Journal of Consumer Psychology, 28*(2), 329-343. https://doi.org/10.1002/jcpy.1036

Ambuehl, S., Niederle, M., & Roth, A. E. (2015). More money, more problems? Can high pay be coercive and repugnant? *American Economic Review, 105*(5), 357-360. https://doi.org/10.1257/aer.p20151034

Ames, E., & Cook, N. (2020). Food becoming compost: encountering and negotiating disgust in household sustainability. *Australian Geographer, 51*(3), 325-339. https://doi.org/10.1080/00049182.2020.1786944

Argo, J. J., Dahl, D. W., & Morales, A. C. (2006). Consumer contamination: How consumers react to products touched by others. *Journal of Marketing, 70*(2), 81-94. https://doi.org/10.1509/jmkg.70.2.81

Baker, M. J., & Churchill, G. A. (1977). The impact of physically attractive models on advertising evaluations. *Journal of Marketing Research, 14*(4), 538-555. https://doi.org/10.2307/3151194

Batres, C., & Perrett, D. I. (2020). Pathogen disgust sensitivity changes according to the perceived harshness of the environment. *Cognition and Emotion, 34*(2), 377-383. https://doi.org/10.1080/02699931.2019.1612735

Cameron, L. D., & Williams, B. (2015). Which images and features in graphic cigarette warnings predict their perceived effectiveness? Findings from an online survey of residents in the UK. *Annals of Behavioral Medicine, 49*(5), 639-649. https://doi.org/10.1007/s12160-015-9693-4

Caro, D., Rugani, B., Pulselli, F. M., & Benetto, E. (2015). Implications of a consumer-based perspective for the estimation of GHG emissions. The illustrative case of Luxembourg. *Science of The Total Environment, 508*, 67-75. https://doi.org/10.1016/j.scitotenv.2014.11.053

Chan, E. Y. (2019). Dirty weekends and personal hygiene products: The embodiment of casual sex in marketing. *Psychology & Marketing, 36*(6), 587-596. https://doi.org/10.1002/mar.21198

Chan, C., Van Boven, L., Andrade, E. B., & Ariely, D. (2014). Moral violations reduce oral consumption. *Journal of Consumer Psychology, 24*(3), 381-386. https://doi.org/10.1016/j.jcps.2013.12.003

Chapman, H. A., Kim, D. A., Susskind, J. M., & Anderson, A. K. (2009). In bad taste: Evidence for the oral origins of moral disgust. *Science, 323*(5918), 1222-1226. https://doi.org/10.1126/science.1165565

Collymore, N. N., & McDermott, M. R. (2016). Evaluating the effects of six alcohol-related message frames on emotions and intentions: The neglected role of disgust. *Journal of Health Psychology, 21*(9), 1907-1917. https://doi.org/10.1177/1359105314567910

Crilly, N., Moultrie, J., & Clarkson, P. J. (2004). Seeing things: Consumer response to the visual domain in product design. *Design Studies, 25*(6), 547-577. https://doi.org/10.1016/j.destud.2004.03.001

Curtis, V. A. (2007). Dirt, disgust and disease: A natural history of hygiene. *Journal of Epidemiology and Community Health, 61*(8), 660-664. https://doi.org/10.1136/jech.2007.062380

Cusimano, C., Royzman, E. B., Leeman, R. F., & Metas, S. (2018). Measurement is the core disgust problem: Response to Inbar and Scott (2018). *Judgment and Decision Making, 13*(6), 639-651.

Deigh, J. (2006). The politics of disgust and shame. *The Journal of Ethics, 10*(4), 383-418. https://doi.org/10.1007/s10892-006-9001-4

Dens, N., De Pelsmacker, P., & Janssens, W. (2008). Exploring consumer reactions to incongruent mild disgust appeals. *Journal of Marketing Communications, 14*(4), 249-269. https://doi.org/10.1080/13527260802141231

Di Muro, F., & Noseworthy, T. J. (2013). Money isn't everything, but it helps if it doesn't look used: How the physical appearance of money influences spending. *Journal of Consumer Research, 39*(6), 1330-1342. https://doi.org/10.1086/668406

Egolf, A., Siegrist, M., & Hartmann, C. (2018). How people's food disgust sensitivity shapes their eating and food behaviour. *Appetite, 127*, 28-36. https://doi.org/10.1016/j.appet.2018.04.014

Egolf, A., Hartmann, C., & Siegrist, M. (2019). When evolution works against the future: Disgust's contributions to the acceptance of new food technologies. *Risk Analysis, 39*(7), 1546-1559. https://doi.org/10.1111/risa.13279

Faraji-Rad, A., & Pham, M. T. (2017). Uncertainty increases the reliance on affect in decisions. *Journal of Consumer Research, 44*(1), 1-21. https://doi.org/10.1093/jcr/ucw073

Fink, J., Pflugradt, E., Stierle, C., & Exner, C. (2018). Changing disgust through imagery rescripting and cognitive reappraisal in contamination-based obsessive-compulsive disorder. *Journal of Anxiety Disorders, 54*, 36-48. https://doi.org/10.1016/j.janxdis.2018.01.002

Fokkinga, S. F., & Desmet, P. M. A. (2013). Ten ways to design for disgust, sadness, and other enjoyments: A design approach to enrich product experiences with negative emotions. *International Journal of Design, 7*(1), 19-36.

Galoni, C., & Noseworthy, T. J. (2015). Does dirty money influence product valuations? *Journal of Consumer Psychology, 25*(2), 304-310. https://doi.org/10.1016/j.jcps.2014.11.002

Gérard, J., & Helme-Guizon, A. (2018). Within a hair's breadth of buying the product: The impact of tangible and intangible bodily cues of contamination: The role of disgust and mental imagery. *Applied Cognitive Psychology, 32*(5), 537-549. https://doi.org/10.1002/acp.3425

Giner-Sorolla, R., & Chapman, H. A. (2017). Beyond purity: Moral disgust toward bad character. *Psychological science, 28*(1), 80-91. https://doi.org/10.1177/0956797616673193

Gmuer, A., Guth, J. N., Hartmann, C., & Siegrist, M. (2016). Effects of the degree of processing of insect ingredients in snacks on expected emotional experiences and willingness to eat. *Food Quality and Preference, 54*, 117-127. https://doi.org/10.1016/j.foodqual.2016.07.003

Guido, G., Pino, G., & Peluso, A. M. (2018). The impact of disgust on consumers' purchase intentions: an empirical assessment. *Journal of Consumer Marketing, 35*(1), 105-115. https://doi.org/10.1108/jcm-04-2016-1786

Hammond, J. (2008). *Branding Your Business: Promoting your business, attracting customers and standing out in the market place*. Brand Halo Ltd.

Harris, L. T., & Fiske, S. T. (2007). Social groups that elicit disgust are differentially processed in mPFC. *Social Cognitive and Affective Neuroscience, 2*(1), 45-51. https://doi.org/10.1093/scan/nsl037

Hazée, S., Van Vaerenbergh, Y., Delcourt, C., & Warlop, L. (2019). Sharing goods? Yuck, no! An investigation of consumers' contamination concerns about access-based services. *Journal of Service Research, 22*(3), 256-271. https://doi.org/10.1177/1094670519838622

Hełka, A. M., & Stefanowicz, I. (2016). The influence of the emotion of disgust on attitudes towards a product and purchase decisions. *Psychologia Ekonomiczna*(10), 5-13. https://doi.org/10.15678/pjoep.2016.10.01

Hemenover, S. H., & Schimmack, U. (2007). That's disgusting! . . ., but very amusing: Mixed feelings of amusement and disgust. *Cognition & Emotion, 21*(5), 1102-1113. https://doi.org/10.1080/02699930601057037

Herbes, C., Beuthner, C., & Ramme, I. (2018). Consumer attitudes towards biobased packaging—A cross-cultural comparative study. *Journal of Cleaner Production, 194*, 203-218. https://doi.org/10.1016/j.jclepro.2018.05.106

Herz, R. S., & Hinds, A. (2013). Stealing is not gross: Language distinguishes visceral disgust from moral violations. *The American journal of psychology, 126*(3), 275-286. https://doi.org/10.5406/amerjpsyc.126.3.0275

Hoefling, A., Likowski, K. U., Deutsch, R., Hafner, M., Seibt, B., Muhlberger, A., Weyers, P., & Strack, F. (2009). When hunger finds no fault with moldy corn: food deprivation reduces food-related disgust. *Emotion, 9*(1), 50-58. https://doi.org/10.1037/a0014449

Hubbard, R. C. (1993). Shock advertising: the Benetton case. *Studies in Popular Culture, 16*(1), 39-51.

Jaeger, S. R., Machin, L., Aschemann-Witzel, J., Antunez, L., Harker, F. R., & Ares, G. (2018). Buy, eat or discard? A case study with apples to explore fruit quality perception and food waste. *Food Quality and Preference, 69*, 10-20. https://doi.org/10.1016/j.foodqual.2018.05.004

Johnson, A. R. (2005). When a celebrity is tied to immoral behavior: Consumer reactions to Michael Jackson and Kobe Bryant. In G. Menon & A. R. Rao (Eds.), *NA—Advances in Consumer Research* (Vol. 32, pp. 100-101). Association for Consumer Research.

Kempen, A. (2017). Corruption exposed One policeman's greed results in 89 child deaths. . . and counting. *Servamus Community-based Safety and Security Magazine, 110*(7), 44-47.

Kraus, M. W., Park, J. W., & Tan, J. J. X. (2017). Signs of social class: The experience of economic inequality in everyday life. *Perspectives on Psychological Science, 12*(3), 422-435. https://doi.org/10.1177/1745691616673192

Kubber⊘d, E., Ueland, Ø., Irene Dingstad, G., Risvik, E., & Jan Henjesand, I. (2008). The effect of animality in the consumption experience—A potential for disgust. *Journal of Food Products Marketing, 14*(3), 103-124. https://doi.org/10.1080/10454440801985985

La Barbera, F., Verneau, F., Amato, M., & Grunert, K. (2018). Understanding Westerners' disgust for the eating of insects: The role of food neophobia and implicit associations. *Food Quality and Preference, 64*, 120-125. https://doi.org/10.1016/j.foodqual.2017.10.002

Lea, S. E., & Webley, P. (2006). Money as tool, money as drug: the biological psychology of a strong incentive. *Behavioral and Brain Sciences, 29*(2), 161-176. https://doi.org/10.1017/S0140525X06009046

Leider, S., & Roth, A. E. (2010). Kidneys for sale: who disapproves, and why? *American Journal of Transplantation, 10*(5), 1221-1227. https://doi.org/10.1111/j.1600-6143.2010.03019.x

Lerner, J. S., Small, D. A., & Loewenstein, G. (2004). Heart strings and purse strings: Carryover effects of emotions on economic decisions. *Psychological science, 15*(5), 337-341. https://doi.org/10.1111/j.0956-7976.2004.00679.x

Lupton, D. (2015). The pedagogy of disgust: the ethical, moral and political implications of using disgust in public health campaigns. *Critical Public Health, 25*(1), 4-14. https://doi.org/10.1080/09581596.2014.885115

Marczak, M., Marchewka, A., Wypych, M., Misiak, M., Droździel, D., Sorokowski, P., & Sorokowska, A. (2019). Wasting food is disgusting: Evidence from behavioral and neuroimaging study of moral judgment of food-wasting behavior. *bioRxiv*. https://doi.org/10.1101/750299

McAteer, O. (2019, September 23). Sadness and disgust among emotions that drive brand value, says biometric study. *Campaign*. https://www.campaignlive.co.uk/article/sadness-disgust-among-emotions-drive-brand-value-says-biometric-study/1660239

Meissner, K., Muth, E. R., & Herbert, B. M. (2011). Bradygastric activity of the stomach predicts disgust sensitivity and perceived disgust intensity. *Biological Psychology, 86*(1), 9-16. https://doi.org/10.1016/j.biopsycho.2010.09.014

Menegaki, A. N., Mellon, R. C., Vrentzou, A., Koumakis, G., & Tsagarakis, K. P. (2009). What's in a name: Framing treated wastewater as recycled water increases willingness to use and willingness to pay. *Journal of Economic Psychology, 30*(3), 285-292. https://doi.org/10.1016/j.joep.2008.08.007

Meng, M. D., & Leary, R. B. (2021). It might be ethical, but I won't buy it: Perceived contamination of, and disgust towards, clothing made from recycled plastic bottles. *Psychology & Marketing, 38*(2), 298-312. https://doi.org/10.1002/mar.21323

Menninghaus, W., Wagner, V., Hanich, J., Wassiliwizky, E., Jacobsen, T., & Koelsch, S. (2017). The Distancing-Embracing model of the enjoyment of negative emotions in art reception. *Behavioral and Brain Sciences, 40*, e347. https://doi.org/10.1017/S0140525X17000309

Morales, A. C., Wu, E. C., & Fitzsimons, G. J. (2012). How disgust enhances the effectiveness of fear appeals. *Journal of Marketing Research, 49*(3), 383-393. https://doi.org/10.1509/jmr.07.0364

Moretti, L., & di Pellegrino, G. (2010). Disgust selectively modulates reciprocal fairness in economic interactions. *Emotion, 10*(2), 169-180. https://doi.org/10.1037/a0017826

Motoki, K., & Sugiura, M. (2018). Disgust, sadness, and appraisal: Disgusted consumers dislike food more than sad ones. *Frontiers in Psychology, 9*, 76. https://doi.org/10.3389/fpsyg.2018.00076

Nabi, R. L. (2002). The theoretical versus the lay meaning of disgust: Implications for emotion research. *Cognition & Emotion, 16*(5), 695-703. https://doi.org/10.1080/02699930143000437

Olatunji, B. O., Forsyth, J. P., & Cherian, A. (2007). Evaluative differential conditioning of disgust: a sticky form of relational learning that is resistant to extinction. *Journal of Anxiety Disorders, 21*(6), 820-834. https://doi.org/10.1016/j.janxdis.2006.11.004

Olatunji, B. O., Berg, H., Cox, R. C., & Billingsley, A. (2017). The effects of cognitive reappraisal on conditioned disgust in contamination-based OCD: An analogue study. *Journal of Anxiety Disorders, 51*, 86-93. https://doi.org/10.1016/j.janxdis.2017.06.005

Oosterwijk, S. (2017). Choosing the negative: A behavioral demonstration of morbid curiosity. *PLoS One, 12*(7), e0178399. https://doi.org/10.1371/journal.pone.0178399

Oosterwijk, S., Lindquist, K. A., Adebayo, M., & Barrett, L. F. (2016). The neural representation of typical and atypical experiences of negative images: comparing fear, disgust and morbid fascination. *Social Cognitive and Affective Neuroscience, 11*(1), 11-22. https://doi.org/10.1093/scan/nsv088

Oppliger, P. A., & Zillmann, D. (1997). Disgust in humor: Its appeal to adolescents. *Humor-International Journal of Humor Research, 10*(4), 421-437. https://doi.org/10.1515/humr.1997.10.4.421

Oum, R. E., Lieberman, D., & Aylward, A. (2011). A feel for disgust: tactile cues to pathogen presence. *Cognition & Emotion, 25*(4), 717-725. https://doi.org/10.1080/02699931.2010.496997

Powell, P. A. (2017, March 24). How disgust stops us from living sustainably—and how to stop it. *The Conversation*. https://theconversation.com/how-disgust-stops-us-from-living-sustainably-and-how-to-stop-it-75110

Powell, P. A., Jones, C. R., & Consedine, N. S. (2019). It's not queasy being green: The role of disgust in willingness-to-pay for more sustainable product alternatives. *Food Quality and Preference, 78*, 103737. https://doi.org/10.1016/j.foodqual.2019.103737

Prokop, P., Ozel, M., Usak, M., & Senay, I. (2013). Disease-threat model explains acceptance of genetically modified products. *Psihologija, 46*(3), 229-243. https://doi.org/10.2298/Psi130416002p

Reynolds, L. M., Powell, P., Lin, Y. S., Ravi, K., Chung, C. K., & Consedine, N. S. (2019). Fighting the flinch: Experimentally induced compassion makes a difference in health care providers. *British Journal of Health Psychology, 24*(4), 982-998. https://doi.org/10.1111/bjhp.12390

Roth, A. E. (2007). Repugnance as a constraint on markets. *Journal of Economic Perspectives, 21* (3), 37-58. https://doi.org/10.1257/jep.21.3.37

Royzman, E., Cusimano, C., & Leeman, R. F. (2017). What lies beneath? Fear vs. disgust as affective predictors of absolutist opposition to genetically modified food and other new technologies. *Judgment and Decision Making, 12*(5), 466.

Rozin, P. (2008). Hedonic "adaptation": Specific habituation to disgust/death elicitors as a result of dissecting a cadaver. *Judgment and Decision Making, 3*(2), 191-194.

Rozin, P., & Nemeroff, C. (1990). The laws of sympathetic magic: A psychological analysis of similarity and contagion. In W. Stigler, R. A. Shweder, & G. Herdt (Eds.), *Cultural psychology: Essays on comparative human development* (pp. 205-232). Cambridge University Press. https://doi.org/10.1017/CBO9781139173728.006

Rozin, P., & Singh, L. (1999). The moralization of cigarette smoking in the United States. *Journal of Consumer Psychology, 8*(3), 321-337. https://doi.org/10.1207/s15327663jcp0803_07

Rozin, P., Haidt, J., & Fincher, K. (2009). From oral to moral. *Science, 323*(5918), 1179-1180. https://doi.org/10.1126/science.1170492

Rozin, P., Haddad, B., Nemeroff, C., & Slovic, P. (2015). Psychological aspects of the rejection of recycled water: Contamination, purification and disgust. *Judgment and Decision Making, 10*(1), 50-63.

Rubenking, B. (2019). Emotion, attitudes, norms and sources: Exploring sharing intent of disgusting online videos. *Computers in Human Behavior, 96*, 63-71. https://doi.org/10.1016/j.chb.2019.02.011

Russell, P. S., & Giner-Sorolla, R. (2011). Social justifications for moral emotions: When reasons for disgust are less elaborated than for anger. *Emotion, 11*(3), 637-646. https://doi.org/10.1037/a0022600

Scanlon, M. (2011). History beyond the academy: Humor and horror in children's history books. *New Review of Children's Literature and Librarianship, 16*(2), 69-91. https://doi.org/10.1080/13614541.2010.540197

Schilke, O., & Rossman, G. (2018). It's only wrong if it's transactional: Moral perceptions of obfuscated exchange. *American Sociological Review, 83*(6), 1079-1107. https://doi.org/10.1177/0003122418806284

Schnall, S. (2017). Disgust as embodied loss aversion. *European Review of Social Psychology, 28* (1), 50-94. https://doi.org/10.1080/10463283.2016.1259844

Scott, S. E., Inbar, Y., & Rozin, P. (2016). Evidence for absolute moral opposition to genetically modified food in the United States. *Perspectives on Psychological Science, 11*(3), 315-324. https://doi.org/10.1177/1745691615621275

Shimp, T. A., & Stuart, E. W. (2004). The role of disgust as an emotional mediator of advertising effects. *Journal of Advertising, 33*(1), 43-53. https://doi.org/10.1080/00913367.2004.10639150

Siegrist, M., Sutterlin, B., & Hartmann, C. (2018). Perceived naturalness and evoked disgust influence acceptance of cultured meat. *Meat Science, 139*, 213-219. https://doi.org/10.1016/j.meatsci.2018.02.007

Simpson, J., Carter, S., Anthony, S. H., & Overton, P. G. (2006). Is disgust a homogeneous emotion? *Motivation and Emotion, 30*(1), 31-41. https://doi.org/10.1007/s11031-006-9005-1

Stellar, J. E., & Willer, R. (2013). The corruption of value. *Social Psychological and Personality Science, 5*(1), 60-66. https://doi.org/10.1177/1948550613484770

Strohminger, N. (2013). *The hedonics of disgust* [Unpublished doctoral dissertation]. University of Michigan.

Strohminger, N. (2014). Disgust talked about. *Philosophy Compass, 9*(7), 478-493. https://doi.org/10.1111/phc3.12137

Tasimi, A., & Gelman, S. A. (2017). Dirty money: The role of moral history in economic judgments. *Cognitive Science, 41 Suppl 3*, 523-544. https://doi.org/10.1111/cogs.12464

Townsend, E., & Campbell, S. (2004). Psychological determinants of willingness to taste and purchase genetically modified food. *Risk Analysis, 24*(5), 1385-1393. https://doi.org/10.1111/j.0272-4332.2004.00533.x

Tybur, J. M., Bryan, A. D., Magnan, R. E., & Hooper, A. E. C. (2011). Smells like safe sex: Olfactory pathogen primes increase intentions to use condoms. *Psychological science, 22*(4), 478-480. https://doi.org/10.1177/0956797611400096

Tybur, J. M., Lieberman, D., Kurzban, R., & DeScioli, P. (2013). Disgust: Evolved function and structure. *Psychological review, 120*(1), 65. https://doi.org/10.1037/a0030778

Ubel, S., Wabnegger, A., Arendasy, M., Zorjan, S., & Schienle, A. (2017). Affective evaluation of one's own and others' body odor: The role of disgust proneness. *Perception, 46*(12), 1427-1433. https://doi.org/10.1177/0301006617721120

van Hooff, J. C., Devue, C., Vieweg, P. E., & Theeuwes, J. (2013). Disgust- and not fear-evoking images hold our attention. *Acta Psychologica, 143*(1), 1-6. https://doi.org/10.1016/j.actpsy.2013.02.001

Vansteelandt, K., Van Mechelen, I., & Nezlek, J. B. (2005). The co-occurrence of emotions in daily life: A multilevel approach. *Journal of Research in Personality, 39*(3), 325-335. https://doi.org/10.1016/j.jrp.2004.05.006

Vrana, S. R. (1993). The psychophysiology of disgust: differentiating negative emotional contexts with facial EMG. *Psychophysiology, 30*(3), 279-286. https://doi.org/10.1111/j.1469-8986.1993.tb03354.x

Wagner, V., Menninghaus, W., Hanich, J., & Jacobsen, T. (2014). Art schema effects on affective experience: The case of disgusting images. *Psychology of Aesthetics Creativity and the Arts, 8* (2), 120-129. https://doi.org/10.1037/a0036126

Winterich, K. P., Mittal, V., & Morales, A. C. (2014). Protect thyself: How affective self-protection increases self-interested, unethical behavior. *Organizational Behavior and Human Decision Processes, 125*(2), 151-161. https://doi.org/10.1016/j.obhdp.2014.07.004

Part IV
Afterword

Chapter 16
An Afterword on Future Directions in Disgust Research: A Global Pandemic, a Divided World, and an Environmental Catastrophe

Philip A. Powell and Nathan S. Consedine

Editorial Reflections During a Global Pandemic

Editing this volume has been a pleasure and a challenge. When we agreed to edit the work back in the summer of 2018, no one could have foreseen the global catastrophe that would occur a year and a half later. As of 10th April 2021, there have been 135 million cases of coronavirus disease 2019 (COVID-19) detected worldwide and around 2.91 million deaths. As geographically disparate editors, our experience of living through the pandemic has been polarised due to the choices of decision makers in our respective countries. In the UK, to date, there have been 4.37 million cases and 127,000 deaths. In New Zealand, these figures are 2571 and 26, respectively. These differences in numbers are reflected in our two Governments' responses to the crisis, with the UK spending a greater proportion of time under stricter restrictions (see Fig. 16.1). Yet, some of the struggles of balancing work and family life and adapting to new ways of operating have been shared by us both, as well as all of the contributors to this volume. With that in mind, it is a joy to see the work finally reach print. This is particularly the case with the quality of the contributions it features.

There is a notable irony in editing a book about an emotion that specifically evolved to help us avoid getting sick during the emergence of the first global pandemic in a century. This irony notwithstanding, here marks a unique point in human history for research into disgust. How the emotion functions, is altered, or utilised, as part of the behavioural immune system (see Bradshaw and Gassen,

P. A. Powell (✉)
School of Health and Related Research, University of Sheffield, Sheffield, UK
e-mail: p.a.powell@sheffield.ac.uk

N. S. Consedine
Department of Psychological Medicine, University of Auckland, Auckland, New Zealand
e-mail: n.consedine@auckland.ac.nz

P. A. Powell, N. S. Consedine (eds.), *The Handbook of Disgust Research*,
https://doi.org/10.1007/978-3-030-84486-8_16

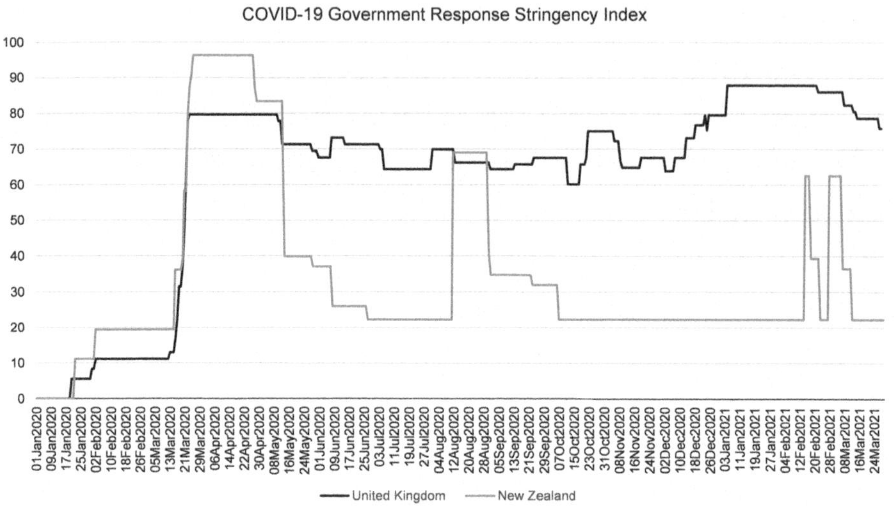

Fig. 16.1 Government response to COVID-19 in the UK and New Zealand based on the Oxford COVID-19 government response stringency index. Higher values mean greater lockdown restrictions. Data from Hale et al. (2020). Data use policy: Creative Commons Attribution CC BY Standard

Chap. 3, this volume) during a once-in-a-lifetime international health crisis will likely be of interest to many. Nonetheless, we are keen that research into COVID-19 does not come to dominate, so that other important knowledge gaps are neglected. As the chapters in this volume attest, there are many unanswered questions in disgust research, many of which relate to significant societal issues that will retain their importance when COVID-19 recedes. In this afterword, we provide an editorial perspective on areas of particular importance for future disgust research related to societal challenges. These challenges include: understanding disgust in the COVID-19 pandemic, exploring disgust's role in an increasingly divided and fragmented world, and regulating and harnessing disgust in the context of pro-environmental behaviour and climate change.

Disgust and COVID-19

The pandemic represents a unique opportunity to learn more about the origins of disgust, its function, and how it responds to a unique health threat in an area highly relevant to its evolved functional origins. It is our opinion that this opportunity should not be overlooked. To the editors' knowledge, only a handful of papers have been published as of April 2021 exploring disgust and COVID-19.

In an early study conducted between February and March 2020, McKay et al. (2020b) demonstrated a significant association between trait disgust and anxiety about COVID-19 in a Chinese sample. A subsequent longitudinal study by Paluszek et al. (2021) between March and May 2020 in over 3000 Canadian and American adults showed that a greater propensity and sensitivity to disgust predicted a higher level of excessive fear and worry ("COVID stress syndrome") in response to COVID-19 a month later. This work also found a significant moderating role of these disgust traits (tested separately) on the effects of anxiety sensitivity on subsequent COVID stress syndrome. Similarly, work by Cox et al. (2020) explored the relationship between pre-pandemic proneness to disgust (measured in 2016) and coronavirus anxiety and safety behaviours (e.g., hygiene, reassurance, and avoidance behaviours) measured in April 2020. The authors found evidence for a diathesis stress model, where the relationship between pre-pandemic disgust proneness and the two coronavirus-related outcomes was stronger when higher levels of perceived current stress were reported. This interesting finding raises the possibility that trait disgust may have a different effect on outcomes depending on contextual factors (in this case stress levels).

Further work by McKay and colleagues explored the consequences of hygiene strategies designed to suppress the transmission of the virus on therapists' willingness to deliver exposure treatment for people living with OCD (involving refraining from hygiene behaviours) (McKay et al. 2020a). The researchers reported that activation of the behavioural immune system (proxied by self-reported physical 'disgust-related' experiences, specifically coldness) was associated with more negative attitudes. While some of the aforementioned studies have evident statistical and measurement limitations, they provide an initial picture of how people's self-reported disgust traits and experiences may be related to their responses to the COVID-19 pandemic.

The above is a snapshot into primary research being conducted into disgust's relationship with this once-in-a-lifetime pandemic and it is likely that more papers are forthcoming. However, while we agree that the current circumstances present a unique opportunity to find out more about how disgust operates, we must issue a word of caution. There is a risk that disgust researchers get caught up in the "hype" of COVID-19 studies (including the rewards of rapid publication and above average citation counts) at the expense of well thought-out studies that will inform a more general understanding of disgust. A recent opinion piece by Ackerman et al. (2020) presented a cautionary tale about how the behavioural immune system's pathogen avoidance mechanisms may not be as useful as people assume in explaining humans' response to the pandemic. Specifically, this work suggests that "mismatches" between modern and ancestral environments (where components of the behaviour immune system, including disgust, evolved), such as non-transmission of the virus through contaminated foods or asymptomatic transmission, may fail to elicit disgust as strongly as other pathogen threats.

This caution notwithstanding, there is clearly a place for understanding disgust's role in the pandemic and reason to believe understanding its relevance may help inform the general understanding of disgust. Several hypotheses are likely to be of interest. First, understanding how much disgust is elicited by pandemic-related cues and how disgust responding may have changed (and/or remain altered) due to the pandemic is an important question. Are people now more disgusted by people coughing in public, handling food in supermarkets, or failing to wash their hands after using the restroom, for example? Will they remain more sensitive to such cues once the pandemic passes? Do we feel moral disgust in response to people refusing to wear masks? Are representational cues sufficient to elicit disgust-mediated avoidance responses, such as knowing that a person (not otherwise exhibiting visible symptoms) has been in a high risk COVID-19 area? How do such dynamics manifest in policy, notably regarding immigration?

Second, work could be conducted into how disgust responses manifest beyond self-report (see also Consedine, Chap. 2, this volume), for example in pandemic-based behaviour. We might profitably consider whether people who are more disgust sensitive are more likely to exhibit and maintain social distancing? Similarly, are disgust prone individuals more likely to follow hygiene advice given during the pandemic? Further still, is there some carryover effect, where COVID-19 safety behaviours lead to maladaptive behaviour patterns in people who are highly disgust sensitive, leading to, for example, disproportionate behavioural responses that persist following our return to "normality"?

Third, it would be interesting to examine how disgust has (successfully and/or unsuccessfully) been leveraged as a tool to help encourage people to adopt safety behaviours throughout the pandemic. There is a history of research on the use of disgust-based campaigns to motivate hygiene behaviours, like handwashing (e.g., Judah et al. 2009; Porzig-Drummond et al. 2009). It would be of interest to explore how such an emotion has or could have been used during an international pandemic. Did advertisements to encourage hand washing, mask wearing, or social distancing feature disgust-based content? How do people respond to disgust-based content designed to encourage more safety behaviours? Were there any unintended effects in the use of disgust to try to encourage particular behaviours, such as stigma?

It is possible that some or all of the questions posed above will have been tackled when this volume comes to print, but we hope they serve to illustrate the unique potential of the current pandemic in facilitating disgust research. As importantly, we hope that researchers will take the opportunity to conduct work that advances our shared understanding of disgust rather than merely using the COVID pandemic as a chance to publish.

Disgust and the Divided World

As an emotion that has special relevance in terms of rejection, disgust has the potential to be of particular significance in our increasingly divided world. Looking to the United States (and Donald Trump's legacy) as a salient example, the "us" and "them" mentality has become ever more entrenched alongside a global rise in nationalism. Many countries have become increasingly conservative in their voting patterns with significant political events, such as the election of Trump in America and the farce that is Brexit in Europe, typifying a movement towards decreased global integration and increased isolationism (see Shook et al., Chap. 14, this volume). Populations tend to become more conservative and nationalistic when threats are present (or created) and questions have been raised over whether the COVID-19 pandemic will compound this dynamic (the stockpiling of vaccines is a clear example of an "us first" approach when faced with limited resources).

Understanding disgust's role in facilitating or representing an increasingly divided world is of great interest. Just as the emotion helps to defend the borders of the self from pathogen, contamination, and moral threats, disgust helps to define group boundaries. A body of research in social psychology has shown, amongst other things, that disgust is lower towards ingroup versus outgroup members (i.e., when smelling their sweat; Reicher et al. 2016), disgust is implicated in prejudice and stigma towards others who differ in group membership from the self (see Vartanian et al., Chap. 10, this volume), and disgust facilitates outgroup dehumanisation (Porzig-Drummond et al. 2009). Outside of the lab, lessons from political rhetoric demonstrate how the language and the purposeful elicitation of disgust is used to promote group discrimination and cement the group hierarchy. Hitler described Jewish people as dirty, disease-ridden rats, and as parasites. More recently, Trump's dehumanising language towards immigrants was, in part, designed to elicit feelings of disgust, facilitating an "us" and "them" mentality.

Much of the research on disgust and division conducted thus far has been based on demonstrating underlying principles of group behaviour, rather than being applied to specific, modern issues and there is considerable research to be done in this area. For example, the relationship of disgust traits and responses to the 2016 Brexit referendum and its aftermath, is, to our knowledge, unexplored. How disgust relates to pressing issues in equality, such as the treatment of women and gender equality only been marginally touched upon. As broadly, considering how disgust responding might either "support" wealth disparities or provide an engine for change might also prove fruitful (see also Powell, Chap. 15, this volume).

Historically, many of the most important insights into human behaviour have arisen following research into events that challenged our preconceptions about ourselves and the world; studies of conformity and obedience in response to Nazism and the Holocaust being the obvious examples. Core global issues the movement of populations, nationalism, resource scarcity, and the distribution of wealth are ripe for consideration through the lens disgust offers. Importantly, from a research point of view, significant nationalistic events are to an extent predictable and likely to emerge over time, such as a potential second Scottish independence referendum in the UK. It is our hope that future investigative opportunities into our divided world will help to make disgust research great again.

Disgust and the Environment

In addition to COVID-19 and a divided world, our species faces another significant threat in the form of climate change and the exploitation of the environment. Record levels of atmospheric CO2 being recorded yearly, hitting a 4 million year peak in

May 2020. Correspondingly, record temperatures are being observed across the globe, with the previous decade the hottest since records began. As discussed in more detail elsewhere in this volume (see Powell, Chap. 15, this volume), consumer behaviour is a significant contributor to greenhouse gas emissions and climate change, with up to a quarter of global emissions originating from food production alone (with a higher carbon footprint from certain food sources, such as meat and dairy vs. plant products). Additional, changeable human factors that contribute to climate change include travel, waste, and a modern societal lack of reuse (i.e. an un-circular economy).

As we have opined elsewhere (Consedine 2019; Powell 2017), disgust seems to underpin the aversion to many of the behavioural changes that would help slow this troubling process—certain features of disgust are antithetical to a shift towards greener living. The disgust system is "conservative" insofar as it responds to cues that act as indicators of contamination and health threat but are, in fact, benign. Empirically, disgust appears to promotes selective consumption of accepted products and the proliferation of waste (e.g., unblemished apples, prime cuts of meat, new rather than second-hand goods) (Jaeger et al. 2018) as well as underpinning the rejection and avoidance to novel, more pro-environmental food products (e.g., foods made from insect proteins or lab-grown meat) (Powell et al. 2019). Disgust has been exploited to encourage over-consumption of "sanitising" products, where the removal of every trace of dirt and germ has been lauded (e.g., in advertisements for home cleaning products) (McAteer 2019).

The core issue, then, in terms of sustainability and climate change, is how best to either leverage disgust or, more likely, *regulate* disgust in the context of environmental behaviours. Little if any research has explored either of these two possibilities. There are, however, clear opportunities in this area including harnessing the effects of disgust reactions towards environmentally damaging behaviours, such as through exposure to the unsustainable and unethical production of meat (e.g., battery farming and mass slaughter). There is also a line of research for understanding how disgust responses influence or are influenced by peoples' interaction with the natural world (and wild conservation) (Bixler and Floyd 1997). Alongside COVID-19 and the divided world, understanding how disgust can both be best altered, circumnavigated, and/or appropriately channelled into pro-environmental behaviour change is a key challenge for the disgust researchers of our time.

Concluding Remarks

It is not without reflection that we choose to close this first edited volume on an emotion that is close to both our hearts with a "call to arms" in the context of the challenges our world faces. Across the development of this project we have variously asked, exhorted, and begged our contributors to avoid lengthy reviews in favour of crisp chapters that succinctly present the core questions and issues in key areas of disgust research. In part, this emphasis reflects changes in norms and preferences regarding the conduct of science. However, it also reflects our belief that the sciences need to become progressively more engaged with the world if we are to exert an informed, evidence-based influence on the direction our societies are taking. Ultimately, if science is for anything, it is for the betterment of our (collective) lives. It is thus apt that disgust, an emotion that evolved as one of the most primitive affective responses we possess should have the potential to prove pivotal in these moments. Yes, the physical sciences behind the Elon Musks of this world may take humanity to the stars. However, it is the science of who we were, who we are, and why we are like this that truly holds the keys to understanding how best we can adapt to surmount the challenges ahead. As we and others throughout this volume have demonstrated, understanding disgust is central to this endeavour.

References

Ackerman, J. M., Tybur, J. M., & Blackwell, A. D. (2020). What role does pathogen-avoidance psychology play in pandemics? *Trends in cognitive sciences*, *25*(3), 177-186. https://doi.org/10.1016/j.tics.2020.11.008

Bixler, R. D., & Floyd, M. F. (1997). Nature is scary, disgusting, and uncomfortable. *Environment and behavior, 29*(4), 443-467. https://doi.org/10.1177/001391659702900401

Consedine, N. S. (2019, September 1). How to get people to eat bugs and drink sewage. *The Conversation*. https://theconversation.com/how-to-get-people-to-eat-bugs-and-drink-sewage-122328

Cox, R. C., Jessup, S. C., Luber, M. J., & Olatunji, B. O. (2020). Pre-pandemic disgust proneness predicts increased coronavirus anxiety and safety behaviors: Evidence for a diathesis-stress model. *Journal of Anxiety Disorders, 76*, 102315. https://doi.org/10.1016/j.janxdis.2020.102315

Hale, T., Webster, S., Petherick, A., Phillips, T., & Kira, B. (2020). Oxford COVID-19 Government Response Tracker. Blavatnik School of Government. Data use policy: Creative Commons Attribution CC BY standard. https://covidtracker.bsg.ox.ac.uk/

Jaeger, S. R., Machín, L., Aschemann-Witzel, J., Antúnez, L., Harker, F. R., & Ares, G. (2018). Buy, eat or discard? A case study with apples to explore fruit quality perception and food waste. *Food Quality and Preference, 69*, 10-20. https://doi.org/10.1016/j.foodqual.2018.05.004

Judah, G., Aunger, R., Schmidt, W.-P., Michie, S., Granger, S., & Curtis, V. (2009). Experimental pretesting of hand-washing interventions in a natural setting. *American journal of public health, 99*(S2), S405-S411. https://doi.org/10.2105/ajph.2009.164160

McAteer, O. (2019, September 23). Sadness and disgust among emotions that drive brand value, says biometric study. *Campaign*. https://www.campaignlive.co.uk/article/sadness-disgust-among-emotions-drive-brand-value-says-biometric-study/1660239

McKay, D., Minaya, C., & Storch, E. A. (2020a). Conducting exposure and response prevention treatment for contamination fears during COVID-19: The behavioral immune system impact on clinician approaches to treatment. *Journal of Anxiety Disorders, 74*, 102270. https://doi.org/10.1016/j.janxdis.2020.102270

McKay, D., Yang, H., Elhai, J., & Asmundson, G. J. (2020b). Anxiety regarding contracting COVID-19 related to interoceptive anxiety sensations: The moderating role of disgust propensity and sensitivity. *Journal of Anxiety Disorders, 73*, 102233. https://doi.org/10.1016/j.janxdis.2020.102233

Paluszek, M. M., Asmundson, A. J., Landry, C. A., McKay, D., Taylor, S., & Asmundson, G. J. (2021). Effects of anxiety sensitivity, disgust, and intolerance of uncertainty on the COVID stress syndrome: a longitudinal assessment of transdiagnostic constructs and the behavioural immune system. *Cognitive Behaviour Therapy*, 1-13. https://doi.org/10.1080/16506073.2021.1877339

Porzig-Drummond, R., Stevenson, R., Case, T., & Oaten, M. (2009). Can the emotion of disgust be harnessed to promote hand hygiene? Experimental and field-based tests. *Social Science & Medicine, 68*(6), 1006-1012. https://doi.org/10.1016/j.socscimed.2009.01.013

Powell, P. A. (2017, March 24). How disgust stops us from living sustainably—and how to stop it. *The Conversation.* https://theconversation.com/how-disgust-stops-us-from-living-sustainably-and-how-to-stop-it-75110

Powell, P. A., Jones, C. R., & Consedine, N. S. (2019). It's not queasy being green: The role of disgust in willingness-to-pay for more sustainable product alternatives. *Food Quality and Preference, 78*, 103737. https://doi.org/10.1016/j.foodqual.2019.103737

Reicher, S. D., Templeton, A., Neville, F., Ferrari, L., & Drury, J. (2016). Core disgust is attenuated by ingroup relations. *Proceedings of the National Academy of Sciences, 113*(10), 2631-2635. https://doi.org/10.1073/pnas.1517027113

Index

P. A. Powell, N. S. Consedine (eds.), *The Handbook of Disgust Research*,
https://doi.org/10.1007/978-3-030-84486-8

The manufacturer's authorised representative in the EU is Springer Nature Customer Service Centre GmbH, Europaplatz 3, 69115 Heidelberg, Germany. If you have any concerns regarding our products, please contact ProductSafety@springernature.com

Printed and bound by CPI Group (UK) Ltd, Croydon, CR0 4YY
30/09/2026
02225690-0011